D1259403

Planck presenting Einstein with the first Planck Gold Medal in 1929. These two titans symbolize the new era in the history of science and humanity.

David Nachmansohn

GERMAN–JEWISH PIONEERS IN SCIENCE
1900–1933

Highlights in Atomic Physics, Chemistry, and Biochemistry

Springer-Verlag   Berlin   Heidelberg   New York

DAVID NACHMANSOHN, M.D.

Professor emeritus of Biochemistry
Departments of Biochemistry and Neurology
Columbia University, New York, N.Y. 10032

With 27 figures.

Library of Congress Cataloging in Publication Data

Nachmansohn, David, 1899-
German-Jewish pioneers in science, 1900-1933.

Bibliography: p.
1. Chemists—Germany—Biography. 2. Biochemists
—Germany—Biography. 3. Physicists—Germany--Biography.
4. Jews in Germany—Biography. I. Title.
QD21.N33   509'2'2 [B]   79-10550

Printed in the United States of America

9 8 7 6 5 4 3 2 1

ISBN 0-387-90402-6 Springer-Verlag New York Heidelberg Berlin
ISBN 3-540-90402-6 Springer-Verlag Berlin Heidelberg New York

This book is dedicated to the memory of Siegfried Moses, president of the Leo Baeck Institute and for two decades one of its most dynamic leaders. The author was invited by the Leo Baeck Institute to give the first Siegfried Moses Memorial Lecture, an honor for which he takes the opportunity to express his gratitude. The topic of the lecture, given in November 1976 in Jerusalem, was the role of German-Jewish pioneers in science in the early twentieth century. This work is an extended version of the lecture.

Preface

The Leo Baeck Institute, to whose late president this book is dedicated, has three branches, located in Jerusalem, London, and New York. Its chief aim is the collection of documents describing the history of Jews in German-speaking countries, the manifold aspects of the association of the two ethnic groups, over a period of about 150 years; that is, from the time of the Enlightenment until the rise to power of the Nazi regime. Twenty-three Year Books (1956–1978) so far and many additional volumes about special fields have been published by the institute. They offer an impressive documentation of the role Jews played in Germany, some of their great achievements, the difficulties they encountered in their struggle for equal rights, as well as its slow but seemingly successful progress. A wealth of interesting material describes the mutual stimulation of the creative forces of the two ethnic groups in a great variety of fields—literature, music, the performing arts, philosophy, humanities, the shaping of public opinion, economy, commerce, and industry. Since the destruction of the Second Temple by the Romans, there have been only a few periods during which Jews played such an eminent role in the history of their host nation. As was forcefully emphasized by Gerson D. Cohen (1975) in his introduction to the twentieth Year Book of the Leo Baeck Institute, the material published so far has already become an invaluable and indispensable source for scholars interested in both Jewish and German history. In addition, it provides a magnificent record of a glorious period that is an integral part of West European civilization.

One special area, which had dramatic effects of tremendous importance for the whole of mankind, has been relatively neglected: the development of science and technology. In the meteoric rise of these two interdependent fields in Germany in the last three decades of the nineteenth centuries and the first three decades of the twentieth, the intimate collaboration of German and Jewish scientists was instrumental, especially after the turn of this century. The mutual influence and inspiration were extremely strong. Many close friendships existed among these scientists. Science is international in character and favors the exchange of ideas and information among scientists all over the world. This exchange sometimes leads to combined efforts in the search to find an answer to a problem; such efforts may be facilitated and encouraged by a variety of factors. The atmosphere and the conditions in Germany prevailing at the turn of the century were particularly favorable for the scientific collaboration between the two ethnic groups.

This book is a limited and—because of the selection of a few topics only—quite restricted attempt at filling the gap that exists in the presentations of the Leo Baeck Institute. Dr. Franz Winkler invited the author to give a lecture on some of the great contributions of German-Jewish scientists. The lecture was given in the spring of 1972 at the Leo Baeck Institute in New York. An article on the same topic was to be written but was postponed because of the death of Aharon Katzir–Katchalsky of the Weizmann Institute during the massacre at Lod in May 1972.* In 1976 the author was invited by the Free University of Berlin to give a historical lecture on the occasion of the bicentennial celebrations honoring the United States. The topic was the great era of the Kaiser Wilhelm Institutes in Berlin–Dahlem in the 1920s. At about the same time the author was invited to give the first Siegfried Moses Memorial Lecture. The topic was the collaboration between German and Jewish scientists in the early twentieth century. The response of many colleagues to the three lectures encouraged the author to write this book.

It must be stressed that this book is not intended to be a systematic presentation of the scientific developments of the fields selected. The author is a biochemist; biochemistry is one of the topics. The history of biochemistry, including the period covered here, has been described in a highly competent, masterful, systematic, and superbly documented way by Marcel Florkin (1977); four volumes have appeared, and two more are in preparation. This history and a great number of other excellent books and publications, such as Joseph Fruton's book *Molecules and Life* (1972), are addressed specifically to biochemists. The present book also includes a chapter on atomic physics. In that field the same situation exists. However, the aims of this book are quite different from

---

* The tragedy, in addition to the personal shock, disrupted a project on which Katchalsky and the author had collaborated: to work out a model integrating the existing biophysical, biochemical and thermodynamic data. Professor Eberhard Neumann, who had spent three years with Katchalsky in the Weizmann Institute, had become greatly interested in the project and in the following years, initially in close collaboration with the author, he succeeded in decisively advancing the project.

those of the other works cited. It tries to convey to the reader some idea of the intellectual atmosphere prevailing during that period, the enthusiasm and excitement about the spectacular achievements, the feeling of a new era in science. The humanistic tradition and the respect for ethical values were very strong. Many of the leaders had been educated in humanistic gymnasiums. The works of Plato, Kant, Goethe, and others were for the scientists of the period described not documents of a forgotten past; their ideas were still very much alive. Many of these scientists were accomplished musicians and loved to play chamber music together. Sharing the same passions and ideals helped forge close ties. The deep devotion to the search for the understanding of nature, the unraveling of its forces and mysteries, was linked to the hope, in fact the firm conviction, that these efforts would contribute to and prepare a better future for mankind. Some of the most significant achievements are mentioned, using language not too technical for many readers. The description of the close collaboration between German and Jewish scientists forms an integral part of the book. Some profiles are given and family backgrounds mentioned, an aspect pertinent for a better understanding of many attitudes, views, psychological factors, and reactions; some of these may be difficult to understand today without such background. There are two reasons for including a few highlights of atomic physics. First, it was the field in which the greatest breakthroughs were made, breakthroughs that have revolutionized all of science in this century. It had and will continue to have the profoundest impact on the fate of society. Second, in no other area was the collaboration of the two ethnic groups equally conspicuous and important.

The author was fortunate to work for several years in the Kaiser Wilhelm Institutes in Berlin–Dahlem in the second part of the 1920s. He was fortunate to know personally several illustrious scientists working there. The smallness of the institutes and the relatively small number of scientists favored personal contacts between young people and the famous leaders, in contrast to today's giant institutes and their huge number of scientists. Some parts of the book are based not on research and reading of the literature, but on personal recollections and reflections. The author's enthusiasm may have affected his "objectivity." The notion of objectivity is in any event questionable, as every historian knows. Even evaluations of scientific contributions are frequently far from being objective. The author has tried to keep his presentation as close to truth and reality as possible. *Ultra posse nemo obligatur.*

The reader should realize that only a few highlights of the various topics selected have been described. The amount of space devoted to personalities or contributions should not be considered an indication of their importance. The familiarity of the author, the sources readily available to him, and his unequal competence in such a wide area played a role. Overall space considerations were also a factor.

The presentation of each of the two main fields has been differently organized: that of chemistry, including biochemistry, is subdivided into the description of personalities. In the case of atomic physics such a subdivision appeared more difficult in view of the frequently inseparable interaction between the various concepts and personalities involved. A subdivision of this field based on personalities would lead to many repetitions.

Finally, the description of the work, the life, and the personality of the greatest German-Jewish scientist, Albert Einstein, whose one-hundredth birthday is being celebrated in 1979, has been omitted for obvious reasons. A vast number of books have been written about him on all levels, and they are readily available. Therefore references to Einstein's work, ideas, or personality have only been made where they seemed essential for the material discussed.

The book is a monument to the illustrious scientists described. At the same time it is a tribute from the author to all those who—either personally or by their writings—influenced his own scientific and cultural formation.

New York, May 1979   D. N.

Taken in 1920 at the Kaiser Wilhelm Institute in Berlin-Dahlem. Standing, left to right: Hugo Grotrian, Wilhelm Westphal, Otto von Baeyer, Peter Pringsheim, Gustav Hertz; Seated, left to right: Hertha Sponer, Albert Einstein, Ingrid Franck, James Franck, Lise Meitner, Fritz Haber, Otto Hahn.

Max von Laue

Max Planck

James Franck

Gustav Hertz

Max Born

Niels Bohr

Werner Heisenberg

Erwin Schrödinger

Walther Nernst

Fritz Haber

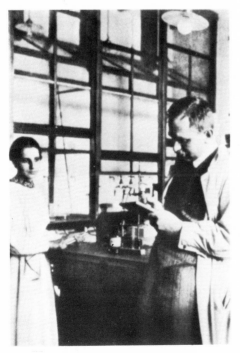

Lise Meitner and Otto Hahn

Manfred Eigen

Adolf von Baeyer

Paul Ehrlich

Richard Willstätter

Heinrich Wieland

Otto Warburg

Otto Meyerhof

Carl Neuberg

Gustav Embden

Hans Krebs

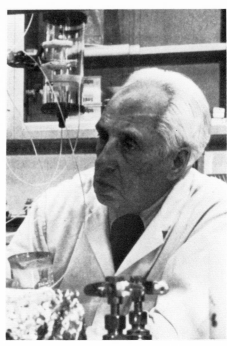

Severo Ochoa

Rudolf Schoenheimer

Ernst Chain

Table of Contents

Acknowledgments

The author would like to acknowledge the help of several friends and colleagues in the preparation of the manuscript. In the first place he is most grateful to Professor Hans Herken for reading the whole manuscript and making a number of most valuable suggestions. He thanks Sir Hans Krebs for giving authorization to use his superbly written biographical notes about Otto H. Warburg. Many of the formulations were so excellent that some sentences were, with his permission, reproduced without change. Professor John T. Edsall made valuable comments on the portrayal of Otto Meyerhof. This description was endorsed by Professor Severo Ochoa, a close friend of both Meyerhof and the author since 1929, when they met in Meyerhof's laboratory. It was also read by Sir Hans Krebs and Dr. Hermann Blaschko; the latter was also an associate of Meyerhof. Professor Daniel Greenberg went over some parts of the chapter on atomic physics. I greatly appreciate the help of Mrs. Elisabeth Heisenberg for providing much important information about her husband's personality, facts which I did not find anywhere else, and for the permission to mention in the text her still unpublished memoirs about her husband. Mrs. Elisabeth Lisco, the daughter of James Franck and wife of Professor Hermann Lisco, Harvard University, kindly provided the author with the family history of her father. Dr. Irene Forrest, the daughter of Carl Neuberg, was kind enough to read the chapter about her father and to make some corrections. Professor Ochoa also read and corrected the pages in which his work is described. Professor Eberhard Neumann kindly helped the author in the discussion of ΔH and ΔG values in the light of non-equilibrium thermodynamics. The author would like to express his appreciation of the help and encouragement of Professor Lewis P. Rowland, chairman of the Department of Neurology, Columbia University, who provided him with the facilities required. The deep devotion and efficiency of Ms. Eva-Renate Busse, and her invaluable help in many respects in preparing the manuscript is gratefully acknowledged.

Without the encouragement of many other scientific friends and colleagues in the United States, Israel, Germany, France and other countries this book would not have been written.

Introduction

Modern science starts at the end of the Renaissance, in the late sixteenth and early seventeenth centuries, with Copernicus, Kepler, Galileo, Harvey, and Newton. What distinguishes modern science from Greek philosophy is the central role of the experiment as the basic element of all notions, theories, and concepts; the interdependence between observed facts and theories, the discoveries starting from theories and their reflection and effects on the progress and rapid growth of scientific knowledge. The complexity of the factors on which scientific progress is based will become evident in the discussion of modern concepts of epistemology. Some of the grandiose speculations of Greek philosophy, such as the theory of continuity and the theory of the atom, still formed part of the thinking of physicists at the turn of the century. This thinking was revolutionized only by the spectacular advances of the physics of the twentieth century.

In the seventeenth and eighteenth centuries, science moved relatively slowly. The number of brilliant scientists was limited. But during the nineteenth century scientific progress began to be made at a rapidly increasing rate. In the last three decades of the nineteenth century, for reasons to be discussed shortly, the center of science moved to Germany. And in the twentieth century, a revolution occurred, one beyond the boldest dreams of earlier scientists. It may be said here that science in the twentieth century revolutionized technology to a degree that completely changed not only the daily life of man, but society as a whole; in fact, the fate of mankind.

Of course, scientific discoveries were used for technical improvements, even in ancient times, and technical inventions have stimulated scientists to explore the associated scientific problems throughout history. But the close working association between fundamental science and its application, the advance in technology, was something more; from its beginning in the nineteenth century it grew to become an integral part of the shaping of society in the twentieth century. Germany played an outstanding role in these developments. This book will relate a few highlights of that time, in particular the fortunate and important effects of the close collaboration betwen German and Jewish scientists, their personalities, and the atmosphere prevailing during that era.

I

Historical Background

A. THE RISE OF SCIENCE IN GERMANY

The monumental rise of German science and technology in the last three decades of the nineteenth century and the first three decades of the twentieth transformed Germany from a relatively destitute and in many respects backward country into one of the great powers of the earth. The result has been a revolution of unprecedented dimensions. We are witnesses to a new era in the history of man. To understand this rapid and unparalleled growth of German science and industry, we must first look at some of the factors that led to these developments.

When the ravages of the Thirty Years' War had been ended by the 1648 Westphalian peace treaty, Germany was a devastated country. It took more than a century for Germany to begin to recover intellectually, politically, and economically. There were of course a few exceptions, such as Johann Sebastian Bach (1685–1750) and his family and Gottfried Wilhelm von Leibnitz (1646–1716). In the eighteenth century Germany was still quite backward compared to the other Western European countries, in particular France and England. During the period of the Enlightenment, in the second half of the century, the country started to recover and to develop again intellectually, as illustrated by the names of Klopstock and Herder, Lessing and Schiller, Goethe and Hölderlin, Kant and Beethoven. But otherwise the country was still in a depressed condition; the middle class was poor and had virtually no influence, compared to that of its counterparts in France and England. Poverty, starvation, and disease were widespread.

An unexpected factor in the rebirth of German intellectual forces was the defeat, occupation, and humiliation of Germany by Napoleon. The

German national pride was deeply hurt. Since the rebuilding of a strong army was, under French occupation, out of the question, many leaders realized that an effective way to rebuild German power would be to create and develop an elite that would provide strong leadership. This aim required the strong support of existing universities and the building of new ones, because the university was the natural place for potential talent to develop, producing leaders in all the fields required for rebuilding a strong nation. Wilhelm von Humboldt, strongly supported by such philosophers as Fichte, Schleiermacher, and Hegel, succeeded in overcoming the reluctance of Friedrich Wilhelm III, the king of Prussia; Humboldt was the driving force in the establishment in 1810 of the University of Berlin (*Friedrich Wilhelm Universität*) in the midst of the Napoleonic wars. Prussia was at that time the only potentially strong military power in Germany proper (not counting Austria), ever since the Hohenzollern King Frederick II had built a strong army, which no other of the about 30 German states was able to match. The emphasis of the new university in Berlin was on humanities—philosophy, history, theology, and art. The first rector was the philosopher Fichte, whose lectures *Reden an die Deutsche Nation* were an ardent appeal for the spiritual renewal of the nation. Only in the 1820s did science begin to develop at the University of Berlin. In addition to the University of Berlin, new universities were established in 1811 in Breslau and in 1818 in Bonn.

But the middle class and the population in general remained in a sad state. They had no influence, no representation. The attempts at creating a National Assembly were crushed when the revolutions in France and Austria in 1848 ended in failure. The revolutionary movement among students in favor of progressive ideas was suppressed by a massacre carried out by troops under the command of Prince Wilhelm, later to become Kaiser Wilhelm I; he belonged to a reactionary group and was forced, after this disgraceful performance, to flee to England.

The dramatic turning point of German destiny was marked by the appearance of Otto von Bismarck, one of the great statesmen in history. He combined a brilliant mind, a superb intellect, extraordinary vision, and broad perspectives with an iron will; he was completely autocratic, the chancellor of "blood and iron" (*von Blut und Eisen*). He was extremely ruthless and brutal; every intrigue, plot, forgery (the famous *Emser Depesche*) were justified to achieve his aims. He belonged to the aristocratic and reactionary Junkers, the dominant class in Prussia. Three victorious wars within seven years (1864–1871) were deliberately provoked by Bismarck and transformed the power distribution of Europe: The war with Denmark led to the annexation of Schleswig–Holstein by Prussia; the war with Austria made Prussia the dominant power in Germany; the war with France ended with, in addition to the annexation of Alsace–Lorraine, the establishment of the German *Kaiserreich* under Wilhelm I.

The unification of Germany under the leadership of Prussia and the strong army were only potential factors in the emergence of the new great power in Europe. They created favorable conditions for transforming a relatively backward population into one of the strongest and most prosperous countries in Europe. Germany was poor in natural resources. Unlike France and England, it had no vast empire to exploit. Bismarck and many other farsighted leaders recognized the vital importance of developing science and technology in order to build a powerful industry and agriculture. Only through economic wealth could Germany become a really great power. The industries had to do more than match the advanced and flourishing industries of France and particularly England; they had to be far superior in quality by greatly improved technology. The universities received strong government financial support, on a scale unprecedented in the history of any nation. Many new institutes were built and new positions created. The size of faculty and student bodies began to increase rapidly. A large-scale expansion took place. Moreover, the number of the more industry-oriented *Technische Hochschulen** was continuously increased. The first *Technische Hochschule* was founded in 1825 in Karlsruhe. By the end of the century there were about a dozen of them. Their standards were extremely high and their graduates were comparable to university Ph.D.s. Many of these graduates became dynamic leaders in industry, with competence in their fields, vision, and perspective. They were fully aware that basic science was the main source of new inventions and improvements in technology, the door to unexpected new possibilities. They established large research laboratories attached to their manufacturing enterprises. The facilities and atmosphere were comparable to those in academic institutions. By the turn of the century Germany had become the leading industrial country. It had a gigantic pharmaceutical and chemical industry; for example, 80% of all dyes were manufactured in Germany. It had an electronic and optical industry of unmatched quality. Close collaboration between industry and universities was frequent and of mutual benefit. The United States was the first country to follow Germany's lead in joining research and industry (evidenced by DuPont and General Electric, among others); France and England soon fell behind. We can see the impact of this remarkable development by noting the changes that took place in Germany; in 1840, with a population of about 35 million, it was a country plagued by poverty, misery, starvation, and disease, the scene of Hauptmann's *Die Weber*. In 1910, with a population of 70 million, it was a rich country with a highly developed middle class and a working class, which in many respects had better living conditions and more advanced social institutions than their counterparts in France and England, although both classes were virtually without political power.

---

* *Technische Hochschulen* are roughly equivalent to American institutes of technology.

B. THE ENTRANCE OF JEWS INTO GERMAN SCIENCE

About German–Jewish history there exists a vast literature. The publications of the Leo Baeck Institute offer a superb source. However, only those features pertinent to the subject of this book may be mentioned here —namely the factors that permitted the development of close German–Jewish scientific collaboration.

The period of the Englightenment in the eighteenth century did not bring about the emancipation of Jews in Germany, but it prepared the way for this step, especially among intellectuals and progressive leaders. A paramount and decisive role in this development was played by Moses Mendelssohn (1729–1786), a brilliant, extraordinary man, by his outstanding contribution to literature and philosophy, by his influence on a number of leading personalities, and by specific actions aimed at improving the conditions of his fellow Jews.

Born in Dessau, the son of a poor teacher in the town's small Jewish ghetto, he started in his youth to learn, with an exceptional fervor, whatever was accessible to him. Through one of his teachers he got hold of *Guide of the Perplexed*, one of the most famous philosophical works of Moses Maimonides, and studied it with great enthusiasm; it had a great and lasting influence on his thinking. When his teacher moved to Berlin, Mendelssohn followed him—quite a courageous act for a 14-year-old Jewish boy at that time. After a few years of great hardship, which did not prevent him from continuing his studies with great devotion and energy, he became the teacher of the sons of a rich Jewish merchant, Isaac Bernhard, and later Bernhard's accountant. This position ensured him and his family a comfortable life and permitted him to continue all his other activities.

The knowledge Mendelssohn acquired—mostly on his own—was truly amazing. He learned not only to speak an elegant German, rare among Jews at that period, but also to speak fluent French and English. In addition, he learned Latin and Greek. He became passionately interested in philosophy and literature and to some extent even in science; only for history had he no interest. When he was 25 years old he became a close friend of Gotthold Ephraim Lessing (1729–1781), who was of the same age and one of the foremost figures in the German Enlightenment. A writer and a poet, Lessing was a vigorous fighter for tolerance and freedom. The hero in one of his most celebrated dramas, *Nathan der Weise*, is a Jewish merchant of great wisdom, for whom Mendelssohn was unequivocally the model. The drama, still performed today, produced a great sensation at its first performances in the 1780s. It was enthusiastically welcomed by liberal and progressive elements, but sharply criticized by anti-Semitic factions in the population.

Mendelssohn wrote many articles and books. He translated parts of some of Shakespeare's plays into German. His first book, *Philosophical*

Dialogues, was the first book in modern times written in German by a Jew. But the contribution that made him famous all over Europe was his book *Phädon or About the Immortality of the Soul*, published in 1767. The first part of the book was a translation of Plato's *Phädon* from the Greek into an excellent German, but the second part was a philosophical discussion of the problem of the title, using the same form of dialogue that Plato had used. The book was a sensational success; it was translated into most European languages and became one of the most widely read books of that era. It was highly praised, by Goethe and Herder, among many others. Kant wrote that Mendelssohn's genius might succeed in introducing a new epoch in philosophy. Lessing referred to him as a second Spinoza. The Royal Academy of Sciences of Berlin elected Mendelssohn to membership, but Frederick II removed him from the list.

Mendelssohn had become an international celebrity. His home became a center where many intellectuals met, freely and frequently, Jews and Christians, Germans as well as people from other countries. All were deeply impressed by his personality. One of them was Johann Kaspar Lavater, a theologian, philosopher, and scientist from Zurich. When a book of Charles Bonnet, professor at Geneva, appeared praising the superiority of the Christian religion, Lavater sent the book to Mendelssohn with the suggestion that Mendelssohn either refute the arguments or convert to Christianity. Mendelssohn had always tried to avoid discussing religion. He considered the Jewish religion not as one requesting certain beliefs, but as one prescribing what to do and how to do it. He was firmly attached to the Jewish religion and its traditions. His answer to Lavater was most forceful, dignified, and impressive. Lavater deeply regretted his action and apologized.

The event had, however, another effect: Mendelssohn started to pay much more attention to the terrible plight of most Jews in Germany and elsewhere. Most of them lived in ghettos in incredible poverty, filth, and ignorance. They spoke Yiddish, and few knew the Bible in the original text. They were completely isolated and ignorant of the revolutionary movements that were sweeping the societies of Western Europe during the era of the Enlightenment. Mendelssohn decided to translate the Bible into German written in Hebrew letters, thereby making German easily accessible to the young Jews in the ghettos. Within a few years several important parts of the Bible appeared in this form. This work marks a turning point in the history of German Jews. The younger generation enthusiastically studied this translation, albeit against the advice of many reactionary elements. They learned German, and that opened the way for them to participate in the cultural life of Europe of that era, decades before the legal emancipation. Many of them went so far, especially in the second and third generations, to become baptized or to have their children baptized, a result that Mendelssohn had not foreseen. Many other Jews initiated reform movements in order to modernize Jewish life and traditions. The translation of the Bible must be considered one of the chief factors in the association between the two ethnic groups.

Mendelssohn was considered the foremost Jewish leader of his time, and repeatedly tried to use his influence in the fight to improve the harsh conditions under which most Jews lived. He had a large family, and several of his six children were baptized. Two of his sons established a banking house that became one of the greatest private banks in Germany, and was closed only in the 1930s by the Nazis. Probably the most famous member of the family was his grandson, Felix Mendelssohn-Bartholdy. He was a great composer, pianist, and conductor, and deeply influenced the musical life of several German cities, as well as Paris and London. A fascinating account of the whole family has been written by Kupferberg (1972).

Mendelssohn's home was not the only one where Jews and Christians met long before the emancipation; there were many distinguished diplomats, artists, and writers who also took part in the interchange. Among the more famous "salons," as they were referred to, were those of Henriette Herz, wife of Marcus Herz, a physician, and Rachel Levin, wife of Karl Varnhagen von Ense. One frequent guest of Henriette Herz was Count Mirabeau, and this experience with distinguished Jews influenced him in his fight for the rights of Jews in France. Wilhelm von Humboldt, a strong supporter of the rights of Jews in Prussia, was a frequent guest of Rachel Varnhagen von Ense. But one must realize that these salons were exceptions and had very little to do with the real situation of the Jews in Germany.

The emancipation and the granting of constitutional rights in Prussia came only after the war of liberation, in the 1812 *Judenedict* of Chancellor Karl August von Hardenberg, an edict strongly supported by Wilhelm von Humboldt. Prussia was the state where the great majority of the Jews in Germany lived (not including Austria). In 1816, there were 124,000 Jews in Prussia, although more than half of them were Polish Jews living in the provinces of Posen and Westprussia. In 1846 the number had increased to about 220,000, or more than two-thirds of all the Jews in Germany. In 1910, 70% of all Jews (not baptized) lived in Prussia, about 400,000.

The constitutional rights did not automatically give genuine equality. Complete assimilation was expected as a prerequisite, even by many liberal elements. The willingness of Germans to accept Jews as equals was bound to another step: baptism, as a sign of complete identification with the rest of the population. This was the "ticket of admission to European civilization," to use an expression of Heine. Many Jews payed the price; many others, reluctant for different reasons to take the step, had their children baptized to spare them the agonies and conflicts of adhering to a religious community that for many had become unimportant or completely irrelevant and meaningless.

Despite all the efforts of the liberal elements and the strenuous fight of the Jews themselves to remove the barriers to their full acceptance by assimilation, anti-Semitic forces remained powerful. They frequently brought about violent outbreaks and riots differing in strength and extent in the cities and states where they took place. This unrest began immediately after the war of liberation and continued until the middle of the nineteenth century. The efforts of Metternich, Hardenberg, and Humboldt at the Congress of Vienna, in 1816, to extend the rights of Jews and to improve their situation had little effect. By 1846 more than one-third of all the Jews in Prussia still did not have the rights of citizens. Whenever adverse events—economic, political, or social—occurred, these feelings of hostility erupted. After the establishment of the *Kaiserreich*, the physical attacks were stopped, but the anti-Semitic movements expressed themselves in other ways.

The two most powerful figures in the *Kaiserreich*, the period from 1871 until World War I, were Bismarck and, after his dismissal in 1890, Kaiser Wilhelm II. Since this was the period of the rapid rise of German science, when Jews began to play an outstanding role in this field (as well as in many others), the attitude of these two men toward the Jews is pertinent. Bismarck was a Prussian Junker, the scion of one of the oldest families. He had great respect for the high qualities, the great talents, and the competence of many Jews, and had not the slightest hesitation to use their services for Germany or for his own interests. The most famous and well-known example is his close association with the Jewish banker Gerson Bleichröder. The history and importance of the help and advice Bleichröder gave Bismarck in the creation of the *Kaiserreich* have been described by Fritz Stern (1977) in his book, *Gold and Iron. Bismarck, Bleichröder and the Building of the German Empire*. It is a superb description of the many facets of the personal relationship between Bismarck and Bleichröder, and it is also a penetrating and illuminating historical analysis of the whole era. It gives an unusually interesting insight into many pertinent factors dominating German society, the economic revolutions that took place in that period, and the successes and failures of the Jews in their struggle for their rights. Bleichröder had helped to finance two of Bismarck's wars; he was his most intimate counselor, even in many political problems during that most critical time. They had frequent consultations and often dined together. Bleichröder stayed for several weeks in Versailles and was the only Jew to participate in the proclamation of the *Kaiserreich* early in 1871. In the following two decades he was one of the most powerful and prominent figures who helped Bismarck in the building of the empire and was frequently referred to as the German Rothschild. He was also Bismarck's adviser—a most successful one—in all his personal financial transactions and investments. In 1872 Bleichröder was elevated to hereditary aristocracy, an extraordinary step in Prussia for a Jew, especially one who still had a Yiddish accent. He was even invited to dine with Kaiser Wilhelm I, and he became one of the most celebrated Jews in Germany. Among the leading ministers of several European countries who sought his advice and help was Benjamin Disraeli; Bleichröder had correspondence with kings. But his contributions to the greatness of Germany did not earn him the acclaim of the population; on the contrary, his success and power made him the target of strong anti-Semitic reactions. Although the nobility was officially forced to accept him, he was detested and despised as an intruder—a fate he shared with many rich Jews. Wealthy Jews became the symbol of capitalism, disliked by those who considered capitalism as a danger that corrupted the German character and life style. A new type of anti-Semitism, which tried to initiate political actions to curb Jewish influence, emerged.

Bismarck and Bleichröder remained intimate friends until the latter's death, even after the fall of Bismarck, despite the resentment their friendship created among Bismarck's other friends and associates. However, Bismarck's attitude toward the Jews in general was essentially influenced by political opportunism. Bismarck used anti-Semitism in his fight against liberals and Social Democrats, since many of their leaders were Jews. It was a convenient political weapon rather than a genuine, deep-seated feeling. He fully recognized the great value of many talented and gifted Jews and their importance for the strength of Germany. He would not tolerate any violent actions against Jews and would not permit their civil rights violated. In principle, the anti-Semitism in Germany did not fundamentally differ from that existing in many other Western countries, such as France, the United States, or even Great Britain, although in the latter it was probably least manifest. But in Germany the percentage of Jews was much higher and their power and influence much stronger than in other Western countries. The combination of these two factors was certainly responsible for a more widespread and more vigorous reaction against Jews.

The attitude of Kaiser Wilhelm II toward Jews was in several respects different from Bismarck's; it was more complex and sometimes more emotional, an important feature of his character in general. During his reign, from 1888 until World War I, the situation of the Jew in Germany continued to improve at the same rate as in the Bismarck era, perhaps even faster. They increasingly penetrated academic professions, and many became recognized and respected leaders, especially in science, but also in many other fields. However, these facts do not reflect Wilhelm's feelings about the Jews. His parents certainly did not have any prejudice against Jews. His mother, a daughter of Queen Victoria and Prince Albert, detested anti-Semitism; she considered the pogroms in Russia a "barbaric tyranny" and strongly influenced her husband, the unfortunate Kaiser Friedrich, who died of cancer after a reign of only 99 days. But the parents had less influence on their son than had the grandfather, Kaiser Wilhelm I. He was very conservative and prejudiced against Jews. He insisted that his grandson, after having finished the gymnasium, join the army for half a year. The prince then went to the University of Bonn, the most preferred university among the ruling families and the nobility. There he joined a student fraternity, the Borussia. He participated with great enthusiasm in the life of the fraternity. After only two years he joined the army in Potsdam. The atmosphere there strengthened his conservative attitude, much to the unhappiness of his parents. At that time, in the 1880s, anti-Jewish feelings, always present, were exacerbated by the economic decline of the ruling Junker class. Since their wealth was essentially based on their large estates, the rising capitalism and industrialization greatly diminished their fortunes, as will be discussed later. Owing to their newly accumulated wealth, Jews began to play a prominent role in society and many Junkers were unable to compete with them. The Junkers consequently attributed their misfortunes, not to the economic trends prevailing in Western Europe, but to the Jews; although unable to compete with the Jews economically, they did succeed, with the full support of Wilhelm, in blocking them from the government and especially from the army, with very few exceptions.

The close friends of Wilhelm, even before he became kaiser in 1888, were all outspoken anti-Semites. Among those friends were the son of Bismarck, Count Herbert von Bismarck, A. von Waldersee, and Philipp von Eulenburg. Through Eulenburg Wilhelm met the famous conductor Hans von Bülow and his wife Cosima. When the latter married Richard Wagner, Wilhelm met the Wagner family through her. The relationship between Cosima and Wilhelm remained quite cool, but in 1901 he met her son-in-law Houston Stewart Chamberlain (1899). Chamberlain's book, *On the Foundations of the Nineteenth Century*, was extremely popular. He considered the Jew the deadly enemy of "Aryan" Germans. The book had a powerful influence on the ideological basis of German anti-Semitism. Wilhelm, greatly impressed by Chamberlain's personality and most enthusiastic about the book, made the book obligatory reading for all high-school teachers. However, although he completely endorsed the praise of Germanism by Chamberlain, he was much more cautious about accepting the claim that the Jews were the most dangerous enemy of Germany. For the kaiser the yellow race (*die gelbe Gefahr*) and the blacks were the real threat to Western civilization.

It is of interest in this context to quote an exchange of views between Lord Balfour and Chaim Weizmann, reported in Weizmann's letters and papers of August 1914 to November 1917, published in 1975. In a discussion that took place in December 1914, Balfour described his visit to Cosima Wagner in Bayreuth and her views about Jews. Weizmann wrote as follows:

> Mrs. Wagner is of the opinion that the Jews in Germany have captured the German stage, Press, Commerce, the Universities, etc.,

that they are putting in their pockets after only a hundred years of emancipation everything for which the Germans have worked for centuries, and that she and people who think like her resent very much having to receive all the moral and material culture at the hands of the Jews.

In this discussion with Balfour, Weizmann formulated his views on the Jewish problem, using the case of German Jews as an example:

> The essential point which most non-Jews overlook and which forms the crux of the Jewish tragedy is that those Jews who are giving their energies and their brains to the Germans are doing it in their capacity as Germans and are enriching Germany and not Jewry, which they are abandoning. . . . The tragedy of it all is that whereas we do not recognize them as Jews, Madame Wagner does not recognize them as Germans, and so we stand there as the most exploited and misunderstood people.*

At the turn of the century these extreme views of Cosima Wagner were shared by a small number of people, but they became widespread in the era of the Weimar Republic. It is remarkable how unaware Jews as well as many non-Jews were of these deep-rooted feelings, almost until the era when the Nazis rose to power.

The anti-Semitic feelings of Wilhelm exploded only occasionally. This was quite in line with his impulsive character. He was unable to accept any kind of criticism. The press, in which several Jews played a prominent role, became a chief target of his hostile outbursts when critical remarks were published against him or his close friends. He even tried, although without success, to limit the freedom of the press. He also considered the Social Democrats, among whose leaders were several Jews, as most dangerous subversive elements. When Wilhelm was visited by Nicholas Murray Butler, the president of Columbia University, he tried to convince Butler of the serious threat that the Social Democrats presented to society.

However, no matter what his feelings and occasional emotional reactions were, his attitude toward Jews was determined by many other factors. The glory of Germany was his primary concern. He was intelligent enough to recognize how much Jews had contributed to the building of the empire, to the power and greatness of Germany. He admired some of the rich Jewish industrial leaders and bankers and became quite friendly with some of them. Albert Ballin, the director general of the Hamburg–American Steamship Line, was generally considered to be a friend of the kaiser, who frequently visited Ballin in his mansion in

---

* I am greatly obliged to Dr. Rolf Michaelis, Jerusalem, who brought this discussion to my attention.

Hamburg and invited him to his palace in Berlin. Other Jews greatly respected by the kaiser were bankers Max Warburg and Karl Fürstenberg, and James Simon, a well-known patron of the arts and the owner of a magnificent art collection.

The kaiser was extremely proud of Germany's increasing leadership in science and technology. He occasionally became interested in specific technical achievements and wanted to be kept informed; sometimes he even asked for a demonstration. He was fully aware of the many outstanding contributions of Jews in this area. He was enthusiastic about the plans of Friedrich Althoff to establish one of the world's great research centers in Dahlem, on a Hohenzollern domain. When Althoff died, Wilhelm personally took the initiative that the plans should be actively pursued and strongly supported the foundation of the Kaiser Wilhelm Society, as will be described later. Thus it was not difficult for the liberal leaders of German science (e.g., Walther Nernst, Emil Fischer, and Max Planck) to convince him that the leaders of these institutes should be the most brilliant scholars Germany had; quite a few Jews were selected. They would provide the best guarantee for maintaining the German leadership in this field and thereby greatly increase the prestige of the Reich. This was for Wilhelm a compelling argument. It is thus not surprising that many people considered the kaiser to be fundamentally liberal and attributed his occasional anti-Semitic outbursts to the bad influence of his entourage, especially to those of his friends who were known to be strongly anti-Semitic. It may be mentioned that even among the members of the nobility many had not the slightest hesitation to have Jews as their personal physicians, bankers, or advisers, despite their anti-Semitism.

Finally, the kaiser would not tolerate any violent eruption of anti-Semitism. Law and order were untouchable, holy principles in his country. Moreover, he knew from his many visits to his grandmother Queen Victoria how barbaric such riots and manifestations were considered in England and how detested they were. He would never permit any activities incompatible with Germany's dignity and reputation as a highly civilized nation. A more detailed description and analysis of the kaiser's attitude toward the Jews and his reactions in exile after the defeat of Germany and his flight to Holland in 1918 may be found in the article by Cecil (1976); the latter period is not pertinent to this work.

The protection of life and property and constitutional rights fostered by Bismarck and Wilhelm II permitted Jews in Germany to improve their position gradually but steadily, especially in the latter part of the nineteenth century. Actually their position became much better than in Austria, where Emperor Joseph II had inaugurated an era of tolerance with an "edict of tolerance" in 1782, long before the emancipation in Germany. In France, which had only a small Jewish population, the emancipation began in 1806 under Napoleon. His attitude toward Jews varied greatly at different periods and he brought as much evil as good to the Jews. Although Jews formed a small fraction of the population, anti-Semitic feelings were strong, especially in the lower middle class. They came to the surface during the famous Dreyfus affair. It was the shock of this eruption that produced a strong reaction in Theodor Herzl.* To him France symbolized real freedom and human dignity and rights; he had believed it to be free of anti-Semitism. Under the impact of the Dreyfus affair he wrote *Der Judenstaat* and in 1897 at the first Zionist Congress in Bâle, initiated the political organization of the Zionist movement for establishing a Jewish homeland in Palestine. He thereby became the uncontested founder of Zionism.

The specific question in the context of this book is, however, that of the factors permitting the Jews in Germany to play an important role in academic life and to become instrumental in the rapid rise of science in the Wilhelmian era and later in the Weimar Republic until Hitler. Even before the emancipation there were a small number of Jews in comfortable and privileged positions. These Jews were under the special protection of the rulers of some of the many little states into which Germany was divided; they were useful to the rulers because of their intelligence, skill, and knowledge. For example, those who had international connections in the banking business sometimes helped rulers get badly needed money in tight situations. The sons of these Jews were permitted to study at universities. An interesting figure is the number of Jewish students in German universities during the eighteenth century: From 1710 to 1800 there were 430 Jewish students registered in a survey covering most German universities (Richarz, 1974). The number of students in all of Germany during the year 1796 was about 6000.

After the emancipation, in the nineteenth century, the economic situation of the Jews and the opportunities open to them began to improve. It has been a frequent experience in Jewish history that, when a more liberal attitude permitted Jews to develop in a country in which they previously had virtually no rights, many of the traditional positions were occupied and therefore closed to them. They had to find new types of activities in which the old social structure had not erected prohibitive barriers. An expanding and increasingly diversified economy, as was the case in Western Europe and particularly in Germany in the second part of the nineteenth century, greatly helped their economic rise. Several factors played an essential role. One such factor was the rapid growth of big cities, which created many new opportunities and new kinds of positions in a variety of fields. Such favorable circumstances rarely existed in small villages, many of them forbidden for Jews. Growing cities offered talented, intelligent Jews, with their willpower and drive for better living conditions, an unexpected chance.

---

* Theodor Herzl (1860–1904): A Viennese journalist and a reporter for the *Neue Freie Presse* in Paris from 1891–1896.

A second factor, perhaps even more important, was the rapid development of the capitalistic and industrial economy around the middle of the century (Mosse, 1976). The whole economy changed completely; it brought new forms in the life of the population, creating a new elite and new ideologies. A new upper middle class in comfortable economic positions began to emerge, whereas the previously dominant nobility, the Junkers, as well as the craftsmen, the lower middle class, the peasants, and the landowners, lost many of their privileges. The assimilation of the Jews, and the baptism of many, was facilitated, although those who had suffered because of the new and inevitable developments became bitter opponents of the Jewish intrusion into the economy. They considered the Jews, rather than the economic situation, responsible for their misfortune and kept alive strong anti-Semitic feelings. It must be stressed that at this time German industry was still far behind that of France and England. But in the many newly created and developing industries of the mid-nineteenth century, Jews played an important role; the number of Jews who became wealthy by their role in these industries was about 5%, a remarkably high percentage compared with their percentage of the total population. When in the *Kaiserreich* German industry started its rapid rise and Germany became the leading industrial power in Europe, the percentage of Jews in high positions rose at a fast rate. In 1910 the number of Jewish directors in the most important industries was about 13% (108 out of a total of 808). In the boards of directors (*Aufsichtsräten*), where bankers played an important role, the percentage was higher, about 25%. Quite a few of the Jews in Berlin, in the era of Wilhelm, belonged to the richest families, with fortunes of about 20 million marks or more.

The third factor was the vast expansion of the universities and *Technischen Hochschulen*. The research activities in these institutions, crucial to the growth of industry and commerce, demanded well-trained and competent people. Jews were well equipped to move into the positions available. Their traditional quest for learning and scholarship goes back to the destruction of the Second Temple, that is, it had been going on for two millenia. The continuous fight for survival in the Diaspora over the centuries forced them to use their mental abilities to obtain high quality in whatever they did. Another important element may have been genetic endowment, favored by a selective process. There was a long tradition of cultured families with highly talented and scholarly sons trying to find equally gifted marriage partners for these sons.

A few figures may be given to illustrate the participation of Jews in the growth of German universities. The expansion was slow until 1870. For example, the number of students in *all* German universities increased from 6000 in 1796 to about 11,000 in 1850. But after 1870 the increase was rapid. In 1810, when the University of Berlin was founded, there were 252 registered students. In 1906 there were 5000 students registered for the winter semester and 4000 registered for the summer semester.

About 10% of the students were Jews. One has to keep in mind that a
large fraction of German Jews, about one-third, lived at that time in
Berlin. When the Nazis came to power, there were 160,000 nonbaptized
Jews living in Berlin, out of a total of 540,000 in all of Germany. There
were large Jewish communities in other big cities, particularly in Breslau,
Frankfort on the Main, and Hamburg. During the *Kaiserreich* Jewish
students made up 7–8% of the total student body in most universities.
This is a remarkably high percentage since Jews constituted not quite 1%
of the total population. Almost 70% of all German Jews lived in Prussia.

It is, therefore, not surprising that the number of Jews in academic
professions was very high. In 1907, 6% of all German physicians and
dentists, 14% of lawyers, and 8% of private scholars, journalists and
writers were Jews (Hamburger, 1968). This high percentage becomes
even more significant in the light of their concentration in a few big cities,
particularly in Berlin. When the Nazis came to power, almost half of the
6000 physicians in Berlin were Jews. In contrast to the large proportion in
free academic professions, the number of Jewish faculty members at the
universities was comparatively small. In 1910, there were about 200
Jewish faculty members, half of them in the faculty of medicine. The
promotion to the rank of *Ordinarius** was for Jews quite difficult, despite
their sometimes extraordinary brilliance. Resistance came primarily from
the faculties themselves. Anti-Jewish feelings were quite strong in some
fields of humanities and in law, not as strong in medicine, and least pronounced in science. The fight of Heinrich von Treitschke against the
"bad" effects of Jewish influence on the German destiny is well known.
His atrocious outbursts against the Jews and his efforts to incite students
met with strong opposition from liberal professors, such as Rudolf
Virchow, Emil Du Bois-Reymond, and Theodor Mommsen. But actually
he expressed the feelings of many conservative and reactionary elements
in the universities. Between 1882 and 1909 there were at any given time
about 20–25 Jews who had the position of *Ordinarius*. There were, however, many alternative opportunities for Jews in medicine and science
(*Naturwissenschaften*) for pursuing creative research activities in private,
industrial, or semiprivate institutions. The foundation of the Kaiser
Wilhelm Society, in 1911, opened a new era. Many Jews became directors
of superbly equipped institutes. The fact that about 1150 Jewish scientists,

---

* *Ordinarius* is the chairman of a department. At the early period of this century he
had a great deal of power to run the department as he wanted. *Privatdozent* was a
title, granted after a dissertation, associated with the *venia legendi* (license to give
students lectures), usually an unpaid position. *Extraordinarius* was sometimes a paid
position, but with limited influence in the department compared to that of the chairman. To compare these ranks with assistant, associate, and full professorship in
American universities, as is sometimes done, is misleading in view of the completely
different German university system and the role of the faculty of the period
described.

many of them of great brilliance and in high positions, left Germany between 1933 and 1935 is an indication of the important role played by Jews in German science before the rise of Hitler.

The acceptance of Jews in the universities was greatly facilitated by their assimilation to German civilization. At the turn of this century many Jews had lost almost all connections with Jewish tradition and the Jewish community. Many of them considered themselves to be *Deutsche Staatsbürger jüdischen Glaubens* (German citizen of Jewish faith). They founded the *Central Verein*, which was actually an expression of resistance to fight discrimination against Jews and to struggle for complete equality. But the Jewish religion actually had little meaning for many of them. Similar education and cultural experiences favored, at least in some groups, close contacts and friendships with non-Jews, especially in the educated middle class. Intermarriages became frequent and accelerated the loosening of the ties to the Jewish community. The friendships developed in the scientific community were so strong that many of them continued even in the Nazi period and led to many tragedies even among Christian scientists. This particular aspect will be discussed later.

Even in the relatively liberal Wilhelmian era, a period of great prosperity and relative affluence, the freedom that Jews enjoyed in the academic and free professions was not extended to all fields. Even baptized Jews were admitted to public office and civil service in very limited and insignificant numbers. They were virtually excluded from the government and the army. The very few exceptions were meaningless in the context of the whole picture. In these fields the situation began to change only in World War I, when many Jews fought in the army; about 2000 became officers, although none were of high rank. Fritz Haber, as will be seen later, played an essential role in mobilizing available materials for the war and producing many new ones, and the General Staff, reluctantly and with great resentment, was obliged to accept his advice. In the short-lived Weimar Republic, seven Jews actually held cabinet-level positions.

To illustrate the role of the Jews in the Weimar Republic at cabinet level, two of these men will be mentioned here. The first state secretary of the interior (*Innenminister*) was Hugo Preuss (1860–1925), appointed by Friedrich Ebert in November 1918 (Hamburger, 1975), although Ebert knew that Preuss was not a Social Democrat. Preuss was an extraordinarily brilliant scholar, with an outstanding record of books published on public law; he also had a keen sense of history. He became *Privatdozent* at the University of Berlin at the age of 29, but the strong anti-Semitism in the faculty of law there kept him from reaching the rank of Professor. In 1906 he was appointed professor at the *Handelshochschule* (college of commerce), where several Jewish scholars, blocked by anti-Semitism at the university, had an opportunity to teach. Preuss belonged to the left wing of the liberals. He is generally referred to as the father of the Weimar Constitution. But he resigned his position in June 1919, in protest

against the Treaty of Versailles. Preuss was married to Else, *née* Liebermann, of a prominent and wealthy Jewish family. Her father was the
outstanding chemist Karl Liebermann, a teacher of Fritz Haber (see IV,
A). Max Liebermann, one of Germany's outstanding painters, belonged
to the same family, which was also related to the Rathenaus. Preuss was
a forceful and courageous fighter for Jewish rights and repeatedly became
personally involved in cases of discrimination against Jews. In the last
years of his life he even took part in Jewish life and as a demonstration
he joined, in 1924, the management of the *Akademie für die Wissenschaft
des Judentums* (Academy of the Science of Judaism). He had a genuine
sense of belonging to the German nation, to which he was deeply and
genuinely devoted. He did not see any contradiction in his attitude because he was a Jew. He considered anti-Semitism, as so many other Jews
did, lunacy and was convinced that it would disappear with the progress
of civilization and in a society in which progressive public laws prevailed.

A second Jewish cabinet member was Walther Rathenau (1867–1922)
who became foreign secretary (*Aussenminister*) in 1922. He was the son
of Emil Rathenau, the founder of the Allgemeine Elektrizitäts-Gesellschaft
(A.E.G.) (Schulin, 1976). Both father and son were extraordinarily gifted
and cultured individuals; both played an important role in the development
of large-scale industry in Germany. Emil had started with the foundation
of a machine factory, influenced apparently by his grandfather, a member
of the famous Liebermann family, who was associated with a machine
factory in Silesia. He was not so much a technical inventor or banker as
a brilliant and efficient organizer. After the foundation of the A.E.G. he
facilitated the introduction of electricity in Germany in many fields; the
role of the A.E.G. was comparable in magnitude to that of the huge
Siemens concern (the firms of Siemens and Halske, and Siemens and
Schuckert). It was a period of rapid expansion of the use of electricity
for a great variety of applications. Thanks to these two giant industrial
empires Germany became at the turn of the century the leading country
in this field, exceeding not only England and France, but even the United
States. The A.E.G. built large power stations in many countries in Europe
and all over the world.

Emil's son Walther, who had studied mathematics, physics, chemistry,
and philosophy and had obtained a Ph.D., was a genuine intellectual with
a great variety of interests. He had many friends among the well-known
writers of that period. His intuition and vision were extraordinary. He
realized very early that capitalism in its then prevailing form would
eventually have to be replaced by a planned economy. These views
created many dilemmas and minor conflicts that led to disagreements with
his father about the management of the A.E.G. and its policy, however
in a critical moment he supported his father strongly. He was greatly
interested in politics. Before the war he had the vision of a European
economic community, many decades before it became a reality.

A few days after the outbreak of the war in August 1914 he became director of the distribution of raw materials for the war effort, and carried out this function with extraordinary ingenuity. He left the war ministry in July 1915, when his father died, and became the president of the A.E.G. In the Weimar Republic he was one of the experts in the negotiations in Spa, in 1921, and later was a delegate to the meetings in London and Cannes. He became foreign secretary in February 1922 and signed the first treaty with the Soviet Union after the war in Rapallo. He was particularly hated by right-wing elements; for them he was one of the worst Jewish traitors. He was assassinated in July 1922.

Walther Rathenau was never baptized. His attitude toward both ethnic groups was torn by many conflicts. He had a deep and genuine love for Germany, but he also recognized many of the weaknesses of German society and of certain characteristics of the German people. He struggled for an answer to his Jewishness and thought that the answer might be either the Zionist idea or total assimilation. When he was foreign secretary he was visited, in April 1922, by one of the prominent Zionist leaders in Germany, Kurt Blumenfeld, together with Albert Einstein. They implored Rathenau to give up his function as foreign secretary; they believed that a Jew should not be the representative of Germany under the humiliating condition in which Germany found itself. Blumenfeld told him: "Postminister dürften Sie sein, aber nicht Aussenminister," meaning that helping Germany as an expert in a field was not comparable to acting, in the exposed position of foreign minister, as the representative of the German nation who was forced to accept continuously the disastrous requests of the Allies, designed to weaken and humiliate Germany. Blumenfeld and Einstein believed that the situation would inevitably lead to a disaster for Rathenau.

But all these aspects are marginal to the main topic of this book. More important is the strong influence of Jews in the Weimar Republic in all cultural activities and in the universities. Their influence was most pronounced in performing arts (theater, opera, music, etc.), but it was also strong in literature, painting, architecture, journalism, and still other fields. Germany experienced a real renaissance in the second part of the 1920s. The lost war, the revolution, the social and economic upheavals apparently stimulated many creative forces in all fields. Berlin became an exciting city for all intellectuals. Many visitors from abroad, particularly from the United States, considered it the spiritual capital of the world. The brilliant contributions of Jews to many of these developments were remarkable. One has to keep in mind that Jews made up 4.3% of Berlin's population; they made up only 0.9% of the population of Germany (Leschnitzer, 1954, 1956).

Unfortunately, this success blinded the Jews and seduced them into ignoring the ominous signs on the wall. The distastrous war, the defeat of 1918, produced a violent anti-Semitism, based on racist grounds rather than on primarily religious grounds as before. Most Jews still considered anti-Semitism a passing phenomenon that would subside, as hostility toward them previously had, with general economic and political improvement. They believed that the high standard of German civilization would not permit a return of the barbaric events of the Middle Ages. The Zionist movement, which had started at the end of the nineteenth century, was joined by a very small fraction of German Jewry; it was vigorously rejected by the vast majority. In the early 1930s, when the strength of the Nazi movement was already quite apparent, there were hardly 10,000 Jews in Germany, of 540,000, who actively supported or even payed lip service to Zionism. But one has to remember that even many highly educated Germans never took the Nazi movement seriously. When they began to realize its disastrous impact, it was too late for any serious opposition. After April 1933, Hitler had absolute dictatorial power and every attempt at opposition was ruthlessly crushed. Few people outside Germany were aware of the brutality of the terror, organized to perfection and applied against the slightest signs of opposition, using assassinations, executions, and concentration camps. It was the tragedy of Europe that even in France, and still less in England, leaders did not recognize the real significance of the events in Germany in time.

II

Atomic Physics in the
Early Twentieth Century

During the eighteenth century France and Great Britain led the world in science. In the nineteenth century marked fluctuations in scientific leadership took place. In the first decades, despite the setbacks related to the French Revolution, France was the leading country in science, particularly in chemistry. In the middle of the century Great Britain took the lead. Then, toward the end of the century, particularly in the last three decades, leadership passed to Germany. Some of the developments in science that took place in Germany, especially in the last part of the century, will be mentioned first. These developments set the stage for the rapid advances made in the twentieth century.

A. DEVELOPMENTS IN PHYSICS IN THE
LATE NINETEENTH CENTURY

The University of Berlin, founded in 1810, was essentially concerned with humanities and philosophy during its first two decades. Physics was not taught until 1834, when Heinrich Gustav Magnus (1802–1870) became *Extraordinarius* of physics; in 1845 he became the first *Ordinarius* of physics. Magnus was primarily interested in ballistics, but he contributed to other fields, such as physiology. There was no institute, and both physics and chemistry were taught in a very rudimentary form. In the 1860s a few laboratories were opened to students. Magnus was a baptized Jew, a member of a distinguished Jewish family. His older brother Eduard was a gifted and popular painter of the Berlin Biedermeier (early Victorian period) and painted many portraits of celebrities. Some other noteworthy members of the family were the outstanding pharmacologist Rudolf Magnus (1873–1927), who was *Ordinarius* in Utrecht at the time of his death; and Paul Wilhelm Magnus, *Extraordinarius* of botany at the University of Berlin; and Adolf Magnus-Levy (1865–1955), a distinguished physician who made many brilliant contributions to medicine, particularly in the field of metabolism. Magnus-Levy left Hitler's Germany for the United States, where he remained until his death.

The situation changed dramatically in 1870, when Hermann von Helmholtz (1821–1894) accepted the chair of physics at the University of Berlin on the condition that the university would build an institute for him. Helmholtz had initially been a physiologist. His main interest had been the problem of sensory perception. His contributions to the physiology of the eye and the ear are classics and represented a major breakthrough. Before going to Berlin he had been professor of physiology in Königsberg, Bonn, and Heidelberg. His interest shifted to physics, to which he made many contributions; his fame in history is particularly associated with his fundamental work on the first law of thermodynamics, the science that relates heat to other forms of energy. This law was first proposed in 1842 by Robert Mayer (1814–1878), a physician from Heilbronn. It was experimentally supported by the eminent British physicist James Prescott Joule (1818–1889) by precise determinations of the mechanical heat equivalent. However, the formulations of Helmholtz surpass by far those of his predecessors in clarity and depth.

Helmholtz had a powerful personality. He had a brilliant mind, and a universality of knowledge that was amazing and covered all fields of science of his time. He was a poor lecturer, but his brilliance attracted some of the most outstanding physicists and their contributions were landmarks in the development of physics. Among them was Ludwig Boltzmann (1844–1906), a native of Vienna, whose interest in mathematics led him to introduce revolutionary methods of statistics into physics. The importance of these methods became increasingly obvious with time; he and James Clerk Maxwell (1831–1879), an eminent British physicist, are considered to be the founders of statistical mechanics. Another brilliant pupil of Helmholtz was Wilhelm Wien (1864–1928), whose work on radiation laws laid the groundwork for Planck's work on heat radiation and thus led to the development of quantum theory (see p. 27). Albert A. Michelson (1852–1931) was another of Helmholtz's pupils; he was a Polish Jew who had emigrated to the United States. His experiments determined with great precision the velocity of light, for which he received the Nobel Prize in 1907. His work (with E. W. Morley) eliminated the notion of the ether and was essential to Einstein's theory of relativity. The work of yet another pupil, Heinrich Rudolf Hertz (1857–1894), was of great theoretical importance and initiated startling developments in technology. Hertz was a member of a prominent Jewish family in Hamburg; his father was a lawyer. Hertz joined Helmholtz in the 1870s and received his Ph.D. with him. Helmholtz was greatly impressed by Hertz's intelligence and suggested that Hertz look for experimental evidence for the existence of electromagnetic waves. Maxwell had suggested that events such as electric discharges generated electromagnetic waves that spread in all directions with the velocity of light, but he did not pursue the idea experimentally. In 1887, after several years of work, Hertz succeeded in demonstrating the existence of these waves experimentally. He was at that time professor at the *Technische Hochschule* in Karlsruhe.

Hertz's discovery had a revolutionary impact on communications. Hertz, who died at the age of 37, had not foreseen it. It took the work of several outstanding physicists and engineers for the new communication systems, including radio, to be developed. One of the chief architects was the Italian physicist Guglielmo Marconi (1874–1937), whose work led to the first wireless communication between Europe and America. He received the Nobel Prize for this work in 1909. Many ingenious technical and engineering problems had to be solved before the complex machines that today permit the transmission of such sounds as human voices and music could be developed. One of the outstanding pioneers was the American physicist and engineer Lee De Forest. But the first major breakthrough opening the whole field was Hertz's discovery.

Helmholtz's success in attracting brilliant pupils to his institute was not the only factor that transformed Berlin into an important center of physics. He also insisted that Gustav Robert Kirchhoff (1824–1887), who had the chair of theoretical physics in Heidelberg at the time when Helmholtz was there, be offered a newly created chair of theoretical physics in Berlin; Kirchhoff moved to Berlin in 1874. One of Kirchhoff's most outstanding contributions, important for many further advances, was the development of spectrum analysis, which permits one to identify substances by measuring their wavelengths. This work was performed in collaboration with the eminent chemist Robert Wilhelm Bunsen (1811–1899) in Heidelberg. After Kirchhoff's death in 1887 Max Planck was chosen, on Helmholtz's suggestion, as his successor; thus, in 1889, one of the titans destined to revolutionize science was brought to Berlin.

Helmholtz and his wife, Anna, *née* von Mohl, had a splendid and elegant official residence and entertained in great style. Their parties brought many intellectuals, scientists, artists, and leaders of both government and industry together. They not only contributed to the splendor and new spirit of the rapidly growing capital of the *Kaiserreich*, but also helped to promote an interest in science and its importance for society. Helmholtz's daughter Ellen married the son of the great electroengineer Werner von Siemens (1816–1892). Siemens gave the government half a million marks for the building of the *Physikalisch-Technische Reichsanstalt* (Federal Institute for Physics and Technology), of which Helmholtz became the director without giving up the chair of physics. He held both positions until his death in 1894.

The field of physics grew rapidly in other German universities. One of the most eminent physicists outside of Berlin was Wilhelm Conrad Röntgen (1845–1923). He made a variety of important and widely recognized contributions, but his worldwide fame is based on his discovery of X-rays in 1895. The importance of X-rays for such fields as physics, chemistry, biochemistry, and medicine does not need any elaboration. In Zurich Röntgen had met August Kundt (1839–1894), who became his teacher in physics. When Kundt became *Ordinarius* in Strasbourg, he arranged for a second chair of physics to be established there for Röntgen a few years later. Röntgen was by that time internationally renowned. He subsequently held the position of *Ordinarius* at Giessen, at Würzburg as successor of Kohlrausch, and in 1900, at Munich. He was the recipient of the first Nobel Prize in physics in 1901.

In the era before the *Kaiserreich*, Munich was actually ahead of Berlin in both physics and chemistry (see p. 30). Georg Simon Ohm (1787–1854) became professor of physics in Munich in 1849. He is well known for his work on electricity and especially for his formulation of the laws on the relationships between resistance, strength of current, and potential difference. Some of his work still forms part of textbook physics. He received the Copley Medal of the Royal Society in 1841, at that time about the equivalent of the Nobel Prize today. Würzburg too had a distinguished physics department, first under Kundt and then under his successor Friedrich Wilhelm Kohlrausch (1840–1910), who joined the university in 1875 and took over the physics chair in 1888. Kohlrausch's main research was devoted to problems of electrical and magnetic measurements. Among his most important achievements are conduction measurements in electrolyte solutions. He attracted many students, quite a few of them from the United States. Among his most outstanding students were Ostwald, van't Hoff, and Nernst, who elaborated several of Kohlrausch's theories.

Wilhelm Ostwald (1853–1932) was a brilliant physicist and chemist; he is considered one of the founders of physical chemistry. He was a dynamic person with an incredible vitality and a wide range of interests: physical chemistry, thermodynamics, color perception, to name a few. He was interested in philosophy; he was a painter and a musician. In 1887 he took over the only existing chair in the field of physical chemistry, in Leipzig; Ostwald soon made his institute the world's leading center. Among the great number of students whom he attracted were Arrhenius, van't Hoff, and Nernst. His contributions in the field of catalysis were vital to the growth of industry and earned him the Nobel Prize in 1909.

Walther Nernst (1864–1941) had been a student of Helmholtz and Boltzmann in thermodynamics. In Kohlrausch's laboratory, where he received his doctorate, he met Ostwald, who deeply impressed him; Nernst later joined Ostwald in Leipzig. In his first paper there Nernst brilliantly combined Boltzmann's atomistic ideas with the laws of thermodynamics developed by Helmholtz and Clausius (see below). Nernst recognized that electrochemistry offered the possibility of integrating the two lines of thought. One of the most important quantities of chemical reactions is their *free energy*, which is the amount that may be used to perform work, as distinct from the *total heat* that can be derived from a reaction. This difference plays a decisive role in many physical and biological reactions and will be discussed later in detail. The quantities involved can be measured as mechanical work, heat, and temperature without taking into consideration the underlying atomic processes. Electric currents create work and heat and thus permit the application of thermodynamic calculations, but they also permit the use of atomic concepts, based on the behavior of ions. Nernst's approach and experimental work resulted in the first theory of the galvanic cell. This was an outstanding achievement of far-reaching significance, as was widely recognized, and he was soon considered one of the leaders in physical chemistry.

In 1890 Nernst moved to Göttingen, attracted by an offer from Riecke, the professor of physics there. By 1894 he had become the head of a new institute of physical chemistry in Göttingen, a chair created especially for him. He therefore declined the chair of physics in Munich offered to him as successor to Boltzmann. This development was brought about by Friedrich Althoff, permanent secretary of the Prussian Ministry of Education from 1882 to 1907. Althoff was extremely autocratic, an enlightened and benevolent dictator who frequently went over the head of faculty members. His primary aim was to create great international centers in Berlin and Göttingen, and because of his determination, his excellent judgment, and his deep devotion, he succeeded beyond all expectations. He kept files on every promising scientist. It was on his initiative that three chairs of mathematics were established in Göttingen in the 1890s—one for Felix Klein, one for David Hilbert, and one for Hermann Minkowski (see II, C)—an action that turned out to be decisive for the glory of Göttingen early in this century. Nernst's new institute offered him superb opportunities to develop his genius. He soon became one of the most influential personalities in German science. Eleven years later he moved to the University of Berlin, as will be described later.

Another outstanding physicist, Rudolf Clausius (1822–1888), was a pupil of Magnus and became professor of physics in Berlin in the Royal School of Engineering. From Berlin he moved to Zurich, where he took over the chair of physics, and then to Bonn, where he held the same position. Clausius proposed, in 1850, a fundamental principle known as the *second law of thermodynamics*. The first law of thermodynamics deals with the conservation of energy and the equivalence of all its forms, but it does not say anything about the transformations from one form into another one. The observations made by French physicist and engineer

Nicolas L. S. Carnot (1796–1832) while working on problems of steam engines made it clear that mechanical energy could be completely transformed into heat, but the opposite process was impossible. According to the second law proposed by Clausius not all types of energy transformations are possible and even the possible ones take place only under certain conditions; the law thus defines the direction of energy transformations. The second law has universal validity and can be applied to many physical and chemical processes, including those taking place in living cells; however its generalized formalism does not reveal the various physical (and chemical) processes involved. Helmholtz and Josiah Willard Gibbs (1839–1903)—independently and actually before Helmholtz—in the United States had already distinguished between free energy, that which can be used, and total heat that can be released, but which is wasted. In 1865, Clausius also introduced the notion of entropy. Boltzmann, in his statistical interpretation of thermodynamics, considered entropy a measure of disorder on the atomic scale. The notion of entropy became increasingly important in the subsequent developments of thermodynamics. Many scientists have greatly contributed to the elaboration of its significance; some of these aspects will be discussed in the following chapters. It may be mentioned that in the last decades a new type of thermodynamics has been developed, referred to as "non-equilibrium thermodynamics." It is particularly important for biological systems, since living cells and structures are never in equilibrium except in death (see, e.g., Katchalsky and Curran, 1965). These developments took place long after the period described in this book.

There were other distinguished physicists in Germany during the nineteenth century, but the few examples mentioned may suffice to give the reader an idea of the events that led to the spectacular development of physics in Germany in the twentieth century.

B. THE PERIOD BEFORE WORLD WAR I

Atomic physics is probably one of the most difficult fields for a scientifically untrained reader to understand. In the 1920s, at a critical phase in the development of atomic physics, David Hilbert (1862–1943), the famous mathematician in Göttingen (Reid, 1970), made the remark that "physics is becoming too difficult for the physicists." This was a time when mathematical formalisms were found, but there was no satisfactory explanation for the physical phenomena of the atom. Despite the difficulty presented by the subject, it is necessary to give a short summary of some highlights and basic principles of atomic physics. There are two reasons. First, as every layman knows, knowledge of the atom has opened a new era in all fields of science and technology; the importance of this knowledge for the future of the human society is well recognized. Second, the field offers a superb illustration of the profound effects of the mutual stimulation between Germans and Jews in science. It was a small but tightly knit group of brilliant physicists, a galaxy of geniuses, that created and advanced this field, making Germany the center of a revolution in physics.

As mentioned before, the existence of electromagnetic waves, postulated by James Clerk Maxwell, had been demonstrated experimentally by Heinrich Hertz (p. 22). The types of electromagnetic waves are distinguished by wavelength. Radio waves are very short, heat waves, or infrared rays, are much longer waves, only slightly longer than the visible-light waves, those that excite the retina of the eye. Light waves are about 400–720 nanometers in length (a nanometer, 10^{-9}, is a billionth of a meter). The shortest light waves are about 4000 times as long as X-rays, and 40,000 times as long as the waves of the gamma particles emitted from radium.

The nature of heat radiation had been explored late in the nineteenth century, by Ludwig Boltzmann (1844–1906); Wilhelm Wien (1864–1928), who received the Nobel Prize in 1911; and John W. S. Rayleigh (1842–1919), among others. Much pertinent information was gleaned, but there were many unexplained contradictions and serious difficulties.

In 1895, following up some observations made by Wien, Max Planck started to reexamine the problem. Much of the preceding work had relied heavily on thermodynamic considerations. Planck realized that it was necessary to study how matter emits and absorbs heat radiation. Thus he shifted his emphasis from heat radiation per se to the radiating atom. He believed that this change of direction might facilitate the interpretation of the empirical facts. During this work, he made some observations that led to a completely new and revolutionary concept: *Matter emits radiant energy only in discrete bursts.* He reported his observations in a famous lecture given in December 1900, which is generally considered to be the beginning of modern atomic physics. He proposed the startling view that matter emits radiant energy in finite quantities proportional to the frequency of the radiation, v, which is the number of waves emitted per second. The proportionality factor is a universal constant, having the dimension of a mechanical action; it is the celebrated constant h of Planck, called the *elementary quantum of action.* The energy is thus equal to hv. The shorter the length of a wave, the higher the frequency. Therefore the quantum of energy grows larger as wavelength decreases. A quantum of ultraviolet light contains more energy than a quantum of red light; a quantum of X-rays has even more. The profoundly revolutionary character expressed by the quantum of action, introduced by Planck in a stroke of genius, is now generally recognized as one of the greatest landmarks in the history of science. It transformed physics in this century perhaps more than any other single notion. The Planck constant—the energy of a single quantum of action—is unimaginably small, 6.6×10^{-27} erg second (an erg is a unit of work). By comparison, the common 100-watt incandescent lamp emits 300 trillion quanta every second.

Planck's observations made obsolete the notion of the continuum in nature, a notion that had dominated science since Aristotle. (*Natura non facit saltus*—"nature makes no leaps.") Isaac Newton had considered this possibility for some time, but finally abandoned it. It has been said that Planck did not recognize the revolutionary character of his experiments. Such a view is incorrect; Planck stressed the importance of his observations for physics and chemistry. Werner Heisenberg (1901–1976) once reported that Planck told his son Erwin, during a walk in the Grunewald (a forest on the outskirts of Berlin), that he had made observations that were either complete nonsense or the greatest discovery since Newton. However, Planck did not pursue the consequences of his discovery. He was a conservative man, and he tried unsuccessfully to reconcile the facts with classical theory for many years.

The open break with classical physics came several years later, in 1905, when Albert Einstein published an article in the *Annals of Physics*. Einstein, at that time an unknown clerk in the patent office in Berne, presented an uncompromising corpuscular theory of light. He regarded light as consisting of quanta, small corpuscles, during both emission and absorption. He based his concept on observations generally called the *photoelectric effect*: Light waves that hit a metal surface eject electrons, which are present in extremely large quantities in every metal. Electrons are very small negatively charged particles present in atoms; they form only 1/1840 of the mass of the smallest atom, that of hydrogen. Their existence was definitely established, in 1897, by the brilliant English physicist Joseph John Thomson (1856–1940) using a cathode-ray tube, a forerunner of today's oscillograph. Electrons are present in large quantities in every metal and move easily and rapidly; they carry, for example, the electricity in a copper wire. Einstein found that red light was not effective; only violet light (or in some metals ultraviolet) was able to eject electrons, since the energy quanta are larger than those of red light. Each quantum of light ejects one electron from the metal. But the most spectacular result was that the constant of the photoelectric effect was exactly the same as the Planck constant. Einstein fully recognized the tremendous implications of his conception of light on the atomic scale, in contrast to Newtonian physics, which is based on large-scale phenomena and remains correct in this restricted sense. He went so far as to describe the energy quanta as individual *Lichtpfeile*, darts of light. He did not hesitate to express his view that this was a break with classical physics, just as he did when he first came out with revolutionary theory of the relativity of time and space, considered to be absolute values in Newtonian physics. There are two phenomena that are not explainable if light is considered to consist of quanta: diffraction* and interference†. Einstein was unable to dispute the contradiction between this wave picture supported by these two phenomena and the idea of the light quanta. He did not even try to explain this inconsistency. He just assumed that an explanation for the contradiction would someday be found.

It was for his work on the photoelectric effect that Einstein received the Nobel Prize in 1921. He had published in the same year, 1905, two other revolutionary articles in the *Annals of Physics*: one on his special theory of relativity, and the other on Brownian movements (movements of small particles due to thermal motion). His postulates in both papers were fully confirmed. However, the work of Einstein, although it forms an integral part of the revolutionary impact of German-Jewish scientists on the twentieth century, will not be discussed here for reasons stated in the Preface. He is only mentioned when essential for the understanding of the progress described.

Planck greatly admired Einstein's audacity in drawing firmly the revolutionary consequences of the notion of the quantum. But he remained silent for quite a long period; he was still struggling to find a compromise with classical physics and for years he maintained objections, hoping to avoid a complete break. In contrast, he and Arnold Sommerfeld (1868–1951), professor of theoretical physics in Munich, immediately recognized the brilliance and revolutionary character of Einstein's theory of relativity. They were the first to endorse it publicly and became its chief promoters. They pronounced it as being probably the greatest event in physics since Isaac Newton. Very soon the theory was also enthusiastically endorsed by Hermann Minkowski (1864–1909), a Russian-Jewish mathematician who had received his education in Germany and was at that time, in 1908, professor at the University of Göttingen (see II, C). He had been one of Einstein's teachers in Zurich. Although many physicists still objected to the theory of relativity, Einstein's reputation began to grow rapidly. In 1909 Planck and Einstein met for the first time at a physics congress in Salzburg. They met again two years later, at a small meeting at a Solvay Conference in Brussels, meetings where about 20 of the world's most illustrious leaders of physics came together every few years. Planck had long discussions with Einstein and was deeply impressed. He became increasingly interested in bringing Einstein to Berlin. He and Walther Nernst made great efforts to create a suitable position for Einstein. In 1913 they traveled to Zurich to persuade Einstein personally to come

---

* Diffraction is the spread of a wave motion into the region behind an obstacle or around the corner of an obstacle in the path of waves.

† If light really has the nature of a wave, two light waves combined properly ought to produce darkness. This is found to be the case. This fact constitutes a strong argument in support of the wave theory of light. Quantum theory would not explain interference phenomena.

to Berlin and accept the position they offered. Although Einstein was hesitant, they were successful and he agreed. With Einstein there, Berlin unquestionably became one of the world's greatest centers of physics.

Planck and Einstein became close personal friends. Their personalities were quite different in many respects and they frequently disagreed in scientific discussions, but they greatly stimulated and influenced each other. Planck was by nature a conservative; Einstein, a rebel. The title of Banesch Hoffmann's (1972) biography, *Albert Einstein, Creator and Rebel*, superbly characterizes Einstein's two most pertinent features. It is not incidental that Einstein was never reluctant to express his views vigorously, even when they were contrary to the well-established laws of Newtonian physics. In addition to fundamental differences in character, Einstein's Jewish origin and some experiences of his youth may have contributed to the differences in viewpoint. Nevertheless, Planck and Einstein shared a great mutual admiration and a genuine and deep affection. They were passionate lovers of music and frequently played chamber music together. Einstein played the violin, and Planck played the piano. Planck was a superb pianist. After his *Abiturientenexamen** he had a difficult time deciding whether to turn to music or to physics.

Einstein's life and personality are well known, but it may be appropriate to describe Planck's background briefly. He was born in Kiel, the scion of a distinguished family that can be traced to the sixteenth century. Among his ancestors were quite a few exceedingly brilliant persons. If one accepts the view that genetic endowment is an important factor in the determination of a person's gifts and creative abilities, as well as other traits, Planck offers a good support for it. It is said that he was strikingly similar to his great-grandfather Gottlieb Jacob Planck (1751–1833). A professor of theology at the University of Göttingen, Gottlieb Jacob was a brilliant scholar and had a wide range of interests. The teachers who had influenced him most were two philosophers who had been pupils of Leibnitz. Gottlieb Jacob's strong interest in mathematics and philosophy was echoed in his great-grandson, as evidenced by Max Planck's passion for theoretical physics and his genuine and deep devotion to the search for understanding all aspects of life in a universal philosophy. Plato had a strong impact on his concepts, theories, and beliefs. His grandfather, Heinrich Ludwig, was a professor of theology at Göttingen; his father, Wilhelm Johann Julius, was a professor of law at the University of Munich, where he died in 1900. There were other outstanding scholars in the family, particularly in law.

---

* Final examinations required when finishing high school. Their passing entitled the student to entrance in a university. There exist no *colleges* on the continent of Europe; the high schools had therefore, of necessity, a higher standard than those in the United States. In the humanistic gymnasium, one spent 9 years learning latin and 6 years learning greek.

The deep religious faith of Max Planck was in line with the family tradition. For him science was no obstacle to strong belief in a divine power controlling the physical world, a power not accessible to any scientific analysis and hidden to the human mind. However, like his great-grandfather, he was far from being rigidly orthodox. He recognized the necessity of complete and absolute freedom in science and research and vigorously rejected all dogmatic views and prejudices. Tolerance, fairness, and justice to opposing views, understanding and kindheartedness, all were integral parts of his personality. He never assumed others to have bad motives before he had evidence. His actions toward Boltzmann reveal much about his character. He at first disliked Boltzmann's statistical approach, and an associate attacked Boltzmann in a paper. When Planck later realized that Boltzmann was correct, he acknowledged his own error and the success of the statistical method. He also immediately wrote a personal letter to Boltzmann. His genius, the lucidity of his mind in the analysis of scientific problems, was thus combined with exceptionally great human qualities.

He was devoted to his country. To him, love of the fatherland was not an abstract notion. It was a deeply felt relationship involving the obligation of all citizens to do their best for the well-being and the future of the country.

Planck had—perhaps because of his background—a great respect for law, in life as well as in science. This may account for his conservative attitude, his reluctance to discard laws that seemed so well established, such as those of Newton. Although it took some time, he did eventually overcome his reluctance, and he accepted new and revolutionary concepts when he recognized the soundness of the facts supporting them.

In the 1930s, when the Nazis seized power, he was horrified. He was shocked by the persecution of the Jews in general and Jewish scientists—including some of his most admired and beloved friends in particular. He foresaw the disaster the Nazi regime would lead to. For a while he remained silent. When he decided to go to see Hitler, and he dared to explain to him the devastating consequences of the persecution of Jews for science and for the future of Germany, he was thrown out.

It may be that his delay was due to the influence of his son Erwin, who was undersecretary of state in the government of Schleicher, which preceded that of Hitler. Like many Germans (and most people in England and France), Erwin did not believe that Hitler could possibly last for more than a few months. He was visited by Kurt Blumenfeld, the leader of the Zionist Organization of Germany, about eight weeks before Hitler became chancellor. Blumenfeld asked Erwin's advice about precautionary measures the Jewish community could take in the event Hitler did take power. Erwin Planck flatly rejected this idea. He assured Blumenfeld that nothing would happen to the Jews, that Germany was after all a civilized country and a man such as Hitler could not possibly remain in power for more than a few months.

Max Planck had the courage to defend and teach the scientific views of his Jewish friends. In 1934 he dared, in a public lecture, to refer to Einstein as the greatest physicist since Isaac Newton. It was probably because of his age—he was 76—that he was not sent to a concentration camp for this statement, an outrage in Nazi views. In 1935 he arranged a memorial celebration for Fritz Haber (see IV, A) as president of the Kaiser Wilhelm Society, a position Planck held from 1930, after the death of Adolf von Harnack, until 1935. The minister of education, Rust, did not permit university professors to participate. Karl-Friedrich Bonhoeffer, a pupil of Haber, was one of those prevented from participating. Planck arranged the celebration against all obstacles and Otto Hahn (see IV, A) presented the memorial lecture prepared by Bonhoeffer. Planck was invited to Copenhagen, in 1935, so he could breathe the air of freedom for a few days. He refused. He could not travel abroad. Once he had been proud to be a representative of German science, now he felt that he would have to hide his face in shame (James Franck's Introduction, in Planck, 1959).

Max Planck had to flee from Berlin in 1943. In 1944 his house and his magnificent and precious library were totally destroyed by fire. He was then 86 years old. He was warmly received by friends near Magdeburg, but soon the region became a battlefield. He and his wife had again to flee; they were robbed of everything they still had. He was trapped for hours in an air-raid shelter that caved in from a hit. Finally he was rescued by American colleagues. He was bent in pain from arthritis, and they brought him to a hospital in Göttingen.

In 1946, when the Royal Society arranged a Newton celebration in London, Planck was the only German scientist officially invited. This gesture was prompted partly by general admiration for one of the greatest scientists in the history of mankind, the spiritual father of the quantum theory. But it was also a tribute to a man who was a symbol of the other, better Germany, a man of great courage and integrity, an active foe of the Nazi regime. His tragic fate was widely known. Triumph and honors, general recognition of his genius, and the inner satisfaction of achievement —all these were associated with bitter personal tragedies. During World War I he lost his first wife, his oldest son, killed in battle, and his two daughters, who died very young. In the cruel and bitter last years of his life, at a very old age, he saw the destruction of his country. The Nazi regime had driven out his most admired and cherished friends, thereby ruining German science and all the spiritual and ethical values to which his life had been devoted. His only remaining son Erwin, his pride, a strong and active anti-Nazi, was executed in an abominable way. In a stirring letter, Planck spoke of his son as "his closest and best friend."

The unbelievable mental and physical sufferings he endured bring to mind the suffering of Job. When Lise Meitner, his faithful pupil and longtime assistant, met him at the celebration of the Royal Society, she wrote in a letter to Margrethe Bohr (the wife of Niels Bohr): "Planck has grown very old; it was grieving to see him. Even his intellectual vigor sometimes deserts him. But when he was at his best—and he often was—and I had him for myself, one could discuss everything with him. And I was happy to see that his human and personal qualities were wonderful as ever" (quoted from Hermann, 1977).

During the celebration some independent parallel meetings took place in which the future of the Kaiser Wilhelm Society was discussed. The name Kaiser Wilhelm had become unacceptable because of its association with the Prussian military spirit. This was later expressed in letters of Sir Henry Dale and A. V. Hill. Lise Meitner, Niels Bohr, and others expressed the same sentiment to Planck and von Laue. It was decided to call the successor society the Max Planck Society. Max Planck accepted. It was his last service to the society. He died in 1947 at the age of 89.

Einstein was greatly distressed about the trials Planck had to go through during the Nazi period and was deeply affected by the news of his death. At the Max Planck Memorial Service, held in 1948 at the National Academy of Sciences, he recalled that Planck's discovery, his great creative idea, became the basis of all twentieth-century research in physics and had many repercussions. He felt that it was fitting for representatives of all who were striving for truth and knowledge to gather, from all over the globe, for the occasion. "They are here to bear witness that even in these times of ours, when political passion and brute force hang like swords over the anguished and fearful heads of men, the standards of our ideal search for truth is being held aloft undimmed. This ideal, a bond forever uniting scientists of all times and all places, was embodied with rare completeness in Max Planck" (Seelig, 1954). History will consider the work of these two personalities as one of the best illustrations of the magnificent results and the mutual benefits of the collaboration between German and Jewish scientists. These two titans will go into history as the founders of a new age in physics.

Rutherford's Theory of the Atom

A revolutionary development laid the foundation for the understanding of the basic structure of the atom and soon permitted the introduction of quantum theory into atomic physics. The rays given off by the element radium, discovered in 1898 by Pierre Curie (1859–1906) and Marie Curie (1867–1934), were found to consist of three types of particles: alpha (α), beta (β), and gamma (γ) particles. Their separation was due in the first place to the work of Lord Rutherford (1871–1937), but also to that of other investigators. The α particles turned out to be the helium nuclei, the β particles to be electrons; the γ particles were found to resemble those of X-rays discovered by Roentgen. The existence of electrons demonstrated by the English physicist J. J. Thomson has already been mentioned. They had been anticipated by Michael Faraday (1791–1867), and this assumption was further elaborated, in the 1890s, by several physicists, in particular by the great Dutch physicist Hendrik A. Lorentz (1853–1928). Observations made on the ejections of the particles indicated a disintegration of radioactive atoms. They thus contradicted the long-prevailing assumption that atoms are unchanging and eternal (*atomos* in Greek means "indivisible"). Rutherford's work had shown that nature transforms uranium into radium and radium into lead.

But what was then the inner structure of the atom? The first theory of the structure was proposed by Rutherford in 1911 and 1912. He and his associates had found, a few years earlier, that α particles were able to pass through a thin film of gold; some particles were deflected, and some particles bounced back. This was for Rutherford an incredible revelation. It soon suggested to him a picture of the inner structure of the atom.

But before discussing Rutherford's picture of the atom, it may be helpful to recall a few elementary facts concerning the atom. Atoms are extremely small in size; they have a diameter of about 2–5 angstroms. One angstrom is one ten-millionths of a millimeter. Atoms are made up of the negatively charged electrons, just mentioned, and the positively charged protons, established by Rutherford. Protons form the nucleus of the atom. Electrons have only 1/1840 the weight of protons.*

---

* There exist about 100 different types of atoms forming the chemical elements. They differ in size, referred to as atomic number according to the number of protons and free electrons. The smallest atom, or element, is hydrogen, formed by 1 proton and 1 electron. It has, therefore, the atomic number 1. Helium has 2 protons and 2 electrons and the atomic number 2; lithium has 3 protons and 3 electrons and the atomic number 3 and so on. Uranium is one of the largest elements; it has 92 protons and 92 electrons and the atomic number 92. The second feature of an atom is its mass or atomic weight. The mass of an atom depends almost entirely on the nucleus, which consists, in addition to the charged protons, of uncharged neutrons, discovered in 1932 by James Chadwick (1891–1974), a pupil of Rutherford. Since electrons weigh only 1/1840 of a proton, they may be disregarded when considering the weight of an atom. Each proton or neutron contributes 1 unit to the weight of an atom. Therefore the weight of an atom is equal to the total number of protons and neutrons in the nucleus. The normally occurring hydrogen, as already mentioned, consists of 1 proton, therefore it has the atomic number 1; helium has 2 protons and an atomic number 2, but it has in addition 2 neutrons, so its atomic weight is 4. Similarly, oxygen has 8 protons and an atomic number of 8, but also 8 neutrons and an atomic weight of 16. Iron has 26 protons and 30 neutrons; uranium has 92 protons and 146 neutrons (in its most widely occurring form). This aspect will be discussed later (see II, E).

On the basis of his observations, Rutherford came to the truly startling conclusion that the atoms are formed mostly by *empty* space! The electrons circle the nucleus, therefore preventing their fall into the oppositely charged nucleus. In this picture the atom is comparable to the solar system, in which the planets circle around the sun. Although our solar system is also mostly empty space, the empty space is relatively greater in the atom. When Rutherford later bombarded nuclei of nitrogen atoms with α particles (helium nuclei), he obtained oxygen, a result in principle the reverse of those of Henri Becquerel (1852–1908) in the degradation of radioactive uranium. Rutherford was the first to bring about the artificial transformation of one atom into another one.

Bohr's Theory of the Atom

Rutherford's model left many questions open and was unable to explain many experimental results. Einstein had already assumed that in contrast to classical views an atom must undergo a change of structure when it emits light. He realized that the change of structure of the atom would be a quantum change and that quantum physics could provide the information for the understanding of the fundamental laws of atomic structure. He postulated that an atom not only emits radiation in quantized steps, but should be able to absorb energy in the form of single quanta.

In 1912 Rutherford was joined by Niels Bohr (1885–1962), a Danish physicist who had just received his Ph.D. in physics in Copenhagen. Bohr had met Rutherford in Cambridge in the Cavendish Laboratory where he spent four months after he received his doctorate. When Rutherford visited Cambridge, Bohr was so fascinated by his personality and by his ideas that he decided to move to Manchester and to collaborate with Rutherford on the many problems that were raised by the picture of the structure of the atom. One of the important unexplained questions concerned the circling of the electron around the nucleus. According to classical Newtonian mechanics, the electron should gradually lose energy; when its energy was dissipated, the electron would collapse into the nucleus. It was thus obvious that the classical principles were not applicable. Bohr became convinced that the electronic constitution of the atom as proposed by Rutherford, particularly the stability of the electrons, could only be explained by Planck's quantum of action. This startling assumption was confirmed by experiment. Bohr was able to determine the dimensions of the atom and the number of electrons in the outer ring.

However, one of the most puzzling observations that remained unexplained by Rutherford's model was the lines in the spectra produced by an atom. A spectrum is a pattern of the frequencies emitted by the radiation of an atom (or a molecule) in its gaseous form. The atoms of different elements are characterized by special colored bands in the spectrum. No two elements have the same spectrum: The number of bands and their colors vary greatly. The method used for measuring spectra is called *spectroscopy*. It was developed in the nineteenth century with a high degree of perfection by Robert W. Bunsen and Gustav R. Kirchhoff in Heidelberg, and permitted the determination of the different frequencies of emission or absorption of heat and light. Since then spectroscopy has been refined and permits the determination of bands formed by various kinds of radiation, including X-rays, and is applicable to atoms, particles, molecules, reactions between components in the living cell, and a great variety of problems in chemistry and biochemistry.

A Swedish physicist, J. J. Balmer, studying the spectrum of the simplest atom, hydrogen, found that it was made up of only three bands of colors, but they exhibited striking regularities in their frequencies. No one could explain these strange regularities. When an electron circling the nucleus radiates light as it circles, it should give off a continuous range of frequencies rather than well-separated ones as observed by Balmer.

During the study of Balmer's data, Bohr developed a whole new concept of the atomic structure, referred to as *Bohr's atom model*. According to Planck and Einstein, energy is given up not in a continuous stream but in distinctly separate particles. The same principle might apply to the atom; that is, one might have to introduce the Planck constant, the elementary quantum of action. Energy is given off by the atom in the form of light quanta. In applying quantum theory to the explanation of atom structure and behavior, Bohr made a number of assumptions: Electrons circle the nucleus only in certain orbits. As long as an electron remains in the same orbit, the atom does not radiate energy. But when the atom absorbs an energy quantum, the electron jumps from a smaller to a larger orbit. When the electron falls back from a larger to a smaller orbit, it emits a quantum of light. The electrons do not take up an arbitrary orbit but are confined to certain ones, determined by whole multiples of the quantum h. Perhaps the most important feature of Bohr's model was its ability to give a satisfactory account of the special lines of the simplest atom, hydrogen. It was this aspect that may explain the immediate, strong response to Bohr's theory. The theory was more comprehensible to many physicists than the preceding theories; in addition, spectroscopy was a widely used method and therefore its application initiated a huge amount of research. Thus the impact of Bohr's model on the progress of atomic physics was tremendous. Some scientists thought that Bohr considered his model as an established fact. Bohr never regarded it this way. It explained some facts, and the introduction of the quantum in explaining atomic structure was an extremely important new principle, but he was fully aware of the many unsolved problems. It was a new starting point for future investigations, and it would be modified and adjusted when new information was obtained. By means of this model and many other great achievements, to be described later, Bohr transformed this century.

Bohr returned to Copenhagen in 1916 and accepted the chair of theoretical physics there. He and his wife Margrethe, whom he married in 1913 and who apparently was an extraordinary person, remained very close personal friends of the Rutherfords for life. After the war the institute was rebuilt. It soon became one of the world's greatest centers of physics, with illustrious visitors and students from all over the world. Its influence on the history of physics can hardly be overestimated. This will become apparent from the events described later.

A few words may be said about Niels Bohr's background. His father, Christian Bohr, was a distinguished professor of physiology at the University of Copenhagen. His mother, Ellen, *née* Adler, came from a prominent Jewish family; her father was a highly respected banker and financier in Copenhagen. Thus Niels Bohr was a half-Jew. He was very well aware and proud of this Jewish origin. When the Nazis invaded Denmark in 1941, he was according to Nazi ideology a Jew, and his life was in constant danger. He nevertheless stayed in Copenhagen, for reasons to be discussed later, but in 1943 the situation became dangerous and he escaped in a small fishing boat to Sweden. After the foundation of the State of Israel and the Weizmann Institute, he actively helped the institute and visited it several times. He had met Weizmann in 1911 in Manchester at one of the dinner parties that Rutherford arranged once a month. He was the guest of honor at the opening of the Weizmann Institute of Physics in 1958.

The Experimental Evidence for Quantum Structure of Atoms by James Franck and Gustav Hertz

Einstein had postulated that atoms were quantum structures. Bohr's model was based on this assumption. But there was no real experimental proof for this concept. Bohr had proposed his model in 1913. In 1914, the proof was provided—by two physicists working at the University of Berlin, James Franck (1882–1964) and Gustav Hertz (1887–1975). In their famous experiments, for which they received the Nobel Prize of 1925 (awarded in 1926), they showed that individual mercury atoms will take up or emit energy in definite quantized steps. These results demonstrated clearly the reality of Planck's energy quanta and put Einstein's postulate and Bohr's atom model on a solid experimental basis. They provided, moreover, a new method for measuring Planck's constant. Furthermore, it became apparent from these experiments that the results of spectroscopy for the identification of chemical elements indicated the quantized nature of the energy steps characteristic of their atoms.

C. THE TWENTIES

Although World War I slowed down most scientific activities to a considerable degree, some work in atomic physics went on, particularly on spectral lines in view of the results of Bohr's work. By the end of the war a considerable amount of work had been done in Sommerfeld's institute. In 1919 he published *Spectral Lines and Atomic Structure*, which, with its many subsequent editions, became a standard work for physicists interested in the new world of quanta and atoms.

It turned out that there were many more spectral lines than those expected from Bohr's model. Sommerfeld suggested that electrons may run not in circles, but in ellipses around the nucleus. The speed of an electron running in an ellipse may vary from point to point and the calculations must take into account relativistic corrections. Sommerfeld traveled to Copenhagen to visit Bohr. When he arrived there, he found that Bohr had come to similar conclusions and was just then ready to publish them. Although the new picture of the atom, referred to as the Bohr–Sommerfeld model, represented a degree of progress, it was far from completely satisfactory. The improved model required a second quantum number for characterizing each energy state. However, when the data were compared with the computations of Sommerfeld, the results indicated that the picture was still incomplete and that it was necessary to introduce a third quantum number.

At this time a new scientist entered the field of atomic physics: Wolfgang Pauli (1900–1958), destined to become one of the great scientific geniuses of the first half of this century. Pauli's contributions to the development and the understanding of the field were monumental. He was born in Vienna, the son of Wolfgang Joseph, professor of physical chemistry at the University of Vienna. Pauli's father was Jewish, but his mother was not. He studied with Sommerfeld in Munich. When he was 20 years old he wrote an article about relativity. Einstein was so amazed about the mature and grandly conceived work that he could hardly believe that it was written by one so young. When Pauli was 21, he suggested the third quantum number, which he referred to as *space quantization*. The Bohr-Sommerfeld model of the atom was basically still a classical structure. The theories proposed for its behavior had introduced the quantum principle, but they were actually still arbitrary, since no one could explain why they worked.

Einstein had already recognized that there was a fundamental difference between large-scale physical phenomena and those taking place on the atomic scale. Pauli extended this idea and his thoughts and notions were based on the realization that the atom was a physical world subject to laws totally unlike those of large-scale observations. The understanding of these laws required entirely new and revolutionary physical and philosophical concepts. Pauli, his teacher and friend Bohr, and his friend

Werner Heisenberg (see below) must be considered as three great geniuses who laid the foundation for the transition from the old to the new physics initiated by Planck and Einstein. A superb mathematician, Pauli also had a profound insight into physics and was never satisfied with mathematical formalism unless it was associated with physically understandable events.

Bohr and Sommerfeld had postulated quantization of the size and shape of the electron orbit. Pauli now suggested that its position in space was only permitted in certain quantized directions. This new idea was far removed from any classical concepts. However, only a year after Pauli had proposed his idea, space quantization was shown to exist by direct experiments. This demonstration was the brilliant achievement of Otto Stern (1888–1969). Born in Sohrau, in Upper Silesia, he was brought up in Breslau, where he received his Ph.D. at the university in 1912. Both parents were Jewish. He joined Einstein at the University of Prague and went with him to Zurich, where he became *Privatdozent* at the *Eidgenössische Technische Hochschule*. In 1914 he became *Privatdozent* at the University of Frankfort on the Main, where he remained until 1921 except for military service during the war. In that year he went as *Extraordinarius* to Rostock. In 1923 he became professor of physical chemistry in Hamburg. There he built up one of the outstanding schools of physics in Germany—to be discussed later—until the arrival of Hitler in 1933.

Stern began his work on molecular beams in Frankfort in 1920. In his experimental arrangement vapor flowed through a tiny hole bored into an electrically heated furnace, and entered a high vacuum region. This geometry produced a thin beam further narrowed by another slit placed in the path of the escaping molecules. The most important result of the use of the molecular beam was obtained when he applied this method to the effect of a magnetic field on the motion of atoms. In collaboration with Walther Gerlach, in Rostock in 1922, he sent a beam of silver atoms between two differently shaped magnetic poles, one in the form of a knife edge, the other in a U-shape. The beam passed broadside-on very close to the knife edge and emerged from the pole region divided into two lines. The separation of the two lines provided a quantitative basis for measuring the atomic magnetic moment. Thus these experiments not only proved the existence of this property but also offered a direct method for the evaluation of its magnitude. Pauli's postulated space quantization had been experimentally demonstrated and explained as the result of the atomic magnetic moment.

However, when spectra of heavy atoms in a magnetic field were analyzed, new difficulties arose. Sommerfeld had already suspected the possibility of a fourth quantum number, but its meaning and nature were completely obscure. Pauli postulated that the fourth quantum number could not possibly be described in any classical form; only something not yet recognized could provide a solution. When he discussed this idea with Niels Bohr in Copenhagen, two young Dutch scientists, Samuel Goudsmit and George Uhlenbeck, listened to these discussions. They started some experiments aimed at testing Pauli's ideas. The position of the electron orbit in space is determined by the third quantum number connected with the magnetic effect produced by the circling motion of the electron's negative charge. In their efforts to find an indication of the postulated additional quantum number, the two Dutch scientists made a startling discovery: They found that in addition to the orbital motion around the nucleus the electron was turning around its own axis. This motion is referred to as electron *spin* and is similar to the spin of the earth around its own axis during its circling around the sun. Although this was difficult to reconcile with classical concepts, Pauli, after some initial skepticism, recognized the correctness of the interpretation. He realized that on the atomic scale electron spin is of great significance. It is an intrinsic property of the electron, just as its mass and charge are. Pauli then postulated that in one atom no more than two electrons can be in the same energy state and that the two electrons must have opposite spins. In the different orbits formed around the nucleus of heavy atoms (see p. 109), each orbit is formed by several pathways each containing only one pair of electrons with opposite spins. This law, proposed in 1925, has become known as *Pauli's exclusion principle*. It had a profound effect on the concepts of the constitution of matter. Although it is an empirical law for which there is no explanation, the exclusion principle is essential for the concept of symmetry and of fundamental importance for the understanding of events in the atom. It also became a key factor in the understanding of the chemical properties of the elements. The influence of the exclusion principle on atomic physics and on our concepts in chemistry will be discussed later.

The Compton Effect

Before continuing the description of the further developments of atomic physics, which was just about to enter its most exciting era, it may be briefly mentioned that a strong support for the corpuscular theory of Planck and Einstein was obtained with X-rays. The wavelength of X-rays is just about the same as the distance between the layers of atoms in the regular arrangements that occur in atom layers of crystals. This means that interference effects (p. 29), such as those known for light, are observed when a beam of X-rays is scattered by falling on a crystal. Instead of going out from the crystal more or less equally in all directions, the X-rays are scattered only in certain specific directions. These directions are determined by the relations between the wavelengths of the X-rays and the spacing of the atoms in the crystal. Thus the effect can be used with a known crystal to determine the wavelengths of X-rays from a variety of sources. On the other hand, the effect can also be used with X-rays of known wavelengths for the analysis of the arrangement of atoms in crystals whose geometry is not known.

The diffraction of X-rays by the orderly arrangement of atoms in crystals was discovered by Max von Laue (1879–1960) in 1912. The scattering is effected by electrons that are relatively tightly bound to the atoms of the crystal. For this work von Laue received the Nobel Prize in 1914. He was a descendant of a prominent Junker family. He had received his Ph.D. with Max Planck in Berlin and became *Privatdozent* there in 1909. He then moved to Munich, where he achieved the analysis of crystals by X-ray diffraction. His discovery had tremendous consequences, not only for physics and chemistry, but also for biochemistry. X-ray diffraction was successfully applied to the analysis of the tridimensional structure of vital macromolecules in living cells, such as nucleic acids and proteins, and initiated a new era in biological sciences, particularly in combination with other physical and chemical methods. Von Laue remained deeply attached and devoted to his teacher Planck. A close relationship developed between them and lasted for life. He moved to Berlin in 1919 and became director of an institute of the Kaiser Wilhelm Society. He was a great admirer of Einstein; both remained close friends until Einstein's death. But he also had close friendships with many other Jewish scientists. During the Nazi period he dared to defy the Nazis on many occasions (see II, E).

In 1922 Arthur H. Compton (1892–1962), an American physicist, found that when a beam of X-rays was scattered from a block of carbon, the wavelength of the scattered beam was longer than that of the incident beam. This meant that the scattered beam had a lower frequency and, therefore, a loss of energy had occurred. Such a loss can be explained by the quantum theory as the result of the impact of individual X-ray particles on electrons: The electrons have gained the energy lost by the particle. This result is referred to as the *Compton effect*. It shows that X-rays eject electrons in the same way as light waves. The result was a new confirmation of the quantum theory. Similar experiments were carried out by the English physicist C. T. R. Wilson (1869–1959). For their work, both scientists shared the Nobel Prize in 1927.

The University of Göttingen

Before turning to the two most important theories in the field of atomic physics of the 1920s, it is necessary to say a few words about the University of Göttingen, which was destined to play an essential role in the developments of that period. It is important to know the historical background that created an extraordinary atmosphere and a fertile ground for developing one of the most glorious and brilliant centers in atomic physics in the 1920s and, in addition, in physics and pure and applied mathematics in general. It strongly influenced the future of these fields.

The University of Göttingen was founded in the 1730s by the Elector of Hannover, who was King George II of England. Göttingen belonged -to the State of Hannover. From the beginning there prevailed in the university a liberal spirit, which it maintained even after the state was annexed by Prussia after the war of 1866. Its international fame during the nineteenth century was due to the extraordinary strength and brilliance of its mathematics faculty, to which belonged such mathematical giants as Karl Friedrich Gauss (1777–1855), P. G. Lejeune Dirichlet (1805–1859), and Bernhard Riemann (1826–1866), among others. In 1886 Felix Klein (1849–1925) became *Ordinarius* of mathematics there. He was one of the most outstanding mathematicians of that era, and an inspiring and superb teacher and lecturer. When he was 23 years old he became *Ordinarius* in Erlangen; he later moved to Leipzig, and from there his close friend Friedrich Althoff (see p. 25) attracted him to Göttingen. He was at that time already an almost legendary figure, a powerful, dynamic and domineering person known to behave and act in a regal way. He was recognized as the leader of mathematics in Germany. When he went to Göttingen his ambition was to create one of the world's greatest centers in mathematics there. He found in his friend Althoff an enthusiastic supporter. When a position became vacant by the move of Heinrich Weber to Strasbourg, his choice was David Hilbert. Klein had attended Hilbert's lectures when he was still in Leipzig, and he had immediately recognized that Hilbert would become one of the world's foremost leaders in the field. Hilbert joined Klein in Göttingen in 1895.

David Hilbert (1862–1943) was born and brought up in Königsberg. His father and grandfather were judges; his mother was apparently an unusual woman, interested in philosophy, astronomy, and mathematics. Hilbert attended the humanistic gymnasium and, after the *Abiturientenexamen*, studied mathematics, most of the time at the University of Königsberg, which had an excellent mathematics faculty. His teacher was Adolf Hurwitz (1859–1919), who later became *Ordinarius* at the *Eidgenössische Technische Hochschule* in Zurich. Hilbert's extraordinary mathematical gift became apparent very early. Moreover, in Königsberg, the city of Kant, it was almost impossible not to become interested in Kant and his philosophy. Kant had been a teacher of philosophy and of mathematics. Reminders of Kant were at many places and there were few students who did not know Kant's famous words—that, for him, the two greatest wonders were "the stars in the sky above me and the moral laws within me." The atmosphere in Königsberg and at Hilbert's parental home could not have failed to attract Hilbert's interest in philosophy, specifically that of Kant.

An important element in Hilbert's life became his close friendship with

Hermann Minkowski. Minkowski was the youngest son of a Russian-Jewish family. He was born in Kovno, Lithuania, and was eight years old when his parents moved to Königsberg. There he attended the Altstadt gymnasium during the same period as the famous physicists Max and Willy Wien and Arthur Sommerfeld. This gymnasium had a much better standard than that attended by Hilbert. Minkowski was an avid reader of Shakespeare, Schiller, and Goethe. He could recite most of Faust by heart. His older brother Oscar Minkowski (1858–1931) became a famous physician and discovered, with Joseph von Mering (1849–1908), in 1889, that the removal of the pancreas produces diabetes. This discovery later led to the development of insulin by Frederick G. Banting and Charles H. Best. When Minkowski was only 17 years old, he submitted a paper in the competition for the *Grand Prix des Sciences Mathématiques* of the Paris Academy, an incredibly bold action. He was awarded the prize, which he shared with the well-known English mathematician Henry Smith (1826–1883), when he was 18 years old. The news, not surprisingly, created a sensation in Königsberg. Hilbert and Minkowski met shortly afterward and soon became inseparable friends (*die "Unzertrennlichen"*); they were referred to as Castor and Pollux.

Hilbert's fame grew by leaps and bounds. He was elected to membership in several academies and awarded the title *Geheimrat* (privy councelor) by the government. But he was not entirely happy in Göttingen. Between him and Klein existed perfect trust and common interests, but no intimate relationship. He missed the vigorous discussions and the close and inspiring friendship with Minkowski. The latter was in Zurich, where he was the mathematics teacher of Einstein, but he too was not happy there. With Klein's approval and support Hilbert succeeded in convincing Althoff to establish a third chair of mathematics for Minkowski, who came to Göttingen in 1902. It was a turning point in the personal life of Hilbert. Now he was completely happy with his dear friend living quite close to him. They continued their interminable discussions and enjoyed many walks, parties, and picnics together.

With Klein, Hilbert, and Minkowski as leaders of the mathematics faculty, Klein's dream had become a reality: Göttingen was now one of the greatest world centers in the field. Hilbert was considered equal in stature to Henri Poincaré, at that time the greatest living mathematician. In addition, all three men were genuinely interested in both pure and applied mathematics. This was completely in line with the tradition of Gauss, who had left an imprint on every part of pure and applied mathematics. As a result of this attitude close relationships were developed with those fields in which mathematics played an essential role, particularly physics and related areas such as applied mechanics, aeronautics, geophysics, and astronomy, fields in which Göttingen had outstanding leaders as department heads. This tendency was greatly strengthened by the appointment in 1904 of Carl Runge (1856–1927), an experimental physicist in Hannover, to full professorship in mathematics, again on Klein's proposal. Minkowski was always greatly interested in physics. When he spent some time in Bonn he had been close to Heinrich Hertz. At a meeting in Cologne, in 1908, of the Society of German Scientists and Physicians (*Naturforschergesellschaft*), he gave a lecture on space and time in which he enthusiastically endorsed Einstein's theory of relativity, although with some reservations on certain mathematical formulations.

Soon after his lecture in Cologne tragedy struck: Minkowski died from a ruptured appendix in January 1909 at the age of 44. It was a shock and a tragedy for Hilbert. A 25-year-long, deep and almost unique friendship, which had enriched their lives in a great many ways and had brought much happiness and joy to both men, had suddenly come to an end. When Hilbert announced the news to the students he wept; it was an unusual experience for students, particularly in Germany, to see an *Ordinarius* weep. In his stirring address in memory of Minkowski, Hilbert declared that his friend was for him a gift of the gods—such a one as would seldom fall to a person's lot—and that he must be grateful that he possessed it for such a long time. Minkowski's successor became Edmund Landau (1877–1938), whose wife Maria was the daughter of Paul Ehrlich (1854–1915).

Although nobody could replace Minkowski for Hilbert, he looked for friends among the younger generation with whom he could discuss the problems that were on his mind. He became quite friendly with the philosopher Leonard Nelson (1882–1927), who belonged to the Kant–Friess school (see IV, D). They had many discussions about problems in areas where philosophy and mathematics meet. Another young friend was Theodor von Kármán (1881–1963), a Hungarian Jew who could trace his ancestry to the famous rabbi Judah Loew (1525–1609) in Prague. Kármán later moved to the United States and became one of the world's leading authorities in aeronautics. He called Hilbert "the greatest mathematician in the history of science." During this period a friendship began between Hilbert and his pupil Richard Courant (1882–1972), who became, in the 1920s, one of the leading figures in shaping the mathematics faculty in Göttingen (see Reid, 1976). Before Minkowski's death Hilbert had become interested in the rapidly growing field of theoretical physics, and between 1910 and 1930 this interest became increasingly stronger. He often lectured on topics of physics and continuously broadened his knowledge of the field. He intensified his contacts with physicists.

A few aspects of Hilbert's personality may be mentioned that are of special interest in the context of this book. He was completely internationally minded and very outspoken in this respect. He was extremely liberal and had left the Protestant church. When his son Franz was asked in school what his religion was, he did not know. The son of the philosopher Edmund Husserl, a baptized Jew, told him that if he did not know his religion, he was certainly a Jew. Hilbert was strongly opposed to World War I; he believed it to be an act of stupidity. In contrast to many colleagues, he dared express his opinion openly. When the famous address, "To the Cultural World," was published in October 1914, denying that Germany caused the war and refuting "the lies and slanders of the enemy," Hilbert was the only one of those asked who refused to sign this declaration. Many of the most famous artists and scientists of Germany, among them many Jews, had signed it. The only name conspicuously absent was Einstein's, but he was a Swiss citizen. In later years, when the Nazis almost destroyed the famous mathematics–physics faculty at Göttingen, Hilbert was desperate, but his age and poor health prevented a vigorous fight. At a meeting in Göttingen in 1934 the Nazi minister of education, Rust, inquired how his faculty was doing. Hilbert asked him what faculty he was speaking of. When Rust said that he meant, of course, the mathematics–physics faculty, Hilbert's answer was that it just did not exist, that he, Rust, had destroyed it by chasing away the most illustrious leaders. (For a detailed description of Hilbert, see Reid, 1970.)

After World War I the physics department in Göttingen attracted two outstanding physicists. There were two departments, one under the direction of Robert Pohl, the other under that of Woldemar Voigt. Voigt died in 1919, and in 1921 Max Born, at that time professor of physics at the University of Frankfort, was offered the vacant position. Visiting the department and giving much thought to its future, he concluded that there was enough space and an urgent need for a third department, devoted to experimental physics, especially atomic physics, since he considered himself more a theoretical physicist. He was a great admirer of James Franck, his friend since their Heidelberg student time, and fully recognized his great accomplishments in atomic physics. On Born's insistence the third department was created and offered to Franck, who accepted. Franck was at that time the director of the physics division in Haber's Kaiser Wilhelm Institute of Physical Chemistry in Berlin–Dahlem.

With Hilbert, Courant, and Runge the interest of the mathematicians in physics was very strong. Between the two faculties there was an extremely active collaboration. This close association between the two faculties did not exist anywhere else. It created an exciting and almost unique atmosphere, which in turn attracted many students and famous scholars in both fields to Göttingen. It became the Mecca of physicists and mathematicians. With the revolutionary developments in atomic physics described before and the challenge of many new and provocative problems for which there were no satisfactory answers, the collaboration between the two fields became of paramount importance. For these reasons Göttingen played an outstanding role in the dynamic and rapidly moving phase of atomic physics in the 1920s.

Wave Mechanics

At the time of the discovery of the Compton effect, the discontinuous character of radiation and the existence of particles (corpuscles, photons) could scarcely be contested. But physicists remained aware of intrinsic contradictions that could not be explained in terms of particles, such as interference and diffraction, as mentioned before (see p. 29). These phenomena seemed to support the assumption of the wave character of radiation, well demonstrated in the case of light and X-ray radiation. Thus the formidable dilemma of waves and particles remained unsolved. Clearly an answer had to be found to the duality of waves and particles in radiation. Einstein's equation $E = mc^2$ states that the energy of a particle is equal to its mass multiplied by the square of the velocity of light.

In 1923 Louis de Broglie (1892–   ) made the first attempt to find an answer to the wave–particle duality. He combined Einstein's equation with Planck's equation (the energy of a photon is equal to its wave frequency multiplied by Planck's constant h). He asked himself this question: Since the existence of stationary states for atoms shows the intervention of the quantum of action in the properties of the electron, should it not be supposed that the electron presents a duality analogous to that of light? At that time no manifestations of the wave properties of the electron had been found; the electron was always considered as an electrically charged particle. Some observations seemed to indicate the insufficiency of this view and suggested that the properties of the electron were not always those of a simple corpuscle; however, these indications were ambiguous.

De Broglie tried, therefore, to attribute to the electron, and more generally to all corpuscles, a dualistic nature and to postulate a wave and corpuscular aspect interrelated by the quantum of action. The problem was to establish a manner of associating the propagation of a certain wave with the motion of any corpuscle, the quantities that specify that the wave is related to the dynamic quantities containing Planck's constant h. This association should be established in such a way that the general rules expressing the connection between the wave and corpuscle would, when applied to the photon, yield the well-known relation demonstrated by Einstein but also include the combination of photons and light waves. When de Broglie analyzed how the waves associated with the electrons might behave in the interior of the atom, he arrived at the following conclusion: The conditions of quantization express the fact that the wave associated with a stationary state of the electron in the atom is a stationary wave in the sense of wave theory. This was a brilliant and revolutionary concept. He received the Nobel Prize in 1929 for his work. In 1925 his concept was taken up and greatly extended and clarified by Erwin Schrödinger (1887–1961).

Schrödinger, like Planck, disliked the discontinuity of the quantum concepts and the idea of electrons jumping from one orbit to another, as proposed in the Bohr model. When he read the paper of de Broglie he immediately became very interested and started to investigate the problem of how the Bohr model could be expressed in terms of waves. Starting with the concept of waves he tried to build a new picture of the atom. He assumed that the atomic orbits were confined waves with certain well-defined shapes and frequencies, just like sound waves produced by violin strings. In this way the waves of each orbit would take on their own distinctive form. With this concept he was able to explain the stability and identity of atoms, properties that were not easily understood on the basis of the planetary model consisting only of particles. Stability exists because the lowest orbital pattern does not change, unless considerable energy is added to it. In the absence of the energy to move the electron to a higher state the orbit is stable. Because the confined wave pattern is always the same, each atom must have its own distinctive identity. One hydrogen atom has to be like all other hydrogen atoms.

On the basis of these ideas Schrödinger went on to find the proper mathematical equations. After some initial difficulties he succeeded in formulating mathematical expressions that laid the foundation of what is called *wave mechanics*.

The basic aspect of de Broglie's postulate was that any material particle was associated with a wave whose length was given by its mass and speed, with Planck's constant providing the order of magnitude. In Schrödinger's concept the orbits of the Bohr model became the diffraction pattern that the waves of the electrons form around the atomic nucleus. Instead of well-defined planetary orbits, wave mechanics postulates a diffuse pattern of standing waves gradually thinning out into space and without clear demarcation.

From the mathematical point of view Schrödinger's theory was more satisfactory than that of Bohr. But the picture of the electron spreading out into a diffuse sphere of electricity inside the atom was not easy for many physicists to accept. While the physicists passionately discussed the physical reality of matter waves, a new and startling discovery was made by two American physicists, Clinton J. Davisson and Lester H. Germer. They were measuring the scattering of electrons directed at a nickel surface. By accident the nickel plate was temporarily heated, causing the metal surface to change to single crystals of nickel. Now the electrons, instead of being scattered in all directions, bounced back at certain angles. Puzzled by these results, Davisson wrote to Max Born, asking him whether he might be able to provide an explanation. Born, discussing the results with James Franck, connected the results with de Broglie's postulate, of which the American scientists were not aware; they asked a young research worker, Walter Elsasser, to work on the problem. His calculations suggested that the angles noted by Davisson were indeed the diffraction pattern of electron waves. They corresponded to the similar pattern that Max von Laue had obtained by applying X-rays to a crystal. Born and Franck wrote to Davisson, explaining to him that his observations presented a great discovery and suggesting some modifications of his experimental arrangements that would still further clarify his data. Almost simultaneously and independently George P. Thomson (1892–1975), the son of J. J. Thompson, in England obtained an electron diffraction pattern with the use of X-rays. He and Davisson shared the Nobel Prize in 1937 for their work.

In addition to the evidence for the scattering of high-energy beams of electrons by crystals, supporting the wave nature of the electron, experiments were performed by Otto Stern in Hamburg demonstrating that protons too were scattered from crystals like waves.

Now the reality of matter waves proposed by de Broglie and elaborated by Schrödinger had been experimentally demonstrated. They were not mere theoretical postulates or a concept based on mathematical equations. Further support of wave mechanics resulted from new calculations of Wolfgang Pauli. He applied what is referred to as *matrix mechanics*, a mathematical procedure introduced by Göttingen mathematicians in the nineteenth century. These matrix mechanics had been applied by Max Born, Werner Heisenberg, and Pascual Jordan to quantum mechanics, which is discussed in the following pages. With the success of de Broglie and Schrödinger the wave–particle dualism required, even more urgently, a satisfactory answer to the dilemma, one that would reconcile two seemingly opposing views.

Schrödinger was born in Vienna in 1882; his father was a chemist who had turned to painting; his mother was a daughter of the chemistry professor at the university. The cultural atmosphere in his home strongly influenced his personality and stimulated his artistic inclinations. He studied physics and mathematics at the University of Vienna and received his Ph.D. there. In 1910 he became *Privatdozent*. After the war he went, in 1920, to the University of Jena. In 1921 he became *Extraordinarius* at the University of Breslau, but soon left for Zurich where he became *Ordinarius* at the *Eidgenössische Technische Hochschule*. In 1927 he became the successor of Max Planck at the University of Berlin. He shared the Nobel Prize in 1933 with Paul A. M. Dirac (1902–   ). When the Nazis took over, he left Germany in 1933 since he was in vigorous opposition to the regime. He had no Jewish ancestors. He went to Oxford, where he became a fellow at Magdalen College. In 1936 he accepted a position in Graz, but left again when the Nazis occupied Austria. He became professor at the Institute of Advanced Studies in Dublin. In 1956, he returned to his native Vienna as professor at the university. He could never accept the interpretations of Born, the indeterminacy principle of Heisenberg, and the notion of complementarity of Bohr. He still hoped that the determinism of classical physics would eventually prevail and become applicable to the physics of atomic structure. In contrast, de Broglie eventually recognized the importance of Bohr's and Heisenberg's interpretations.

Quantum Mechanics

Almost parallel with the startling development of wave mechanics was the development of another theory, referred to as *quantum mechanics*. During 1924–1927, when these two theories were developed, vigorous discussions took place, not only between those holding opposing views but also between the proponents themselves, particularly in quantum mechanics. It was a struggle between scientific giants, rarely encountered with such intensity and passion. It created an atmosphere of great intellectual excitement. The concepts that emerged were revolutionary, producing a greater and more complete break with classical physics than the theory of relativity. The key figures in the development of quantum mechanics were Max Born, Werner Heisenberg, and Niels Bohr; the events took place primarily in Göttingen and Copenhagen, as will be discussed.

In Bohr's original theory there was a discrepancy between the orbital frequency of the electrons and the frequency of emitted radiation. This discrepancy was interpreted as a limitation to the concept of the electronic orbit, which seemed somehow doubtful from the beginning. The lowest orbits do not change without additional energy. In the higher orbits, however, the electrons should move at a larger distance from the nucleus. There one should speak about electronic orbits. In these higher orbits the frequencies of emitted radiation approach the orbital frequencies and their higher harmonics. In 1916 Bohr had already suggested that in this case the intensities of the emitted spectral lines approach the intensities of the corresponding harmonics. This *principle of correspondence* had proved useful for the approximate calculation of the intensities of spectral lines. This theory appeared to give at least a qualitative, but not yet a quantitative, description of what happens inside the atom. Some new feature was expressed by the quantum conditions; it was thought that they might be somehow connected with the dualism between waves and particles.

Before we turn to the emergence of quantum mechanics, it may be useful to portray a scientist who played a key role in the elaboration of this theory: Werner Heisenberg (1901–1976). He was born in Würzburg. In 1910 his father August became *Ordinarius* of Greek philology in Munich; thus Werner was actually brought up in Munich and always considered Munich as his home. His extraordinary brilliance became evident at the gymnasium, in physics, mathematics, and Greek philosophy. He was deeply influenced by Plato in his scientific and philosophical thinking; this is apparent in his late work about the unity of the theory of matter, which he first presented in 1958 on the occasion of the centennial celebration of Max Planck's birthday. After the *Abiturientenexamen* in 1920, slightly delayed because of the war, he joined Arnold Sommerfeld, who was a most inspiring teacher and whose institute attracted many brilliant students; it was an ideal place for Heisenberg's passionate interest in the new world of atomic physics of Planck, Einstein, and Bohr. There he met Wolfgang Pauli who, still a student and only about a year older than Heisenberg, was an assistant to Sommerfeld. Almost from the beginning of Heisenberg's studies he developed a close friendship with Pauli. This was a most fortunate event. The two were congenial personalities, and the friendship was a great stimulus, and of mutual benefit for their scientific development and thinking. Heisenberg was a more enthusiastic type and more driven by his imagination than Pauli. The latter was famous for his sharp and incisive critical judgment and his extraordinary insight into the problems of atomic physics. Many eminent physicists, when they proposed a new theory, wondered what Pauli's reaction would be. Nobody was safe from his biting comments. When Einstein tried for years to propose a unified field theory and failed, Pauli did not hesitate to remark that man should not try to put together what God had put apart. He became Heisenberg's mentor and regularly criticized him, frequently very harshly. Pauli and Heisenberg spent only one year (two semesters) in Sommerfeld's institute. After receiving his Ph.D., Pauli moved to Born in Göttingen, and from there to Otto Stern in Hamburg, where he stayed until his appointment, in 1928, as professor of theoretical physics at the *Eidgenössische Technische Hochschule* in Zurich. But Pauli and Heisenberg kept in touch. They frequently exchanged their views in letters about the essential mathematical, physical, and philosophical problems and met often. The close friendship, so useful for science, lasted until the death of Pauli in 1958.

One of the great events for atomic physicists of the year 1922 was a series of lectures about quantum theory given by Bohr in Göttingen (nicknamed the *Bohr-Festspiele*). Not only Born and Franck and their associates, but many of the famous mathematicians were present. In addition, scientists from many different places came there to attend these lectures. Pauli came from Hamburg, Sommerfeld and Heisenberg from Munich. Heisenberg was profoundly stimulated and excited, particularly about Bohr's philosophical attitude. Lively discussions followed the lectures. To the great amazement of all present, Heisenberg, then a 21-year-old student, challenged Bohr after the first lecture and a vigorous discussion arose. Bohr was apparently so impressed that he proposed that Heisenberg join him for a walk. For Heisenberg it was one of the most intensive discussions he had ever had about the physical and philosophical problems of atomic physics. It decisively influenced his thoughts. He realized that

Bohr was more skeptical about his own theory than many other physicists, including Sommerfeld. Bohr derived the insight into the processes within the atom not so much from mathematical formalism but from intensive efforts to recognize the physical reality, based more on intuition than on precise analyses. He believed that the physical reality should form the basis of our concepts, and only then should an attempt to formalize the concept in mathematical terms be made. Heisenberg was fascinated by Bohr, both as a person and as a scientist. But he too made a deep impression on Bohr, who said to some friends after their return from their walk: "He understands everything." This was an extraordinary compliment from a man accustomed to discussions with the most brilliant contemporaries and known to be extremely critical toward ideas proposed. During the following weeks many long discussions took place following the lectures, in which James Franck and Max Born took very active parts. For Heisenberg, Bohr had opened a new world. When they later became friends, he wrote to Bohr that it was these lectures that enabled him to grasp the true nature of atomic physics.

The following winter, 1922–1923, Heisenberg spent with Max Born in Göttingen, since Sommerfeld had accepted a visiting professorship in Wisconsin. Heisenberg was still a student, yet Born included him in a seminar on Bohr's theory in which only eight physicists and mathematicians participated. The seminars took place in Born's home. When Pauli left Göttingen for Hamburg, Born thought that he would never again have such a brilliant collaborator. Now Heisenberg turned out to be equally brilliant. In addition, Born was fond of Heisenberg's personality, his modesty and enthusiasm, and delighted that Heisenberg was, as he was himself, a superb piano player. He was anxious to get Heisenberg in his institute. After Heisenberg received his Ph.D. in Munich, he joined Born in the fall of 1923.

Although Born's greatest strength was mathematics, he had always followed with great passion the development of atomic physics, and his work was deeply involved in this field. In 1915 he had published a book on the dynamics of crystal lattices (*Dynamik der Kristallgitter*), which was widely acclaimed as an outstanding contribution to the understanding of atomic structure. It was most fortunate that he and his friend James Franck, with his great insight and experience in the experimental aspects of atomic physics, were in the same institute. Since they complemented each other almost perfectly, their frequent, almost daily discussions were a great stimulus to both. When, in the early 1920s, Franck worked in Göttingen with a large and talented group of associates on quantum effects in the interactions between atoms and molecules, Born followed their work very closely and his many discussions with Franck resulted in interesting common work on the application of quantum theory to the kinetics of chemical reactions, which they published together.

One of the brilliant contributions of Born to quantum theory was his interpretation of diffraction. Among the unsolved problems in Einstein's description of the photoelectric effect was the explanation of the diffraction spots on a photographic plate when a beam of electrons is directed on a crystal. This phenomenon seemed to favor the assumption of a wave rather than a particle structure. In Schrödinger's and de Broglie's concepts, the particle was endowed with some wave properties—not yet discovered—that caused the observed diffraction pattern. This concept leads, however, to a dilemma. An electron beam, on undergoing diffraction by the crystal, produces locally separated spots. But where does the charge of the electron go, since it carries a unit charge that cannot be subdivided? An ingenious solution was proposed by Born. In his paper in 1905 Einstein had considered diffraction as a large-scale effect. In Born's interpretation the diffraction effect indicates that not one but many particles are involved. A single electron could never produce a whole diffraction pattern; it would settle on the photographic plate, after being deflected by the crystal, in a region that corresponded to one of the recorded spots. Some angles of deflection are statistically more favored than others, and these are therefore the directions in space followed by most electrons. The dark areas on the photographic plate correspond to these most probable directions, whereas the white areas separating them indicate regions where it is less probable for an electron to arrive. In respect to Schrödinger's wave picture, Born's proposal indicated that the particle wave is just a measure of probability of finding the particles in a particular region of space. The diffuse diffraction rings in Schrödinger's theory had acquired a physical meaning that had replaced the planetary orbits of Bohr. Born's concept was strongly supported and assumed an even more concrete form by Heisenberg's principle of indeterminacy, to be discussed subsequently.

Born's main interest turned then to the quantum theory of the structure of the atom. The question was how far Bohr's theory was compatible with the observed facts and where its limitations were. According to Bohr, the atom in its stationary state may be described by equations of ordinary mechanics, determined by certain quantum rules. No radiation is emitted from the atom in such a stationary state, whereas emission or absorption of radiation takes place in transitory states, in which the difference of energy equals the energy $h\nu$ of the quantum of light. This principle worked well for the hydrogen atom, but how far could it be generalized? To find an answer to this problem was the first aim of Born and his associates. Using a new approach, their results, applied to the helium atom, did not agree with those Bohr obtained with his methods. Born and his associates became increasingly convinced that a radical change of the fundamental basis of physics had become necessary, a new type of mechanics to which the term *quantum mechanics*, first introduced by Born in an article in the *Zeitschrift für Physik* in 1924, was applied.

In the following three years, 1924–1927, quantum mechanics developed at a very rapid rate. It did not actually begin in Göttingen; pertinent observations had been made by others, particularly Hendrik A. Kramers (1894–1932), who recognized the interactions between the electrons in the atom and the electromagnetic field of the light wave. His observations were actually the first steps away from classical mechanics to the still obscure and unexplored world of quantum mechanics.

While at Born's institute, Heisenberg tried to make a transition from the qualitative model mechanics to real quantum mechanics. He summarized his ideas in a paper on a change of formal rules of quantum theory in regard to the problem of the abnormal Zeeman effects. Born was so deeply impressed by the extraordinary quality of the paper and the ingenious interpretations offered that he accepted it for the promotion to *Privatdozent*, although Heisenberg was only 22 years old and a year had not yet passed since he became a Ph.D.

In the fall of that year, 1924, Bohr invited Heisenberg to join him as a research associate for half a year in Copenhagen, at that time one of the great intellectual centers in atomic physics. To the great regret of Born, Heisenberg accepted the invitation, not surprising in view of his great admiration for Bohr. These six months in Copenhagen and the subsequent visits there were highlights of Heisenberg's life and career. As he once stated, he learned from Sommerfeld optimism, in Göttingen mathematics, from Bohr physics.

In the spring of 1925 Born, stimulated by Kramers' observations, raised the question of whether comparable interactions might be found between systems with two electrons as a function of transition quantities. He believed that by appropriate new interpretations of classical perturbation theory one might construct the corresponding quantum formula. Born's assumption was later completely borne out by quantum theory. Born approached the problem with his associate Pascual Jordan. When they reexamined Planck's work on radiation theory, which had led to the discovery of the quantum of action, they found that he had applied classical mechanics to the interaction between matter and light. Introducing transition quantities instead of the classical ones they were able to translate Planck's calculations into terms of quantum theory. Their work appeared in the *Zeitschrift für Physik* in 1925.

In the meantime Heisenberg had prepared a manuscript with some new ideas. He gave it to Born for comments in the summer of 1925, before leaving for Cambridge, where he was to deliver a series of lectures in the Cavendish Laboratory. Heisenberg was not fully satisfied with his analyses and hoped that Born would critically review them, especially some of the mathematical formulations proposed in view of Born's great competence in that field.

When Born read Heisenberg's manuscript, he was fascinated. Heisenberg had developed a series of a new type of mathematical formulations. He had succeeded in expressing Bohr's quantum theories in a new mathematical formalism. Born recognized that the ideas presented an important new step and submitted the paper for publication. He was extremely excited about the new calculations and devoted much thought to them. One day it suddenly dawned on him that Heisenberg's formulations were in fact quite similar to matrix calculations, with which Born had been familiar since his student days in Breslau thanks to his teacher Rosanes. In Göttingen's mathematical atmosphere, matrix calculus was a familiar ground. Although Heisenberg's formulations were satisfactory in some essential parts, others required improvements and refinements.

When Heisenberg returned from Cambridge, Born, Heisenberg, and Jordan immediately started to work, at a hectic pace and with great intensity, on a final and comprehensive presentation. This paper was destined to become the fundamental mathematical basis of quantum mechanics, a milestone in its development. It was written in a rather short time, since Born was anxious to have it completed before his trip to the United States, where he was to deliver a series of lectures at the Massachusetts Institute of Technology. The paper contained the most essential mathematical elements of quantum mechanics, among them, the presentation of physical quantities by noncommuting symbols, the generalization of Hamilton's mechanics, the theory of perturbation, the treatment of time-dependent disturbances with an application to the optical dispersion theory (derived from Kramers' equation), and theorems of the impulse and angular momentum (quantization of direction, abnormal Zeeman effects). The paper appeared in the spring of 1926 in the *Zeitschrift für Physik*. As usual, it is impossible to separate the individual contributions to this complex, profound, and fundamental work. It was the result of the combined efforts of three brilliant minds. But it may be stressed that since matrix calculus forms the basis of the work, Born of necessity played a key role as the uncontested master in the field. Whereas Born was intimately familiar with matrices, Heisenberg was just starting to learn this special field, although his genius enabled him to grasp it very fast and to use it efficiently.

When Pauli, at about the same time, applied matrix formalism of wave mechanics to the hydrogen atom, he obtained the correct result. A year later he also successfully applied it to the helium atom and even to some problems of heavier atoms. But the paradox of the dualism of wave and particle was not solved. It remained hidden in the mathematical scheme. The basis of the mathematical approach to quantum mechanics was a philosophical postulate that Heisenberg had worked out during a stay in Helgoland, where he had gone for a cure for a strong attack of hayfever in Göttingen. He had stressed that only observable quantities should be used in the calculations. Atomic particles could not be regarded as the usual objects in large-scale physics, being only very small ones. No effort was made to introduce physical models in the mathematical formalism.

The pure mathematical formalism does not explain the events inside the atom in physical terms. This lack of physical interpretation was unsatisfactory to many physicists, including Heisenberg and Pauli, who were anxious to understand the reality of the events taking place in the atom. In Göttingen, where mathematics played a dominant role, the new development was accepted with great satisfaction. Only Heisenberg remained unhappy; as he wrote to Pauli, without an insight into the under-lying physical events the whole mathematical formalism would remain for him meaningless. Pauli was in full agreement. Both looked for a new kind of physical notion that would help them understand the nucleus of the mathematical structure.

In the following year Bohr invited Heisenberg to join him in his insti-tute in Copenhagen as a *Dozent* for theoretical physics. Heisenberg ac-cepted the invitation and went to Bohr in May 1926. Within less than a year Bohr and Heisenberg succeeded in working out a theory that pro-vided a basis for the missing physical concepts.

A short time before, in 1924, N. H. Bohr, H. A. Kramers, and J. C. Slater had published a paper in which they introduced the concept of the *probability wave*, something entirely new in theoretical physics. This notion was something standing in the middle between the idea of an event and the actual event, a strange kind of physical reality to be conceived as being between possibility and reality, in a certain way related to the old version of *potentia* in the philosophy of Aristotle. Born took up this idea of the probability wave and gave a clear definition of the mathematical quantity in the formalism, which was to be interpreted as the probability wave.

During the time that Heisenberg spent in Copenhagen, intensive and passionate discussions took place between him and Bohr. Both were ob-sessed with the problem of how to elaborate the comprehensive physical picture underlying the mathematical formalism and find a satisfactory answer to the dilemma of wave and particle. They were occasionally joined by other physicists. In the fall of 1926 Schrödinger spent a few weeks at the institute and participated in the discussions, which went on until the participants were near exhaustion. Schrödinger had shown dur-ing the summer that his formalism of wave mechanics was mathematically equivalent to that of quantum mechanics. He made a great effort to aban-don the idea of quanta and quantum jumps altogether and replace the electrons in the atom by his three-dimensional matter waves. But during the discussions in Copenhagen it soon became apparent that such inter-pretation led to great difficulties. It would not even be sufficient to explain Planck's formula of heat radiation; this was obviously unacceptable.

After Schrödinger had left, Bohr and Heisenberg again took up their efforts to elaborate a satisfactory physical interpretation of the informa-tion available. There was no easy solution. Their discussions went on day after day, week after week, and frequently ended in despair. Both were strong fighters and these intellectual controversies were so exhausting that Bohr apparently became tired. He decided, in February 1927, to go on a skiing vacation in Norway. He wanted to have an opportunity to clarify his ideas for himself, relaxed and in isolation. While he was on vacation he formulated his ideas in a paper. During the same time Heisenberg wrote his ideas down in a manuscript.

A solution was proposed in two different formulations, since there was a subtle difference between the approach of Bohr and that of Heisenberg. Bohr tried to make the dualism between wave and particle the starting point of his formulations of a new physical interpretation. Heisenberg, on the other hand, put the primary emphasis on quantum mechanics without stressing too strongly wave mechanics. But when they compared their ideas, these turned out to be much closer to each other than they had previously realized. Both formulations led to decisive limitations of concepts that had been the basis of classical physics.

In order to know the value of a physical quantity exactly, it is necessary to measure it in a precise way. In classical physics such precise measurements are possible because the observations are made without appreciably disturbing the state of the objects. The measurement acts only to determine the existing state; it does not introduce a new element. Since on the macroscopic level the requirement of the absence of any significant perturbation is fulfilled, it is possible to study a phenomenon quantitatively and to predict subsequent events in an infallible way. This possibility of precise prediction of a system when a certain amount of data is known about its present state, for instance in astronomy, constitutes the determinism of classical mechanics. This determinism of natural phenomena, implying their complete predictability, at least in principle, has become a scientific dogma.

In contrast, on the microscopic level, as in atomic physics, the existence of the quantum of action has as a consequence the fact that the perturbation arising in the operation of measurement cannot be diminished beyond a certain degree; as a result the measurement materially disturbs the phenomenon under study. In Heisenberg's theory, a somewhat more radical one than that of Bohr, the position of an electron can never be defined with absolute precision with simultaneous determination of its velocity. The most delicate way available to localize a corpuscle with precision is the use of radiation of short wavelength. This method permits the distinction of two points in space whose distance is at least of the order of this wavelength. To determine the exact position of an electron, for example, we must use gamma rays. But here the existence of the quantum of action enters, in the form of radiation quanta. The shorter the wavelength the greater the frequency of the radiation and consequently the energy of its photons; they would therefore disrupt the movement of the electron, preventing the measurement of its velocity. If, on the other hand, one reduces the energy of the photon by using photons of lower frequency, the weaker one encountering the stronger electron would be diffracted or deflected. There would be a blur around the electron, precluding the measurement of the exact position at any given time. If the uncertainty about one of the two conjugate quantities, position and velocity, is very small, the uncertainty about the others is very large.

However, there could be a middle course. The photon might be adjusted so that it would not disrupt the electron and force it off its course, but would show the electron's general path with a certain degree of approximation. Even if the exact position could not be determined, and only a general idea was obtained, it would be possible to calculate from these data the probability of its presence in a certain area. When the probability function has been determined at the initial time of the observation, one can calculate, from the laws of quantum theory, the probability function at any later time and can thereby determine the probability of a measurement giving a specified value of the measured quantity. We can predict, for example, the probability of finding the electron at a later time at a given point. However, the probability function does not in itself represent a course of events in the course of time; it represents a *tendency* for events and our knowledge of events. The probability function can be connected with reality only if a new measurement is made to determine a certain property of the system. Only then does the probability function permit the calculation of the probable result of the new measurement. This "uncertainty relation" is the direct consequence of obtaining an accurate measurement of the electron's position and its velocity simultaneously. Actually, the product of these two inaccuracies turned out to be not less than Planck's constant divided by the mass of the particle. Profound analyses of the procedure of measurement permitted Heisenberg and Bohr to show that no measurements can effectively lead to results in disagreement with the relation of uncertainty.

This is the nucleus of Heisenberg's famous *relation of uncertainty* or *principle of indeterminacy*. The limitation of determinism, as it was established in classical physics, and its substitution in quantum theory by the principle of indeterminacy and by the general principles of probability interpretations represent a profound change, as will be discussed later. It may be mentioned that the notion of probability was also used in classical physics, but it had an entirely different significance. Where it was applied, the elementary processes were considered to be controlled by rigorous laws. The notion was used to describe large-scale phenomena relating to an immense number of elementary processes. The basic difference between this notion and that introduced by Heisenberg into quantum theory is the direct application to the course of the elementary processes in the atom.

Immediately after Heisenberg had finished his manuscript, he sent it to his friend Pauli. The reaction was enthusiastic. In his answer Pauli said he considered the new concept as the dawn (*Morgenröte*) of the new era; ". . . the day starts in quantum theory . . .", he wrote.

Bohr's ideas, which he had worked out simultaneously during his vacation in Norway, resulted in the concept of *complementarity*. Schrödinger's concept of the atom as a system not of a nucleus and electrons but of a nucleus and matter waves contained an element of truth. Bohr considered the particle and wave picture as two complementary descriptions of the same reality. Any one of these descriptions can be only partially true. There are severe limitations to both descriptions, be it the wave or the particle concept, otherwise one could not avoid contradictions. He also stressed, in full agreement with Heisenberg, the difficulty of determining simultaneously the position of the electron and its velocity, and the resulting consequences. The knowledge of the position of the particle might be considered as complementary to the knowledge of its velocity, just as the knowledge of a particle to that of a wave. By knowing both, a more complete description, a new synthesis, was possible. Predictions could be made, not of exactly what would happen at any instant, but of the probabilities that certain things would happen. Here again his views are close to those of Heisenberg.

In Bohr's view complementarity is a requirement of the laws of nature and an indispensable logical tool for their understanding. According to this notion atomic systems may be described in different pictures that are entirely appropriate for certain experiments, but not for others; they are mutually exclusive. One may, for example, consider the atom as a small planetary system with a nucleus around which electrons are circling. But in other systems it may be useful to imagine that the nucleus is surrounded by a wave system, in which the frequency of waves determines the radiation emanating from the atom. But one must pay attention to the limitations given by the uncertainty relations. The use of matter waves is convenient for dealing with the radiation emitted by the atom. By means of the frequencies and intensities of the radiation one obtains information about the oscillating charge distributions in the atom. There the wave picture comes nearer to the truth than the particle picture. Since the two pictures are mutually exclusive, because a certain thing cannot at the same time be a particle, a substance confined to a very small volume, and a wave, a field spread out over a large space, Bohr postulated the two pictures as complementary to each other.

The dualism between the two different descriptions of the same reality is actually no longer a difficulty since we know from the mathematical formalism of the theory that contradictions cannot arise. The dualism between the two complementary pictures—waves and particles—is also clearly brought out in the flexibility of the mathematical scheme. The formalism is normally written to resemble Newtonian mechanics, with equations for the parameters of the particles. But by a simple transformation it can be rewritten in a way that resembles a wave equation for an ordinary three-dimensional wave. Therefore, the possibility of playing with different complementary pictures has its analogy in the different transformations of the mathematical scheme. It does not lead to serious difficulties of the interpretation given by the quantum theory.

Thus in the spring of 1927 a consistent interpretation of quantum theory had emerged, frequently called the *Copenhagen interpretation*. It had taken more than a quarter of a century from the first idea of the existence of the quantum of action to a broader understanding of the quantum theoretical laws. Great changes of fundamental concepts had to take place before one could understand the newly developed world of atomic physics. One of the most incisive and revolutionary changes was the necessary modification of the law of causality; it is taken for granted in Kant's philosophy as one of the a priori notions, just as time and space. Einstein's theory of relativity made inevitable a new and more precise formulation of the last two notions. Now the principle of indeterminacy and the uncertainty relations of Heisenberg and the notion of complementarity of Bohr forced a reformulation of the law of causality. Determinism has only a limited applicability; in macroscopic events classical physics still prevails. In contrast, as to the events in the atom, it is impossible to determine the existing state in a precise way. Consequently, one cannot predict subsequent events with certainty.

Many physicists were confused by the new quantum theory; others, particularly those of the older generation, rejected some of the basic new concepts. However, others recognized the new theory as one of the greatest events in science. Robert Oppenheimer calld it "the inauguration of a new phase in the evolution of human thinking." Paul Dirac said that the theory "led to a drastic change in the physicist's view of the world." John A. Wheeler, professor of physics at Princeton University, described the theory as "the most revolutionary concept of this century." Pauli's enthusiastic comments have already been mentioned.

In the midst of further intensive discussions of the new theory between Bohr and Heisenberg, Bohr received an invitation to attend an International Congress of Physics, to be held in Como, Italy, in September 1927 in commemoration of the hundredth anniversary of the death of Alessandro Volta, who had been professor of physics in Como for a few years in the eighteenth century. Enrico Fermi (1901–1954) had been the first professor of theoretical physics in Rome; he had first spent some time with Born in Göttingen. He was the first physicist in Italy to introduce the theory of relativity and the new atomic physics, wave and quantum mechanics. He was anxious to have Bohr at the meeting. Bohr chose as his title, "The Quantum Postulate and the Recent Development of Atomic Theory." There he presented the new concepts of quantum mechanics. One can easily imagine the excitement created in the audience. Many objections were raised.

Einstein was not in Como. Although some brilliant physicists vigorously rejected the theory, among them von Laue and Schrödinger, the question all physicists asked was this: What would Einstein say? The opportunity to learn his reaction came very soon. In October, the Fifth Solvay Conference was held in Brussels. Einstein was there, as were Bohr,

Heisenberg, and Pauli. The topic of the meeting was "Electrons and Photons." Bohr was invited to give a full report about his concept of complementarity.

After the presentation of the new quantum theory Einstein rose immediately and rejected the new notions. He was opposed to uncertainty relations. He did not consider complementarity as an acceptable solution. The new theory did not offer, in his opinion, a closer connection with the wave concept; moreover, it left to chance the time and direction of the elementary processes. Heated discussions followed. In the evenings Einstein and Bohr had personal discussions. Both had the greatest admiration for each other. The exchange of views took place in a friendly way, but the differences were fundamental and both vigorously defended their views. Bohr reminded Einstein of the ideas he himself had ingeniously raised. He had shown, in 1905, that the photon is a corpuscle as well as a wave. He had formulated, in 1917, rules indicating that the atom may spontaneously emit radiation at a rate corresponding to a certain probability. Only the probability of disintegration could be calculated, not the exact moment at which the radioactive material would give off another particle in its long or short decay. Having thus laid the foundations of atomic physics, he should not object to the further drastic changes of quantum theory arising from new observations.

Einstein then came up with several "ideal" experiments, that means imagined experimental setups, designed to show that uncertainty relations would be unnecessary. These proposed experiments led to further intensive discussions and in the end Bohr was always able to show that these experiments would not be able to go beyond the interpretations given. But Bohr was unable to convince Einstein. Einstein said that he did not believe that God would play dice, an expression he used later on other occasions. Bohr's answer was that caution was needed in ascribing attributes to Providence in ordinary language. But Einstein firmly maintained a deterministic attitude and refused to admit any modification of the law of causality. An admirable account of Bohr's ideas presented at the conference—and on later occasions—may be found in his article, "Discussions with Einstein on Epistemological Problems in Atomic Physics" (Bohr, 1949), published in a volume that appeared on the occasion of Einstein's seventieth birthday. He began the article by paying tribute to the epoch-making contributions of Einstein to the progress of physics and philosophy and by acknowledging the indebtedness of the whole generation for the guidance his genius has given. But then Bohr summarized in a sharp and lucid way the arguments in favor of the new interpretation, taking into account the objections raised by Einstein.

Einstein remained adamant in his attitude. He never accepted the new ideas. Born, who was not only a great admirer of Einstein, but had a deep and genuine affection for him, tried in many letters to influence his atti-

tude. This exchange of letters between Einstein and Max—and his wife
Hedwig—Born (1969) over a period of nearly four decades offers a fascinating documentation of the personalities and the dramatic events of the
era involved, with many human, political, philosophical, and scientific
aspects. As Heisenberg states in his Preface to Born's book, the extraordinarily great difficulties encountered in the understanding of atomic phenomena are well illustrated by the fact that two such illustrious scientists,
close personal friends, were unable to agree about the meaning of quantum
theory. Einstein accepted the mathematical formalism of quantum theory
in which Born had played an essential role. But he was not willing to
accept the interpretation of quantum mechanics as a definite or complete
presentation of the physical events taking place in the atom.

At the end of the Solvay Conference, there was little doubt that Bohr's
and Heisenberg's interpretations had triumphed. This conference of 1927
is widely considered as a major turning point in the history of atomic
physics. Several illustrious physicists, especially of the older generation,
in addition to Einstein, von Laue, and Schrödinger, remained unconvinced.
Here again, as frequently in the history of science, one is reminded of
Planck's statement that scientific opposition never dies except with the
death of the opponents. However, many others recognized that the new
interpretation was the starting point of a new era. In his book, *The
Revolution in Physics*, de Broglie (1953) fully discusses and emphasizes
the importance of the concepts of Bohr and Heisenberg. Some physicists
took a more cautious and neutral attitude. They recognized the strength
of the arguments presented. However, they reasoned that one day new
types of analytic methods might emerge that might overcome the present
difficulty leading to the uncertainty relations, due to the impossibility of
precise determination of position and velocity simultaneously. This kind
of reasoning appears not pertinent. Every scientific concept must be based
on experimental facts. If new facts become available, the theory must be
adjusted. But considering a theory as not yet the last word because one
day in the future new observations might require a change is a reservation
that applies to every theory.

Quantum theory and quantum mechanics and the theory of relativity
had a profound impact not only on physics but on many fields of science
and technology. Before turning to these aspects, the family backgrounds
and personalities of Franck and Born will be described. A few remarks
will be added about Gustav Hertz, about whom relatively little is known.
All three played an eminent role in the growth of atomic physics in Germany until 1933. Moreover, Franck and Born in particular influenced the
later developments in the field by their work outside Germany after 1933
and by the schools they had developed. Many of their pupils became
famous and prominent leaders in many countries, particularly in the
United States.

D. THE LIVES OF FRANCK AND BORN; GUSTAV HERTZ

James Franck

James Franck was born in Hamburg, the son of a Jewish banker. His family has been traced to about 1760, when a Jacob Franck was born in Cuxhaven, a son of Moses Franck. Jacob Franck was the great-grandfather of James. James' father, also named Jacob, married his cousin, Rebecka Nachum Drucker: One illustrious forebear of Rebecka's was Rabbi Lase of Berlin; it seems there were many rabbis in her family. Jacob Frank Jr. founded a banking firm in Hamburg (J. Franck and Company), but he moved to Berlin in 1910. James Franck married Ingrid, daughter of Hendrick and Ellen, (*née* Delemonte) Josephson in Göteborg, Sweden; she was one of six children. Henrick was a businessman, a partner in a wholesale house for yard goods. Apparently the Delemonte family had come originally from Portugal to Sweden. Ingrid lived from 1882 to 1942. Franck was deeply devoted to her. She became sick soon after their marriage and died in Chicago. Four years later Franck married Hertha Sponer, a physicist associated with him in Berlin and Göttingen. She became professor of physics at Duke University. Franck had two daughters: Dagmar, married to Arthur von Hippel, a physicist, who joined the Massachusetts Institute of Technology and became professor emeritus in 1961, and Elizabeth, married to Hermann Sisco, professor of anatomy at Harvard University.

Franck's father was deeply religious. He observed Jewish holidays with fasting and chanting. He had gone to the Hamburg *Talmud Thora Realschule* and had a strongly religious upbringing. He was immensely generous; but he had no patience with anyone who was inexact in money matters or dishonest in any way. He was a voracious reader and impressed his grandchildren with his ability to read at least one book a day. In contrast, his son James was not only not orthodox but very liberal. As he said, science was his God and nature his religion. He did not insist that his daughters attend religious instruction classes (*Religionsunterricht*) in school. But he was very proud of his Jewish heritage and never considered being baptized. He was, like most of his contemporaries, a patriotic German, a German Jew. And like many German university professors, he had no interest in politics. When the war broke out in 1914, he voluntarily joined the army. In spite of being a Jew he became an officer, a sign of his courage and the quality of his military activities.

Franck had attended the gymnasium in Hamburg. He studied chemistry in Heidelberg for a year, where he met Born. The two became friends and, as Born once wrote, neither the professors nor the romantic atmosphere of the town were the most important things in his life, but his friendship with Franck. Franck then went to Berlin and under the influence of Emil Warburg he became interested in physics. He measured mobilities of ions by Rutherford's method; he received his Ph.D. at the University of Berlin in 1906. He continued his work on similar lines, particularly on the forces between electrons and atoms. He spent a few years in Frankfort on the Main, but returned to the University of Berlin in 1911. When he was joined by his younger colleague and friend Gustav Hertz, their collaboration led to their great discovery, demonstrating the quantized energy transfer from kinetic energy to light energy. Their work proved the reality of the energy quantum, postulated by Planck, even more convincingly than the observations of Einstein. It was a milestone in the history of quantum theory and they shared the Nobel Prize for their work.

After the war Franck returned, in 1917, to Berlin and joined Fritz Haber (see IV, A). He became the head of the physics division of the Kaiser Wilhelm Institute of Physical Chemistry. It was there that he met Niels Bohr for the first time. Bohr, in his discussions with some younger physicists, stressed the provisional nature and the philosophical inconsistencies of his quantum theory. Bohr's wisdom and personality deeply impressed Franck, who became a great admirer; he visited Bohr in Copenhagen. These contacts established a lasting personal friendship between them.

In 1921 Franck was offered, through the efforts of his friend Born, the chair of a Second Institute of Physics at the University of Göttingen. He and Born moved to their new positions at about the same time. The position was in many respects ideal for Franck. The Institute of Physics was divided into three independent departments. Born's department of theoretical physics consisted of a few rooms. The smallness of the departments favored a very close and intimate collaboration. Franck did not have to give the general lecture on physics and was able to spend his time in a way that suited his interests. He did not easily accept students, but those who were accepted immediately became part of the family. Contacts were completely informal, a feature rare at German universities where stiffness and formality usually prevailed between professors and students.

But the most important aspect was his friendship and collaboration with Born. Their two fields, experimental physics and theoretical physics, supplemented each other in an ideal way, particularly important during that era. He and Born discussed every one of their problems and publications during the 12 years in Göttingen, although they published only a few papers together. During the years when quantum mechanics was developed and when mathematical formalism, especially matrix calculations, was dominant, Franck's competence and deep insight into experimental atomic physics, his views and criticisms were invaluable for Born.

The weekly colloquia were held jointly by the three physics professors. Close contacts existed between the three departments. When distinguished visitors gave seminars or spent some time in any one of the departments, all members shared the great benefits derived from contacts with them.

The years in Göttingen were one of the happiest and most creative periods in Franck's life. His activities covered a broad range of phenomena relevant to atomic physics. Collisions between atoms, formation of molecules and their dissociation, fluorescence, and chemical processes were in the center of his field of interest. In the first years he was still particularly interested in the problems of collisions of electrons with atoms, a problem he was concerned with in Berlin. He summarized his work in a book with P. Jordan, which was published by Springer-Verlag in Berlin in 1926. He and Born separated the motions of electrons and nuclei; he was very interested in the analysis of the elementary process of photochemical reactions. By combining a series of observations he was able to obtain a clear picture of the connection between electronic transition and the motion of the nuclei. Later the problem was approached by the American physicist Edward U. Condon on a quantum mechanical basis (see p. 117). This brought additional information to what is now known as the *Franck–Condon principle*, which has since been applied to a large number of phenomena in chemistry and spectroscopy. Having established many of the basic relationships that determine the interactions between simple molecules and light quanta, Franck became interested, while still in Göttingen, in the fundamental photochemical biological process of photosynthesis, in which the energy of sunlight is used and transformed by plants. At this point his activities in Göttingen came to an end, because of the rise to power of the Nazis. He took this work up after he moved to the United States in 1935.

In the early 1930s Franck had become one of the leading experimental physicists in Germany. In 1932 he was the foremost candidate for the chair of physics at the University of Berlin as successor of Walther Nernst, due to retire in the spring of 1933. Nernst had taken over the chair in 1924 after the death of Heinrich Rubens in 1922. Moreover, Haber's plan was to have Franck take over the Institute of Physics in the Kaiser Wilhelm Institute in Dahlem, for which funds had finally become available, partly from the Rockefeller Foundation and partly from the German government. Berlin was a great attraction as the scientific and cultural center of Germany. But the situation changed drastically when, in January 1933, Hitler became chancellor.

In the first few weeks there was considerable confusion. Officially the Nazi party had only 40% of the seats in the Reichstag. There were serious doubts in the minds of many people, Jews and non-Jews, about how long the regime would last, and especially about whether the Nazis would go ahead with their stated aims to deprive all Jews of citizenship eventually, and, in particular, to eliminate all Jews from civil service immediately. Judges, university professors, schoolteachers, and many others belonged in this category. By the end of March and the beginning of

April there was no doubt that the Nazi regime was going to adhere to the anti-Semitic principles proclaimed by Hitler. According to a new civil service law proclaimed in April 1933, all "non-Aryans"—all those who were Jews or had a Jewish parent or grandparent—were to be removed from their positions in the civil service. As a concession to President Hindenburg those Jews who were actively engaged in World War I were exempted—at least for the time being.

The University of Göttingen was particularly hard hit. The number of brilliant Jews in prominent positions was extremely high, especially in the faculties of physics and mathematics. In Franck's department were Hertha Sponer, his chief assistant; Arthur von Hippel, Franck's son-in-law; and Eugene Rabinowitch. In Born's department were Walter Heitler and Edward Teller. In the mathematics faculty were Richard Courant, Edmund Landau, Emmy Noether, and Hermann Weyl, who was not Jewish but had a Jewish wife. There were intensive, continual discussions about how to react to the new situation. Franck, who would be able to stay on as a war veteran, was determined not to accept the situation. He decided to resign on April 17 as an expression of active protest. He was one of the first among the famous and prominent Jewish scientists to take this action. Einstein was in the United States that winter and on his return to Europe stayed in Belgium; he never entered Germany again. In his letter of resignation Franck wrote to the minister of education that this action was an inner necessity for him because of the attitude of the government toward German Jewry. In a letter to the rector of the university he stressed that it was intolerable that Germans of Jewish descent should be treated as aliens and enemies of the Fatherland. If only for the sake of the children this attitude was unacceptable to him.

Franck did not intend to leave Germany immediately. He hoped he would be able to continue useful work outside the university and possibly help to fight the Nazis from inside. But the conditions for the Jews deteriorated rapidly and no resistance was possible. In the fall of 1933 he left Germany. He spent a year with Niels Bohr in Copenhagen and then left for the United States at the invitation of Johns Hopkins University. He spent three years at Johns Hopkins and then moved to the University of Chicago.

During World War II, Franck, as one of the leading atomic physicists, participated in the Manhattan Project, aimed at the production of an atom bomb, because he and other physicists were worried that this weapon could be produced by Nazi Germany (see II, I). However, Franck and his colleagues believed that this weapon should only be used if Germany became able to produce it. Franck was the chairman of the Committee on Social and Political Implications of Atomic Energy. When Germany was totally defeated and the war ended, in May 1945, he and other colleagues were greatly concerned that the weapon—not yet completed—should not be used in the war still going on in Japan. A report was prepared stressing the revolution in warfare that the use of the atom bomb was bound to bring about. It emphasized the great dangers that the use of this destructive weapon of completely new dimensions would create for the future security of all nations, including the United States. Such an action would be a fateful political decision. The Government should pay attention to the danger of beginning a nuclear arms race (Rabinowitch, 1964). This famous letter, generally referred to as the Franck Report, was addressed to Secretary of War Henry Stimson. Leo Szilard, who in 1939 had frantically tried to promote the production of the atom bomb, was one of the chief contributors to the Franck Report. Frightened that the bomb would be used, Szilard wanted to approach Truman directly, bypassing the Secretary of War. Franck disagreed.

In any event, the Franck Report never reached Truman. Hiroshima and Nagasaki were destroyed. The predictions of the Franck Report have been borne out by the subsequent events: The threat to mankind has become a reality. But it is an everlasting credit to James Franck that he, as one of the world's brilliant leaders in atomic physics, took the initiative in the attempt to prevent the catastrophe. For his vision and action he deserves the gratitude not only of scientists, but of all mankind. The voice of reason was forcefully expressed. It was not the fault of scientists that statesmen did not listen. The Franck Report stands out as a document that should always be remembered. Franck's life was completely devoted to science. But with this action he became a symbol and a pioneer of a new generation of scientists who recognized the great responsibility their work imposed on them and their obligation to weigh carefully the possibly dangerous consequences of their research.

After the war Franck returned to his work on photosynthesis at the University of Chicago. He probably had underestimated, having lived in the world of atoms and molecules, the infinitely greater complexity of living cells. It was many years before the rapid advances of molecular biology were able to unravel some of the most important steps of this complex process. Franck's controversy with Otto Warburg will be described later (see IV, C). There is no doubt that Franck was correct as to the quantum number required, the specific and central topic of the controversy.

Franck was a great and admirable personality. His high ethical standards, his integrity, his warmth, and his broad cultural background were apparent in all discussions. The nobility of his character revealed itself in the expressive and beautiful features of his face. He was loved and admired by students and colleagues, and it is not surprising that he attracted a great number of excellent pupils, many of whom became leaders in the field in many countries.

Franck was always proud of his Jewish heritage, as his action in April 1933 has shown. He was not a Zionist. But, as he told the author in a discussion, when Weizmann asked him in 1934 if he would be willing to continue his work in Israel (at that time Palestine), he strongly expressed his great interest. He never heard from Weizmann again. Many of Weizmann's plans for building up science in Israel were unfeasible, because of the lack of funds and facilities.

In 1964 Franck and his wife returned to Göttingen for a visit. The city had made him an honorary citizen in 1953. He had spent some of the most creative and glorious years of his life in Göttingen, and it was there that he died, quite suddenly, at the age of 83.

Max Born

Max Born (1882–1970) was born in Breslau. He came from a very interesting family. Born's (1975) book *Mein Leben* (being translated at present into English) is a fascinating presentation that describes many interesting personalities, views, and attitudes of the large paternal and maternal family. It gives a vivid account of the family background and the many factors that influenced his own life from early childhood on. It is an absorbing and illuminating documentation of that era, covering many economic, scientific, artistic, and political features of the life of an affluent (Jewish and non-Jewish) society in Germany over a period of half a century before Hitler.

Born's grandfather was Marcus Born, a physician, married to Fanny, *née* Ebstein, the daughter of a rich industrialist. Marcus Born practiced first in Kempen, Province of Posen, and moved later to Görlitz. There he became a successful and highly respected physician, with many clients among the nobility. He was appointed by the Prussian government as a *Kreisarzt* (a kind of official position with special functions). This was a rare distinction for a Jewish physician. His oldest son was Gustav, the father of Max. Gustav had several brothers and sisters. Close relationships were maintained among all family members. Gustav married Margarete, *née* Kauffmann, a member of a large, wealthy, and prominent family. Margarete's grandfather was Meyer Kauffmann, who was the founder of a large and extremely prosperous textile business. Meyer had five sons: Salomon, Julius, Robert, Wilhelm, and Adolf. Four of them became extremely capable and prosperous industrialists in the family business; only Adolf became a physician. He was a passionate music lover and the founder of the Breslau Orchestra.

Salomon was the father of Max's mother. His wife, *née* Joachimsthal, came from a very rich merchant family in Berlin. One of her ancestors had greatly helped Frederick the Great in the Seven Years War and had received full civil rights long before the general emancipation. Salomon and his wife lived in the grand style, commensurate with their wealth and tastes. They were fond of great parties, to which they attracted many artists and especially musicians. Max's mother was a brilliant piano player;

at the parties given by her parents she had met many famous musicians, among them Johannes Brahms, Clara Schumann, and Max Bruch. One of her brothers, Max, had married Luise Helfft, a daughter of one of the greatest bankers of Berlin, owner of a large private bank, like that of Bleichröder. His bank survived the crisis in the 1890s, when the rapid expansion of the economy required huge banks and forced most of the smaller banks to merge with large banking concerns. The Helffts had a magnificent mansion at one of the most elegant squares of Berlin, the Pariser Platz near the Brandenburger Tor. Max and Luise had a magnificent and superbly decorated house in Breslau. They were great patrons of artists and musicians and attracted many celebrities to the concerts in their home.

Gustav was professor of anatomy at the University of Breslau. He had two children, Max and Käthe. Max's mother died when he was four years old, a great tragedy. His father was very busy at the university and the education of the two children raised many serious problems. Grandmother Kauffmann was a domineering lady and tried to maintain nearly complete control over the children's upbringing. Her ideas and concepts frequently clashed with those of Gustav. The disagreements did not prevent the children from visiting the homes of the great number of relatives on both sides of the family, where they were received with great warmth and every effort was made to make them feel at home. The atmospheres in the homes of the two branches differed greatly. The Kauffmann family loved the grand style; the Born family was more for simplicity and stressed intellectual values more strongly. Few relationships existed between the two branches. The increasing problems with the education of the children led Gustav to think of a second marriage. His second wife was Bertha, *née* Lipstein. Her father had been a very rich merchant in Russia. In view of the strong anti-Semitism there and the frequent pogroms, he had sent his sons and daughters to Germany or Switzerland for their education. He finally emigrated and settled in Königsberg, but died soon afterward. His widow Ida moved to Berlin, where two of her daughters were married. One of them, Helene, was the wife of Ismar Mühsam, a highly educated man and a passionate music lover. In his elegant house near the Tiergarten he arranged many concerts of chamber music. Max and his wife Hedwig were frequent guests of the Mühsams when they lived in Berlin, and they met many famous musicians there.

The new stepmother was an extremely kind and warmhearted person and made great efforts to be a true mother to Max and Käthe. The marriage was a very happy one and Gustav could again devote all his time and energy to his duties at the university and especially to his research work. A son Wolfgang was born who was 10 years younger than Max. Max attended a humanistic gymnasium. He did not feel attracted to Latin, but was fond of Greek literature, especially Homer. In retrospect he wondered about the value of his humanistic education, but he concluded that it was after all an important contribution to his general outlook. Especially after he had traveled to Italy and Greece did he realize the great value of the knowledge of these two civilizations that had profoundly influenced the cultural history of Europe.

When Max was 16 years old his father became seriously ill with a heart condition. He could go out only for very short periods of time and spent the evenings at home. This gave him an opportunity to devote much time to his children, who greatly benefited from the frequent discussions with their father, a highly educated and intelligent man and a strong personality. He utterly disliked wars and the Prussian military. His brother Julius had participated in the wars of 1866 and 1870–71 and had advanced to the high rank of a physician in the army: *Oberstabsarzt*. Gustav profoundly impressed his children with his descriptions of the horrors of war, its cruel reality with its mass killings, the unimaginable sufferings and misery; he showed them how the reality contrasted to the glowing, heroic stories glorifying wars that they heard and read in school. These discussions had a deep impact on Max, who remained a strong opponent of war throughout his life. This attitude played a considerable role in his life during the first world war, and also in his friendship with Albert Einstein, whose deeply pacifist convictions are well known. Gustav was frequently visited by distinguished colleagues from the university, among them Albert Neisser, the discoverer of gonococci, who was married to a cousin of Max, Antonia Kauffmann. Listening to the discussions with the visitors was most stimulating to the children. The discussions covered not only medical problems but also literature, philosophy, and politics, and they were full of amusing stories and anecdotes. In the last few years of school, Max developed a considerable skill for mechanical work and seriously considered becoming an engineer. His father advised him against making a hasty decision and urged him to spend a year or two attending a broad range of lectures on topics that interested him before making a final decision. Half a year before Max's *Abiturientenexamen*, Gustav died.

At the University of Breslau Max attended a great variety of lectures, but most of them did not stimulate his interest. He learned much mathematics; his teachers were Rosanes and London, the father of Fritz London, who later became a famous physicist and a pupil of Max. It became apparent that physics, mathematics, and astronomy attracted him most. On the advice of his friend Dr. Lachmann, a physician and former assistant of his father's, who was about 20 years older and acted as a kind of mentor, he decided to study physics and mathematics. He spent some semesters in Heidelberg, Zurich, and Göttingen. In Heidelberg he became a close friend of James Franck. Both men could not foresee how important their friendship would become later in their lives. Another close friend was the physicist Rudolf Ladenburg, the son of Albert Ladenburg, professor of chemistry at the University of Breslau. They spent many vacations together and maintained their close friendship for life. In 1921, when James Franck moved to Göttingen from Haber's institute, Ladenburg became his successor, and later professor of physics at Princeton (IV, A). In Zurich Max was fascinated by the lectures of Adolf Hurwitz, who had come from Königsberg, where he had been the teacher and friend of David Hilbert and Hermann Minkowski. When he returned to Breslau, he had no teacher who really interested him, and few friends. The house of the Neissers was one great attraction; he met many interesting and famous people there, among them Gerhart Hauptmann, Richard Strauss, Artur Schnabel, and Edwin Fischer. But as much as he enjoyed these parties, they could not compensate for the lack of inspiration at the university. Following the advice of a friend he decided to spend a semester in Göttingen.

In Göttingen, Born's brilliance attracted the interest of two of the greatest mathematicians there, Hilbert and Klein (see II, C). Through a letter of recommendation from his stepmother he also met Minkowski, who invited him for lunch and took him later for a walk on the famous Hainberg, where they were joined by Hilbert and Klein. To the great surprise of Minkowski, Hilbert greeted Born very warmly and immediately started a scientific discussion with him. Born attended many lectures in mathematics. Klein was so impressed that he proposed that Born write a paper on a certain topic, to be awarded a prize. Born declined the offer because he was not interested in the topic. This was quite naive, an incredible affront for any *Ordinarius* in Germany, but particularly for Klein, who was accustomed to being treated as a king, even by his senior colleagues. Klein was furious and Born suffered for this mistake for a couple of years. He finally managed to get his Ph.D., in 1906, under the sponsorship of Carl Runge.

In 1906 his sister Käthe married an architect, Georg Königsberger. They lived in Grünau, near Berlin. After getting his doctorate, Born learned—to his surprise and anger—that he had been accepted for military service. Earlier in his career, after the *Abiturientenexamen*, when most people served their term, his service had been postponed because of chronic asthma, so he did not expect to have to join the army at this stage in his life. Although he reached an old age—he died at 88—his health was always fragile.

Born was reluctant to join the army for other reasons. He was strongly antimilitaristic and pacifistic; he had listened well to his father's stories of the horrors of war. He also faced, for the first time in his life, anti-Semitism. A strong anti-Semitic spirit had permeated the army; it was very difficult for Jews to advance, whereas in the preceding generation many Jews had reached the rank of reserve officer. For Born, the Jewish problem had not existed before. In elementary school he had attended Christian religious teaching, since no other was available. His father wanted him to know the Bible, since it had greatly influenced European civilization, but by whom it was taught seemed to him irrelevant. In the gymnasium there was Jewish religious teaching, but Born disliked it and his father did not object when Born stopped attending it. Both Born and his father considered themselves, as most other assimilated Jews did, German Jews. Actually, they considered themselves Germans, since Jewishness had no meaning or value except that their ancestors were Jews.

In spite of all his antagonistic feelings, Born was now forced to join the army. On the advice and with the help of his stepmother's brother, he joined a cavalry regiment that was stationed in Berlin. The regiment attracted many young men of rich families and aristocrats who did not take their duties too seriously. The sergeants were accustomed to accepting bribes for giving lots of free time. Born liked riding and he thought he would be able to visit his sister in Grünau, and perhaps have the time to attend the great theaters, operas, and concerts in Berlin. His friend James Franck was serving in the army at the same time in Berlin, although in another regiment.

The reality of military life was, however, quite different than Born's expectations. He suffered almost continuously from asthma attacks. He was much too tired to enjoy anything. There were other complications resulting from the famous story of the "Hauptmann von Köpenick"* which took place at that time and led to extremely strict controls. During a parade in honor of Kaiser Wilhelm's birthday, Born had to march for hours. He was not accustomed to such strenuous exercise, and he considered the whole event humiliating, incompatible with human dignity. On his return from the parade he fell ill and spent some time in the hospital. To his great relief he was soon dismissed as "unsuitable for military duty."

Born returned to Breslau, unsure of the direction he should take. A relative was going to England and Born decided to go along. He was influenced by James Franck's great admiration of English physicists, such as J. J. Thomson and Rutherford. A serious difficulty was that he did not speak English, but he decided to learn it. In 1907 he went to Cambridge, but his English was poor, a great handicap in trying to get into a college. By accident he met a group of German students, among them H. O. L. Fischer, the son of Emil Fischer, and they tried to help him. It was not easy, since all the colleges were crowded, but he was finally accepted. He attended many lectures, including those of J. J. Thomson. He was fascinated by Thomson's brilliant experiments, but was unable to follow the theoretical explanations because of his poor English and his unfamiliarity

---

* A shoemaker, having spent a long time in prison, bought an old officer's uniform and ordered a group of soldiers to go with him to the small village of Köpenick. He arrested the mayor there and seized the cash box of the city with all the money. This affair created a great sensation and ridiculed the respect for the army uniform in Germany. The army was furious and strengthened the discipline.

with the field. He attended a rather elementary course in experimental physics. When his studies had ended, he traveled through England, visiting many famous cathedrals. He also visited Edinburgh. He could not foresee that one day fate would bring him there to live for many years. Perhaps the most important result of the trip was his decision to become a physicist and to abandon pure mathematics, except as it was necessary for theoretical physics. In Göttingen he had attended the physics lectures of Voigt, who was a poor teacher, and Born's interest in the field had not been stimulated.

Before he could start with his studies he had to spend a few more agonizing weeks in the army; he received a draft order. On the advice of a physician, a pupil of his father, he joined one of the most aristocratic and exclusive regiments of the army, stationed in Breslau. The physician there also was a pupil of his father and a friendly man. Born noticed that the physician immediately considered him totally physically unfit for the army, and after a month he was discharged as not being physically suitable for service. Now he was free to proceed with his studies.

At the University of Breslau there were two chairs of physics. O. von Lummer was the professor of experimental physics, A. Pringsheim, the professor of theoretical physics. Both had been at the *Physikalisch-Technische Reichsanstalt* in Berlin, where they had become very close friends. Their personalities were very different. Lummer was full of vitality, dynamic and impulsive; he was a brilliant experimenter. Pringsheim was a quiet thinker, extremely reserved, cautious in his judgments, and very modest. Pringsheim was Jewish; Lummer was not. Lummer had accepted the chair only on the condition that Pringsheim would get the chair of theoretical physics. Their work greatly contributed to the final recognition of Planck's notion of the quantum of action. Pringsheim became Born's sponsor and put him on a research problem related to Planck's work on heat radiation.

Born had three good friends in the laboratory. He had known Rudolf Ladenburg from early childhood. They shared many common interests and tastes in science, music, the arts, and love of nature. They took many trips together, and often went skiing. Born admired Ladenburg's ability to make rapid decisions, his elegance, his well-trained body. Another friend was Fritz Reiche, who came from Berlin and was a pupil of Planck's. He was Jewish and had the typical Berliner wit and humor, but was prone to depressions and pessimism. Born learned much from him about quantum theory. Reiche later became professor of theoretical physics at the University of Breslau, the chair once held by Pringsheim. A third friend was Stanislas Loria, a Pole from Krakow, at that time under Austrian rule. He was highly educated, elegant, and proud of his Polish nationality; he hated Austria. Born learned much later that Loria's mother was Jewish, although Loria was the prototype of a Polish aristocrat. During World War II Loria was in a concentration camp for a long time, but he survived and became professor of physics at Breslau, at that time a Polish city. Born had a very active social life. He enjoyed the many parties given by various members of his vast family, where he met many famous artists and musicians. He played much piano, took part in chamber music, and went frequently to the theater, opera, and concerts.

But a terrible accident happened in the laboratory. In his work on heat radiation he used a black box kept in a metal holder through which water was continuously running for cooling. It was a complex system of glass and rubber tubing. At the end of the experiment he had to wait for an hour before turning off the water. One day the work dragged on; he was late for some appointment and decided to let the water run overnight. The next morning he found not only his laboratory but the laboratory on the floor below flooded; the ceiling was near collapse. The rubber tubing had gotten loose from the faucet. Lummer was furious; Pringsheim was very angry. Born had to pay all the expenses for the repair, a considerable sum. Born was not very skillful and did not have much experience with experimental work. Lummer, who frequently visited him, was dissatisfied with what he saw and also with Born's knowledge in other fields of physics. After this catastrophe it was clear that he had very little chance for an academic career. He continued his life as usual. He even went on a trip to Greece. But he was unhappy. More important than anything else was for him to have an opportunity to continue his work in his chosen profession. In this respect there seemed to be little hope. He was at a loss as to what to do. Just at that time Reiche asked him whether he had heard about Einstein's theory of relativity. He had not, but because he had been fascinated by electrodynamics and optics since he had attended lectures of Hilbert and Minkowski on these topics, he became most interested in the problem. Both he and Reiche became increasingly enthusiastic, discussing the theory, and their interest was shared by Ladenburg and Loria. Born gave much thought to the problem and when he encountered some difficulties, he wrote a letter to Minkowski asking for advice. Minkowski did not answer his questions, but wrote that he was working on similar problems and was looking for a collaborator. Would Born be willing to join him? This would be the beginning of an academic career. He suggested that Born attend a meeting in Cologne where he would give a lecture on space and time (see II, C). There they could discuss everything.

Born was fascinated by the lecture and joined Minkowski in Göttingen. A few weeks later, in January 1909, Minkowski died. Born was in a state of deep despair and depression. Shortly afterward he was asked to go over Minkowski's manuscripts on mathematical physics. It turned out that most of them were already published. Only one research project was sufficiently advanced and Born finished it and the paper was published. He now started to study the field with great intensity and became completely absorbed in the problems that were at that period in the center of interest.

He worked on a specific problem during the whole winter and then asked permission to present a paper on his results to the Mathematical Society. The lecture turned out to be a catastrophe. It was given before a galaxy of stars, among others Klein, Hilbert, Landau, Runge, Voigt, and Hermann Weyl. Klein was opposed to some aspects of the theory of relativity and probably still had some resentment against Born. He raised quite a few objections and questions. Born was inexperienced in giving lectures and soon became totally confused. Finally Klein said that he had never in his life heard such a poor lecture. Born was in a state of collapse. When he was leaving the room, Runge told him that he found the work interesting and that he should come the next morning to explain his ideas to him. The next morning they had a long discussion. Runge seemed convinced. He said he would take up the matter with Hilbert. He and Hilbert then tried to convince Klein that he had been both unfair and wrong and that the situation should be repaired. All this Born learned only later. He felt so miserable and depressed during the following period that he decided to leave Göttingen, to give up his academic career, to go to a *Technische Hochschule* and become an engineer. These were the worst days of his life. Other difficult times in his life—the wars and exile, for example—could be attributed to general events, which he shared with others. In this case he felt that he just lacked the ability to become a scientist, a blow that was very hard to take.

One day, however, he got a letter from the secretary of the society saying that his lecture apparently had not been correctly understood. Would he be willing to repeat it? This was a great event. "Felix the Great," as Klein was called, the uncrowned king of mathematics of that era in Germany, thus publicly admitted that he may have been wrong and the young fellow right. What a testimony to the spirit of that era, to the scientific atmosphere in Göttingen, to the prevailing magnanimity and fair-mindedness. Born was full of anxiety and reluctant to accept, but he was finally persuaded by his friends. The second lecture was a triumph. Voigt approached him right after the lecture and offered him a position as *Privatdozent* in his institute. He could use the lecture as basis for his *Habilitationsschrift* (a paper—when accepted—by which the author became *Privatdozent* with the license to lecture to students). That summer, in 1909, he became *Privatdozent*.

Five happy years followed until the outbreak of the war. In addition to his lectures, which attracted many brilliant students, he was very active in research. The most outstanding and brilliant work of that period was that on crystal lattices, which he carried out together with Kármán. It was an ideal collaboration. Although Kármán was fascinated by the beauty of pure mathematics, his real passion and interest were applied mathematics. His fame in the field of aeronautics is well known; he became one of the greatest pioneers in this field in the United States, where he went long before Hitler's rise to power. From him Born learned to recognize the essentials in mathematical physics and to see problems in the right perspective. The extraordinary importance of their work on crystal lattices found wide recognition, although only after a considerable delay. Born was repeatedly honored for this work for which he always retained great interest and to which he frequently returned throughout his life. Peter Debye had simultaneously and independently done some similar work on crystal lattices, but the work of Born and Kármán turned out to be in some respects more satisfactory. This shows the correctness of the Latin proverb: *Si duo faciunt idem, non est idem.*

In addition to his work, Born enjoyed these years in many ways. As *Privatdozent* he was frequently invited to the parties of the professors and met many interesting and stimulating people. He devoted much time to music. He went on interesting trips, some of them with Ladenburg to Italy. They visited Florence and the many enchanting little cities in Toscana. One summer they spent in Southern Tyrol, with which Born remained in love for the rest of his life. Although he loved Wagner operas, he disliked the snobbism of Bayreuth. But he could not refuse the invitation from Toni Neisser, a relative of his, to go by car to Bayreuth and visit many lovely places in southern Germany on the way. Albert Neisser was a great enthusiast of Wagner operas. He had strongly promoted the building of the opera house in Bayreuth and had contributed large sums of money. When a huge party was given for the guests, the Neissers were not invited. Jews were absolutely excluded. Neisser was deeply hurt, although the topic could not be mentioned. Cosima Wagner's violent anti-Semitism, comparable to that of the Nazis, and that of her daughter and son-in-law, Houston Stewart Chamberlain, have been discussed before (see I, B).

In those years many famous mathematicians and physicists were attracted to Göttingen and spent some time there as guest professors. Born became very friendly with some of them, especially H. A. Lorentz and A. Michelson. The Michelsons invited him to spend a few months in Chicago. He accepted their invitation and visited the United States in the summer of 1912. He gave many lectures there, all well received, on the theory of relativity. The Michelsons also saw to it that he had a very pleasant time. At the end of his stay in Chicago he went on an extended tour and saw many parts of the country, including California, the Rocky Mountains, and the Grand Canyon.

In the following year, in the spring of 1913, he became engaged to Hedwig, the daughter of Professor Victor Ehrenberg. There was some difficulty with Mrs. Ehrenberg. Hedwig's father was Jewish, her mother was not. Her mother was the daughter of a famous lawyer, Rudolf von Ikering, a direct descendant of Martin Luther. Although she was very fond of Born, he was a Jew and refused to be baptized. She wanted a wedding in a church. Since this was impossible, a compromise was found. The wedding took place in the house of Born's sister, Käthe, in Grünau.

But there was in addition a kind of half-religious ceremony in the presence of a pastor. Many guests of both families were invited. It was a wedding in the grand old style. Nevertheless, Hedwig's mother continued to pressure him to get baptized and he finally gave in. He had never practiced the Jewish religion and he was completely unfamiliar with Jewish rites and traditions. He remained a Jew, just as his father did, because what was good for their ancestors should be good for them. In fact, he was prejudiced against Christian churches, since the history of Christianity had shown how many cruel persecutions and abominable crimes had been committed in the name of Christ. There were other reasons. But in essence he felt that religious confessions and churches were irrelevant. Although he received a few lessons in the Christian faith, baptism did not change him. He never regretted his decision. He did not live in the Jewish world and felt it difficult to live in a Christian world as an outsider. But he never tried to hide that he was a Jew. Those who tried were found out by the Nazis anyhow. When the Nazi persecution started, he felt in spite of his baptism that he belonged to the Jewish people and suffered with them. As he once confessed to Einstein, after the rise of the Nazis he felt even stronger than ever before that he belonged to the Jewish people. This reaction was not uncommon among his Jewish colleagues. Obviously there were great differences as to the degree of these feelings and the strength of the reactions.

Shortly after Born's wedding Voigt decided to retire. Peter Debye became his successor. Born started some work with his friend Richard Courant on quantum theory, which just had taken such a startling turn because of Bohr's atom model. But his interest centered on the problems of crystals. He worked out a theory, but had no means of testing it. He had the great satisfaction of having his theory confirmed three decades later, by an Indian physicist. However, the era of the splendor and glory of Europe was soon approaching its end. Quite unexpectedly to many people, among them Born, World War I broke out. Few people knew the actual background of the outbreak of the catastrophe. Nevertheless, everywhere a spirit of patriotic enthusiasm prevailed. To Born, with his deep hatred of war and his strong pacifist attitude, the war seemed utter foolishness, an explosion of madness and insanity. For him the main nightmare was the mass killings of young, talented people, the flower of many nations. Few people realize even today that it was not the physical or economic destruction, but the loss of a whole generation, among them the elite and the many potential geniuses in all fields, the hope of the future of all nations involved in the war, that was the key element of the disaster that befell Europe. France alone lost almost 2 million (of a population of 38 million); Germany lost more than 2 million (of a population of 70 million). Churchill stated shortly after the war that it would take two generations—60 years—for the consequences of this large-scale slaughter to become manifest. Born's thoughts were continuously preoccupied with these killings, of which he was constantly reminded by hearing and reading of the unbelievable death toll. Completely unaware of what had really happened, Born, like most other people, thought that Germany was the victim of a ruthless and unprovoked attack. Physically unfit for the army, he volunteered to help harvesting crops at a farm, a job for which people were urgently needed. Even this soon turned out to be too much for him and he was forced to give it up. Since most of his students had been drafted, he had much time and used it to write a book on the dynamics of crystals. Writing the book was a great relief from the tension, despair, and exasperation he felt about the ongoing war.

On the day the war broke out, Born received a letter from Planck offering him a professorship of theoretical physics in Berlin. Planck was unable to carry out all his obligations and had asked the ministry for another professorship; the request had been approved. Born had met Planck at a few meetings but did not know him very well. But he felt honored and gratefully accepted. An embarrassing situation developed, however, because von Laue, who was already very famous, was anxious to move from Würzburg to Berlin to be close to his beloved teacher. The difficulty found an unexpected solution. The ministry insisted that von Laue take over the chair of physics at the newly created University of Frankfort on the Main. So the Borns moved to Berlin in May 1915 and took an apartment in the Grunewald, a lovely suburb of Berlin, although quite far away from the university.

In Berlin too the number of students was small. To compensate for the terrible losses on the battlefields, the standard of health required for army service was lowered and more and more students were drafted. The book on crystals was finished. Born worried that even he, in spite of his physical weakness, would be drafted. At that time a group of physicists and engineers was formed to help improve the communications system between airplanes and the ground forces. Born became a member. The group was stationed near Spandau, quite near Berlin. One day they had to march in pouring rain, and Born came down with severe bronchitis and had to stay in bed. When he recovered he did not know what to do, still worrying about the draft. Ladenburg heard about his predicament and asked him to join his group, which tested all inventions concerning the artillery. The job was easy, the working hours only from 9 A.M. to 4 P.M. Most of his assistants were students of physics and mathematics, and Born was happy that they thereby escaped the battlefield and possibly death. One student was nevertheless drafted; Born considered him a real genius and tried desperately to save him. When he learned that the student had been killed, he was shattered.

Einstein was one of the first people whom Born visited in Berlin, although the two had previously met only briefly. But soon Einstein appeared at Born's home with his violin. Hedwig was delighted. From then on they played music quite frequently together. All three became good friends, and after the Borns left Berlin for Frankfort they maintained their friendship through correspondence until Einstein's death in 1956. Their exchange has been published by Born (1969) and is most interesting; it reflects their ideas about many aspects of that era, for almost four decades.

Born and Einstein had many discussions about physics. Born had read Einstein's latest papers on general relativity, and he was struck by Einstein's way of thinking, his faith in the simplicity of fundamental laws. Although all Einstein's theories were based on experimental facts, he admired Einstein's ability to interpret well-known but inconspicuous facts that had escaped everybody else. Perhaps the most important example is the equivalence of gravitation and acceleration, known since Newton's day but not recognized as the key to the understanding of the cosmos until Einstein. Extraordinary intuition was the clue to Einstein's genius, not his mathematical knowledge, as is shown by his interpretation of the photoelectric effects in terms of Planck's energy quanta of action, as mentioned before (see II, B). At the time they met, Einstein was trying to enlarge his mathematical knowledge, but thereby removed himself from physics. When Born, Heisenberg, and Niels Bohr developed quantum mechanics, Einstein disagreed with their interpretation. He and Born became scientific opponents, but this did not affect their friendship in the least. In his letters Born repeatedly tried to convince Einstein, but had no success.

In the following years (1916–1919) the Borns met the Blaschkos, who became their best friends during their Berlin years. Dr. Alfred Blaschko was an outstanding dermatologist. The two families met very frequently. They talked much about politics. Blaschko was a convinced socialist, although of the moderate wing. He was less interested in research than in social problems and public health affairs. He was very active in fighting for proper legislation on venereal diseases. Born listened with great interest to the passionate discussions about politics between Blaschko and Eduard Bernstein, a leading socialist, although he could not get really excited about economic theories and problems. Born was, however, quite impressed by the ethical and religious ideas of Blaschko. The Blaschkos were relatives of the Ullsteins, owners of a huge publishing house. The Borns were frequently invited to the Ullsteins and met many interesting people, especially writers and artists. Alfred Blaschko's son is Hermann Blaschko (see V, D) who became a leading scientist in England.

Born's work in Ladenburg's group did not prevent him from being very active scientifically. His book on the dynamics of crystals had been published and he turned to other problems, such as the properties of optically active molecules (those able to change the polarization of light passing through them), the electromagnetism of crystals, and Sommerfeld's work on quantum theory (based on Bohr's atom model). He was deeply impressed by Sommerfeld's work. He believed that the great importance of this work was not fully recognized and appreciated, and that he had richly deserved the Nobel Prize.

The war was finally drawing to an end. Germany surrendered under the conditions of President Wilson's Fourteen Points. As is well known, the Versailles treaty disregarded Wilson's conditions. Einstein and Born were appalled by the harsh conditions of the Versailles treaty. They were very worried that the great number of incredible humiliations—such as paying fantastically high reparations for four generations (120 years)—would sooner or later lead to terrible reactions among the German population and disastrous consequences for all of Europe. The postwar conditions in Germany were frightful, with continuous fighting between the different parties from the right and left, frequent assassinations, economic disintegraton, and the like. In some cities, among them Berlin, street fighting made life frequently dangerous by the outbreak of unexpected shooting.

At that time James Franck was already in Berlin in Haber's institute in Berlin–Dahlem. Born recognized the outstanding contributions of Haber to the war effort. But he disliked his work on chemical warfare, although he realized Haber's intention. Haber's hope was to end the trench warfare with this new weapon and thereby make a rapid decision possible and in the end save innumerable lives. During a visit in Franck's laboratory, Haber entered and Born met him for the first time. He was fascinated by Haber's personality, his charm, his energy and dynamism, the brilliance of his mind, and his familiarity with a great number of different fields of science. Born received from Haber much invaluable information that was of decisive help for his work. Haber also improved Born's calculations of chemical energies involved and proposed a graphic presentation known later as the *Born–Haber cycle*. The results found widespread interest. Born was frequently invited to present this work at chemical meetings. In spite of some objections, particularly by Nernst, the importance of his work was fully recognized and brought him his first award, an honorary doctoral degree in Bristol, which he received in 1927, together with such celebrities as Rutherford, William Bragg, and Paul Langevin.

In 1919, von Laue again tried to join his beloved teacher Planck and suggested to Born an exchange of professorships, quite an unusual procedure. This time Born felt that he should not resist, so he accepted the position in Frankfort. The move also meant that he would become an *Ordinarius*, whereas in Berlin he was *Extraordinarius*.

In his institute he had Otto Stern as *Privatdozent*. Stern soon became a good friend. During Born's time in Frankfort, Stern started the measurements of the magnetic properties of the atom by the deflection of an atom beam in an heterogeneous magnetic field, which led to his famous work (see below). The professor of experimental physics, Wachsmut, was a charming man, but not too much interested in research. His assistant, Walther Gerlach, was much more attracted to Born and they collaborated on several problems. Gerlach later joined Stern when he left Frankfort. But soon the inflation became noticeable, the instruments became too expensive for the available budget. At that time Einstein's relativity theory was confirmed and became *the* great sensation. Born used his familiarity with the theory to give a whole series of lectures, for which an entrance fee of a few marks was charged, and he used the money to buy instruments. But this source soon ran out. When a friend left for the United States, Born told him jokingly that if he found a German-American interested in Germany, he should tell him of the urgent needs for money to continue important experiments. Soon Born got a letter from his friend, telling him to write to Henry Goldman, the grandson of a Jewish immigrant who had left Germany because he could no longer stand the anti-Semitism there. Born did write Goldman and got a charming letter with a check for several hundred dollars enclosed, at that time a large sum that solved all his problems. Goldman was the director of one of the biggest private banks in New York. He later visited Germany and extended his support to Born and also to many other scientists and artists. Many others in the United States generously offered their help; the financial support tendered by industry, foundations, and other sources became a decisive factor in the striking development of German science in the 1920s. Later it turned out to be an excellent investment; when the Nazis drove so many brilliant scientists from the country, most of them settled in the United States.

Born's chair had been established by a gift from a very rich Jewish merchant, Oppenheim. His wife was a Viennese and an enthusiastic piano player, trained by Clara Schumann. The Borns were frequently invited to their home and he and Mrs. Oppenheim frequently played music together. The Borns were also taken many times to the opera and saw many wonderful performances. During the Nazi period the Oppenheims committed suicide.

Born enthusiastically supported the socialist regime. He became very upset about the Kapp putsch, when a military group tried to overthrow the government, and he was greatly relieved when the attempt failed because of a general strike. Born was also disturbed by the French occupation, which was deeply resented by the whole population and considered an unnecessary and unprovoked humiliation. It led to some violent reactions of the population and as a result quite a few people were killed or wounded.

When Hedwig's parents visited them, her mother became a victim of a worldwide influenza epidemic, which killed many people. It was a great shock for Hedwig as well as for Born, since she was like a real mother to him. They decided to spend their summer vacation in the southern Tyrol and Hedwig fell in love with the country. They also visited Venice. It was during this period that Born published a book presenting Einstein's relativity theory without the use of higher mathematics. The book was a great success. But Einstein's fame triggered much hostility and hatred. The leaders of this anti-Einstein movement were the strongly anti-Semitic physicists Philipp Lenard and Johannes Stark, the leaders of the "Aryan" physics during the Nazi period. At a meeting in Nauheim, in 1920, they vigorously attacked Einstein, while Planck, Wilhelm Wien, and von Laue defended him. During that time Einstein lived in Born's house in Frankfort, which is not far from Nauheim.

In 1921 Born was offered the chair of physics in Göttingen. Both Riecke, professor of theoretical physics, and Voigt, of experimental physics, had died. Peter Debye had taken over for a short time, but then decided to accept the chair in Zurich, in view of the difficult conditions of life in Germany. Pohl directed the department of experimental physics, and Born was to take over theoretical physics. As mentioned before, Born realized immediately that there was space and the need for a second chair of experimental physics, and on his insistence James Franck was offered the chair, which Franck accepted. At that time Felix Klein retired, and Richard Courant, a friend from the time in Breslau, became his successor. Hedwig was a close friend of Nina, the wife of Courant and the daughter of Carl Runge. Nina was an excellent violinist, and the Courants' house soon became a center not only of mathematics but also of music.

The next decade was the period when Göttingen was one of the greatest world centers of physics and mathematics, as already mentioned, but it was also Born's most creative and happiest time. His contribution was just as essential for quantum mechanisms as were the contributions of Heisenberg and Bohr. When Heisenberg received the Nobel Prize in 1932, he wrote a letter to Born, saying that his conscience was uneasy because the prize had not also been offered to Born. He added that every good physicist knew how great Born's (and Jordan's) part in this achievement was. He went on to say that unfortunately he could do nothing about it but be ashamed. This letter is a great testimony to Heisenberg's personality. There are few examples of such magnanimity in any field of human endeavor. When Born received the letter, in 1933, he was already in exile in Cambridge. It took 28 years for the Nobel Prize committee to recognize its mistake: Born received the Nobel Prize in 1954.

Quantum mechanics, particularly its mathematical foundation, is perhaps the greatest of Born's achievements, but it is by no means the only one. He was incredibly creative in many other fields of theoretical physics during this period. A superb documentation of his scientific stature, the brilliance of his teaching, and his great human qualities is the amazing number of extraordinarily outstanding pupils whom he attracted and who were trained by him. Wolfgang Pauli, Heisenberg, and Pascual Jordan have been mentioned; among his other pupils, who either received their doctorates with him or spent some time for training and research with him were Robert Oppenheimer, Max Delbrück, Maria Göppert Mayer,

Paul Dirac, Edward U. Condon, Edward Teller, Eugene P. Wigner, John von Neumann, Fritz London, Linus Pauling, E. Hueckel, and W. Heitler. Even if some of them were not actually his pupils, they all were greatly inspired by Born. About 10 of them became Nobel laureates, and all of them became great leaders. But there were, of course, many others from many countries.

In the summer of 1928 Born was invited by the Russian physicist Joffe to attend a physics congress in Leningrad. He accepted. Following the meeting they were taken to quite a few cities, but Born was so exhausted that he left the group and returned home. In the winter of that year he became tired and weak; he could not sleep. He was forced to interrupt his work for almost a year. He went to a sanatorium in Konstanz on the Bodensee. Since it was a very cold winter, people had to stay inside. Most of the patients were middle-aged merchants, lawyers, officials, and the like. Their discussions were a great shock to Born. All spoke about politics and all were Nazis. Born had until then no idea of how this fanatic nationalism, militarism, and anti-Semitism had permeated ordinary citizens. They all believed the lies spread by Hitler: that the war was lost only by a stab in the back (*Dolchstoss-Legende*), that military leaders were not to be blamed for this defeat, that socialism and liberalism had to be suppressed, that Germany had to fight again to regain the lost territories, and that the Jews were to blame for the defeat. It was for him an almost incredible revelation; like so many of his colleagues, he had no idea what was going on in Germany. It was about that time (1930–1933) that the fight of the Nazis to take over began to intensify, although in Göttingen, at least on the surface, life went on as usual. In fact, in 1932 Born was elected to the deanship of the science faculty, his first administrative job. But the economic crisis grew rapidly, the number of unemployed rose, and finally Hitler became chancellor. There were frequent discussions between Franck, Courant, Weyl, and others. Friends in Switzerland advised them to leave Germany as early as possible. After the proclamation of the civil service laws, Franck resigned on April 17 (p. 65). A week later, the newspapers published the names of discharged professors, among them Born. All that he had built up in Göttingen was ruined. Although the news was not unexpected, Born was shattered. Helwig suffered even more. They left early in May for Wolkenstein in the southern Tyrol, where they had rented an apartment. By the end of May invitations started to pour in from many parts of the world, one of them from Cambridge. Just at that time a physics congress was being held in Zurich. Blackett was there, whom Born knew well. He was from the Cavendish Laboratory, and Born discussed the conditions with him and accepted. In Zurich he met several of his German-Jewish colleagues, among others Otto Stern, who went to the United States, and Franz Simon who, after some hesitation, accepted an invitation from F. A. Lindemann in Oxford. Early in July the Borns left for Cambridge. Lindemann was traveling all over Germany, trying to get as many of the excellent German-Jewish scientists as possible to England, since he believed that England would greatly benefit. He visited Born, who had, however, accepted a position in Cambridge. Lindemann had many connections with the British aristocracy. He was a good friend of Churchill and became his chief scientific adviser during the war. Born had no idea until after the war that Lindemann was responsible for the indiscriminate bombing and the destruction of German cities, which was designed to break the will of the German population to resist. Born found this idea abominable, as one could expect from his whole attitude toward war from his youth. That the Nazis were the first to use this kind of warfare, destroying Warsaw, Rotterdam, Coventry, and countless other cities, seemed to him no excuse for such retaliation. Like his friend Einstein, Born was at that time no longer a pacifist, however. He considered the Nazi regime as the greatest evil to befall mankind, and he believed waging the war against it until its destruction was a necessity.

In Cambridge they met Haber. He was sick, depressed, and lonely, a shadow of himself. Hedwig too was most unhappy in Cambridge. She felt suddenly cut off from her beloved country, her friends, her language. She concealed it from her husband and children not to complicate the situation and confessed it only much later. Born had at least the compensation of his work and close contact with brilliant colleagues in the Cavendish Laboratory. He learned much about nuclear physics, a field he had previously neglected. He met the astronomer Sir Arthur Eddington, whose observations on the deflection of light by the sun had confirmed Einstein's predictions and were decisive for the great triumph of the theory of relativity; overnight they had made Einstein the greatest celebrity of the century. Among Born's students in Cambridge was Maurice Pryce, who later married Born's daughter Margaret. Born was particularly pleased by frequent visits with Erwin Schrödinger and Franz Simon, who lived in Oxford. Being in the same situation fostered a quite intimate relationship. Through Simon, Kurt Mendelssohn, Kurti, and other German-Jewish refugees, the Clarendon Laboratory in Oxford became one of the leading institutions, especially in the field of thermodynamics at low temperatures. Born traveled quite a bit; he was frequently invited to give lectures, several of them in the *Institut Henri Poincaré* in Paris. Much of his time was devoted to correspondence to save friends and colleagues from Nazi Germany. In many cases he was successful, but not always. He was particularly shattered that he was unable to save a brilliant violinist, Alfred Wittenberg, a member of the famous Klingler Quartet, who was later killed by the Nazis.

One day he was invited by Sir Chandrasekhara V. Raman to spend half a year in Bangalore at the Indian Institute of Science. Raman was famous for his discovery of light deflection by the change of light frequency, the so-called *Raman effect*, for which he received the Nobel Prize. Born's life in India was a new and quite interesting experience. They met many Indians, but also Englishmen. Once they were invited by a maharaja and saw the luxury of palace life. Both enjoyed their life in India, Hedwig even more so than Max. Raman tried to get a permanent position for Born. Born was not too enthusiastic. He disliked the great contrast between the luxury of some, and the incredible poverty and misery of the great majority of the people, the many prejudices and obsolete customs. But he was willing to accept. This decision depended, however, on the approval of the council of the institute. At the decisive meeting, at which Born was present, an Englishman, Professor Aston, vigorously objected to the nomination, saying that the council should not accept a second-rate foreigner, one who was driven from his own country. Born was deeply hurt. The Borns returned to Cambridge in the spring of 1936. He was worried, since his contract was for a limited time. He again got many invitations for lectures, one of them from his friend and colleague Charles G. Darwin, professor of theoretical physics in Edinburgh.

Soon afterward Darwin became headmaster of a college in Cambridge, and his chair was offered to Born, probably on Darwin's advice. He was happy to accept the offer. For the next 17 years the Borns lived in Edinburgh, the longest time they had spent in any place. They were warmly received. The laboratory facilities were quite limited and primitive, but he had many excellent pupils there. One of them was Klaus Fuchs. He and his father Emil, a protestant pastor, were German refugees because they were convinced Communists. The father lived in Oxford and became a Quaker. Fuchs was a brilliant student and went into nuclear physics. During the war he became involved in the atom bomb project. After the war he became involved in the famous spy story for having transmitted secrets of the atom bomb to the Russians. It was a surprise to the Borns.

The Borns had many friends in Edinburgh. They became British citizens before the outbreak of the war. This saved them from the hardships other German refugees, including many scientists, were exposed to during the war. Nevertheless they passed through many difficult years. Long before the outbreak of the war Born was greatly concerned because the British government did not take the German war threat seriously, and it was therefore totally inadequately prepared when the war broke out. Even politically naive Germans, such as Born, took the war threat very seriously. Churchill himself was at first a great admirer of Hitler. He had always considered the Versailles treaty as a terrible blunder; he called it "a sad and complicated idiocy." He hoped Hitler would correct the impossible situation. It was not until the end of 1934 that a German journalist, Leopold Schwartzschild, opened Churchill's eyes about what was really going on in Germany. But very few Englishmen took Churchill seriously until many years later, as the author knows from personal experience. Then came the terrible war and postwar years that brought so much suffering to Great Britain.

In 1953—he was 71 years old—Born reached retirement age. In 1954 the Borns decided to return to Germany. They lived in Pyrmont, a small and quiet spa near Göttingen. It was an extremely difficult decision. They were torn between many contradictory feelings about returning to Germany. As Born said, he was unable to explain the decision. Germany had been dominated by the most criminal gang that ever came to power. Millions of Germans had died fighting for Hitler. Many of Born's own close relatives and friends were killed by the Nazis. Millions of Jews perished in the Holocaust in the concentration camps. Despite all these horrors, the Borns still had a strong nostalgic feeling for the *Heimat*, for the German language, for the countryside they loved. Scotland had invited them; they had been received with great kindness and warmth. They had experienced the meaning of a real democracy and of fair-mindedness. But they were not Scots and would never be Scots; they would remain strangers for the rest of their lives. It was a struggle between love and hatred on the one hand, and between love and strangeness on the other. Probably the final choice in this dilemma was forced upon them by financial problems. The pension of a retired professor in Scotland amounted at that time to almost nothing. In Germany they could live comfortably on their restitution money, which at that time could not be transferred to England. They had always been accustomed to a comfortable life, and it would have been a great hardship to live at that age and in poor health in real poverty. After his return to Germany, Born wrote several interesting books, one of them particularly remarkable and of great relevance on the responsibility of scientists (Born, 1965). Born died in 1970, at the age of 88. His son Gustav is a distinguished pharmacologist and chairman of the department of pharmacology in Cambridge.

Franck and Born will go down in history as two of the great architects of the atomic age. These two giants had a decisive impact on the development of atomic physics. They were central figures in the great and glorious era of the University of Göttingen in the 1920s. It was a great loss for German science that they were forced to leave their country. But their activities in research and teaching continued and were of great benefit to the countries in which they settled. These two men are among the finest representatives of German-Jewish pioneers in science, not only because of their outstanding scientific stature, but because of their great human qualities and their high cultural standards. They lived in a unique era that resulted from the close association and genuine friendship between the scientists of the two ethnic groups. They represent an exceptionally superb combination of their Jewish heritage with its high ethical, spiritual, and intellectual values, its zest for generations for learning and study, and the finest traditions of German culture.

Gustav Hertz

Gustav Hertz was born in Hamburg. His father was a lawyer and belonged to a prominent Jewish family; his uncle was Heinrich Hertz, the physicist who had demonstrated the existence of electromagnetic waves. Gustav's mother was not Jewish, so he was a half-Jew, or for the Nazis a Jew. After his *Abiturientenexamen* in 1906 he studied mathematics and mathematical physics in Göttingen with David Hilbert and Carl Runge. He then went to Munich, where he studied with Arnold Sommerfeld and was introduced to the new world of theoretical physics. From Munich he moved to Berlin. Under the influence of James Franck and Robert Pohl he became greatly interested in experimental physics. He received his Ph.D. in physics in 1911 in Berlin. He then collaborated with Franck on the interactions between electron and atoms. Their work led to the classical observations for which they received the Nobel Prize.

During World War I Hertz was gravely wounded and returned, in 1917, to Berlin. There seemed little hope for an academic career and Hertz went into industry. He spent a few years in Eindhoven, Holland, at the Philip's Laboratories. In 1925 he returned to academic life, accepting a position at the University of Halle. He became professor of physics there and, in 1928, professor of physics at the *Technische Hochschule* in Berlin. By this time he had become a famous physicist and Nobel laureate. He built one of the most modern institutes of physics in Berlin. He had many excellent students and collaborators. He devised a method for separating isotopes of neon by means of diffusion techniques. When the Nazis came to power he refused to take the loyalty oath and resigned in 1934 from his position. His wife was regarded as strongly anti-Nazi and pro-Allies. He became the chief physicist of the Siemens concern and a special laboratory was built for him.

After World War II, he and about 200 scientists went to the Soviet Union. Stalin built a special laboratory for him. He was probably one of the most prominent physicists responsible for the success of the Soviet Union in the building of the atom bomb. He returned to East Germany in 1954 and taught physics at the University of Leipzig. There he built a modern physics institute to replace the one that had been destroyed during the war. It is today one of the best physics institutes in East Germany. He received many awards for his postwar work, including the Planck Medal and the Lenin Prize. After his retirement in 1961 he went to Berlin, and died there at the age of 88 in 1975. He was a quiet and modest man; little is known about his feelings and views, especially after 1933.

E. EFFECTS ON PHILOSOPHY AND EPISTEMOLOGY

Science, the systematic effort of man to explain the structure and origin of the universe, cosmic events and the forces driving them, the matter of which our world is formed, the nature of man, biological phenomena, and so on, started in the sixth century B.C. in Greece. Thales—sometimes referred to as the father of science—and his school in Miletus, Aanaximander, and Anaximenes represent the beginning of these efforts. They reached their pinnacle roughly in the years 400–200 B.C., the period of Democritus, Leucippus, Socrates, Plato, Aristotle, and Chrysippus. It is refered to as the Classical period of Greek philosophy. It was followed by the Hellenistic period, which lasted up to the second century A.D. Then a decline began, although the efforts continued to some extent up to the sixth century A.D. Except in a very few fields, such as astronomy and geometry, Greek philosophy was based on speculations, hypotheses, and theories. However, the extraordinary intuition of Greek philosophers, their perspicacity, their brilliant visions produced notions and concepts that are in many respects different from, but in other respects amazingly close to the ideas that are the results of modern science, particularly physics. A fascinating presentation and a competent evaluation of the similarities and differences between the notions developed by Greek philosophers and those of modern physics may be found in the superb and masterfully written book of S. Sambursky (1965), *Das physikalische Weltbild der Antike*. For a millenium few significant developments took place in science.

Since the beginning of modern science in the sixteenth and seventeenth century, there was a continuous parallel development of new philosophical ideas and systems stimulated by the concepts emerging from the new scientific knowledge. These movements became particularly strong in the twentieth century; quantum theory, quantum mechanics, and the theory of relativity in particular, had a profound impact on philosophical thinking. Many of the great figures who created modern atomic physics were at the same time philosophers who spent much time and effort reevaluating the prevailing philosophical thoughts in the light of the newly emerging scientific concepts and theories. This characteristic applies to, among others, Einstein, Planck, Bohr, Born, Heisenberg, and Pauli. In their writings and lectures they frequently tried to analyze the philosophical implications of the newly gained knowledge and experience and to adjust or modify them, especially the theory of cognition. They were all passionately interested in the philosophical aspects as well as in their scientific problems. Adolf von Harnack once stated, "People complain that our generation has no philosophers. Quite unjustly: it is merely that today's philosophers sit in another department, their names are Planck and Einstein."

Among the outstanding philosophers of the early period were Francis

Bacon (1561–1626) in England and René Descartes (1596–1650) in France. Bacon was primarily a philosopher; he underestimated the importance of mathematics and stressed empiricism and the observation of nature. His views had a strong influence in England. On the Continent the philosophy of Descartes prevailed. He is generally considered as the first great philosopher of the new era. Two of his books, *Discours de la Méthode* and *Principia Philosophiae*, were the first classical works in which serious attempts were made to build bridges between the exciting new information resulting from scientific observations and knowledge, and philosophical thinking, until then still strongly influenced by Greek and particularly Aristotelian philosophy. Descartes stressed the fundamental importance of mathematics and made important contributions to geometry. He firmly believed in the power of human reason, which he considered to be the basis of cognition of nature and man. Although he recognized the necessity of observation and experiment, he was convinced that science must rely heavily on deductive methods. Some parts of his philosophy continued to influence scientific thinking for more than two centuries. He considered the physical world as a kind of machine or parts of a machine; therefore physical phenomena could be explained by mechanical models and analogies. Although several of his theories were discarded, some sooner, some later, in particular by Newtonian mechanics, a few of them persisted to dominate scientific thinking. This is particularly true of his dualistic concept in which he divided the world into two basically different categories, one the *res extensa*, the physical world that may be explained on a purely mechanical basis, and the second one the *res cogitans*, the human mind; its analysis is not open to a mechanistic approach. The division between mind and matter or between soul and body was actually not so much a totally new idea as a sharp and concise formulation of similar thoughts previously considered in Plato's philosophy and then again in the era of the Renaissance. Descartes' dualistic concept, the antithesis between mind and body, prevailed in philosophical thinking actually until the turn of this century, when it became increasingly apparent that many aspects of his philosophy required considerable modification. The sharp distinction looks different in the light of quantum theory: Science does not simply describe and explain nature; it is part of an interplay between nature and man. The new theory forced us to think in different areas of connections not recognized before. There exists between them a relationship that Niels Bohr called complementary. The areas considered may strongly differ, but they also complement each other. Only through the interplay between them does their full meaning become apparent. However, it should be mentioned that some of Descartes' theories met with strong objections almost from the beginning; among the critics was Gottfried Wilhelm von Leibnitz (1646–1716), a universal genius equally brilliant as a scientist and as a philosopher, who introduced a completely new mathematical formalism into philosophical thinking. The philosophy of Leibnitz was soon followed by many other philosophical schools, such as empiricism and positivism, most of them influenced by the continuous growth of scientific knowledge.

Turning now specifically to the role of concepts of the atom in the history of philosophy, it is necessary to recall briefly a few notions of Greek philosophy. There are in few fields so many strikingly similar ideas, despite basic differences, in modern or Greek philosophy as in the theories of the smallest building stones of matter. The idea of the smallest individual building blocks of matter came up as early as the first period of Greek philosophy with Thales and his Milesian school, in particular Anaximander and Anaximenes. In their speculations about the properties and the character of the basic elements they proposed different ideas. It is remarkable, however, that Anaximander was the first to envisage the possibility of the transformation of one primary substance into another one. Still more outstanding is the notion of continuous change in the philosophy of Heraclitus. Even if one starts with the assumption that matter is composed of one fundamental principle, it would be difficult to derive from it the infinite variety of things. Therefore, for Heraclitus, the foremost fundamental principle was the notion that everything is in constant flux (the famous πάντα ῥεῖ). He assumed that fire is the basic element, which is both matter and moving force. As Heisenberg (1954) pointed out, if the word *fire* is replaced by *energy*, modern atomic physics is in some ways extremely close to the doctrines of Heraclitus. Energy is in fact the substance from which all things are made, and energy is what moves. From developments in the field of elementary particles we know that they can actually be created by energy. Energy can be changed into motion, heat, light, and tension. Thus energy may be called the fundamental cause for all change in the world.

The concept of the atom was first proposed by Democritus and Leucippus. They considered the atom the smallest indivisible unit of matter, eternal and indestructible. The atoms of Democritus were all of the same substance, but had different sizes and different shapes. Although the idea of the atom prevailed in principle until the turn of our century, it must be stressed that the idea of the unit of matter was greatly modified by later Greek philosophers. The most prominent figure in these developments was Plato. In his dialogue *Timaeus* he gave an account of his ideas about the creation of the universe; he discussed his philosophy of science and his ideas of the cosmos. For Plato every theory was of necessity based on ideas, on the searching intellect. Science is never able to reach any conclusion with absolute certainty. All the phenomena that we can perceive with our senses, the objects of physical science, provide only a picture of the transcendental world of ideas, which represented for Plato the real world. Whereas the physical world is subject to continuous changes and consists of passing phenomena, the actual reality of the world of ideas is permanent. Plato thus refused to accept the absolute character of science.

Plato admitted in *Timaeus* that science may have a certain amount of precision. But in that case it must use mathematics, which was for him the natural language of science. Deeply influenced by the Pythagorean school, Plato realized the powerful creative force of mathematical formalism. Mathematics is somehow in the middle between the world of ideas and the observable world. His emphasis on the paramount importance of mathematics for science is widely considered one of the most important and far-reaching contributions to the development of human thought. His ideas were greatly extended and elaborated, several centuries later, by Neoplatonism, particularly by its most outstanding representative, Proclus.

Plato flatly rejected the idea of the atom as proposed by Democritus and Leucippus. For him the smallest parts were geometrical forms: Those of the earth he compared with cubes, of fire with tetrahedrons, of air with octahedrons, and of water with icosahedrons (forms with a surface of 4, 8, and 20 triangles respectively). Moreover, he did not consider these smallest parts as indivisible. The elements can be transformed into each other; they can be taken apart and new regular solids can be formed by them. It is quite evident that for Plato the form was more important than the substance of which it was the form. Although the atom as conceived by Democritus and Leucippus was revived in classical physics in the nineteenth century, in modern physics the elementary particles such as protons, neutrons, electrons, mesons, quarks, and the like, are more comparable to the original notion of the atom than to the atom itself in modern physics.

Planck, Heisenberg, Pauli, and other leading atomic physicists were greatly influenced and inspired in their philosophy by Plato and Neoplatonism. They considered themselves to be much nearer to Plato and the Pythagoreans than to the materialistic view of Democritus. The elementary particles are not eternal and indestructible units of matter. As modern physics has indeed shown, they can be transformed into each other. If two such particles moving with a very high kinetic energy collide, many new particles may be created from the available energy and the old particles may disappear in the collision. Such events have been frequently observed and support the idea that all particles are made from the same substance, namely energy. These modern views resemble in a remarkable way those of Plato as expressed in *Timaeus*. In the last analysis, the elementary particles are for Plato not substance but mathematical forms. In modern atomic physics elementary particles are also considered as forms, although of a much more complicated nature. In Greek philosophy these forms were considered to be static; modern physics stresses their dynamic nature.

Although some statements of ancient philosophy are amazingly close to those of modern science and testify to the ingenuity and amazing intuition of Greek philosophers, there is this fundamental difference: Modern science, from its beginning in the sixteenth and seventeenth centuries, has been based on detailed experimental studies of nature and on the postulate that only such statements should be made that have been or can be verified by experiment.

These few remarks about the ideas of Greek philosophers concerning the nature of the unit of matter appear useful for the discussion of the effect of modern physics on philosophy. A detailed discussion cannot be included here. Readers interested in this subject are referred to the vast literature of philosophy and, in regard to the special problem of the relationship between ancient and modern concepts of the atom, to the literature dealing extensively with this subject (see, for example, Sambursky, 1965, 1975, 1977; Heisenberg, 1958, 1971, 1973, 1974; Planck, 1959).

Returning to the impact of modern physics on philosophy, it is necessary to discuss some of the concepts of Immanuel Kant (1724–1804). Among several of Kant's books pertinent to the philosophy of science and the theory of cognition, one of the most important is the *Critique of Pure Reason (Kritik der reinen Vernunft)*, which first appeared in 1781. There Kant analyzed the question of whether knowledge is derived only from experience or may come from other sources. His conclusion was, like that of other philosophers before him, that only part of our knowledge is based on experience. He realized that even if all our cognition starts with experience, it is by no means derived exclusively from experience. The world we observe is only a part of the reality that we are able to conceive. Another part is not inferred inductively from experience. It is open to empirical observations by our senses; they are, however, filtered and elaborated by our reason, our thinking and intelligence. These processes of the mind permit a rational empirical cognition. It is reason that brings laws, order, and regularity into the phenomena observed. Thus, the laws conceived are the result of the process of reason, which does not derive them from nature, but prescribes them to it by the contributions of our mind. We understand these phenomena because we approach them with certain notions and concepts for which Kant used the term a priori. Among such necessary notions, or categories, are space and time. They are prerequisite and basic structures into which we must fit all our perceptions. We cannot imagine that there should be no space, even if we can imagine that there should be nothing in the space. The same reflections apply to time. Without these two a priori notions we would be unable to perceive a well-ordered universe. The law of causality is another a priori notion. When an event is observed, it must be determined by a preceding event. For Kant the a priori category of the law of causality forms an absolute necessity of all science; it is not an empirical assertion that can be proved or disproved by experiment. Rather, it forms the basis of all experience.

These a priori concepts are based on Newtonian mechanics, which strongly influenced not only Kant's philosophical thinking but that of the nineteenth century. These laws of physics had absolute validity for Kant and were not subject to any question. Their limited applicability only became apparent through the results of modern physics. Moreover, in Newton's physics, the geometry that formed an essential basis in his concepts was that of Euclid. Not till the nineteenth century was a new geometry developed, particularly by the pioneer work of Gauss (1777–1855), which then greatly influenced thinking in physics and philosophy.

Einstein's theory of relativity has, as is well known, changed and extended the concepts of space and time. Since, as outlined in the Preface, Einstein's theory and his concepts have been described in a huge number of books on all levels, popular and scientific, they need not be discussed here. But quantum theory has in addition changed the a priori character of the law of causality for the behavior of elementary particles in the atom. This change is even more far-reaching and fundamental than that of time and space, just because Kant had considered this law as the necessary basis of any future scientific analysis. It seemed simply impossible to observe any effect without having a definite cause.

Kant obviously could not have foreseen the startling developments of modern physics, neither the theory of relativity nor the quantum theory. The former forced the change of the a priori concept of time and space, and the latter demonstrated that the law of causality is not applicable to the events in the atom. Let us take for example radium atoms capable of emitting a particles. The precise time of the emission of the a particles cannot be predicted. Only the average time of emission is determinable, but each single emission may be very much shorter or very much longer than the average time. In observing the emission one does not know the preceding event that determined the emission at a certain moment. Theoretically one could ask whether it would not be possible to look for such a preceding event causing the emission at a given moment; one should not be discouraged by the fact that so far none has been found. In classical physics the answer would be that the failure to find a cause of a certain effect does not mean that such a cause does not exist.

Now the laws of quantum theory have shown that this argument does not hold. The preceding event is known, but it cannot be determined with accuracy, as discussed in describing the law of uncertainty. Since the forces in the atom cannot be determined accurately, it follows that the forces in the atom responsible for the emission cannot be determined accurately. The principle of indeterminacy of the initial state only permits the calculation of a probability function. If it is intrinsically impossible to know the cause in a precise way, there can be no absolute prediction of the effect. In quantum theory a new method of objectifying perception must be used, one that just could not be anticipated in Kant's philosophy. Every perception refers to an observational situation that must be specified if experience is to result. This includes the instruments for the observation used as well as the observer. This subjective element did not exist in classical physics. In other words, in the case of atomic events a perception can no longer be objectified in the way of classical physics. If two observational situations are in a relationship that Bohr calls complementary, then complete knowledge of one necessarily means incomplete knowledge of the other. Thus the law of causality is no longer contained in the system of atomic events; there, the a priori causality concept of Kant has no absolute validity.

Nevertheless, it must be stressed that the use of the concepts of time, space, and causality is in fact also the condition for observing atomic events. They are therefore by no means eliminated; only their relative character becomes apparent. In a limited sense of the word they are still a priori. These concepts are still the conditions for scientific observation, but at the same time they have been demonstrated to have a limited applicability. The paradox of quantum theory could not be foreseen in a philosophy based on classical physics. Quantum theory made it necessary to realize that the a priori concepts have the character of a relative truth; they are even today indispensable tools of all scientific work. But what has become evident by the advances of science is their limited applicability, and further limitations may well be found in the future. Thus, in the light of modern physics, Kant's philosophy can be considered only as an incomplete approximation of reality.

Although Newton's physics has been greatly extended and modified by the achievements of physics particularly in this century, it has not lost its fundamental value and is still the basis of a wide range of physical reality. In all fairness it must be emphasized that despite the necessary modifications and adjustments of Kant's philosophy by modern physics, the fundamental importance of many of his basic concepts remains and forms an integral part of philosophy. The most important result of the developments in science since Kant is the realization that there is no sharp distinction between cognition based on empirical observation and that based on pure reason, as Kant believed. The a priori notions of Kant were for a long time an important, and perhaps necessary, support for the progress of the theory of cognition, but in the light of the new knowledge such a distinction cannot be maintained.

Heisenberg was not only one of the founders of quantum theory, he was also one of the most prolific writers in philosophy. Many of his philosophical writings appeared in three volumes of *World Perspectives: Physics and Philosophy, Physics and Beyond,* and *Across the Frontiers* (Heisenberg 1958, 1971, 1974). In his essays he discussed not only the strong influence of quantum mechanics on physics and on our thinking, but a great variety of philosophical notions and concepts. There are many fascinating discussions, but only two of them are pertinent to the problems raised by atomic physics.

In his article, "Planck's Discovery and the Problems of Atomic Theory," Heisenberg (1974) recalled, in analyzing the effects of the notion of the quantum of action and of atomic theory, the impact of Newton's *Principia* on the thought of the following centuries. The influence was not simply limited to permitting a decision between contending philosophies. Similarly, the effect of atomic or quantum theory did not result in a position taken for or against one of the earlier or present-day systems of philosophy. The scientist is primarily interested in raising questions. The way in which these questions are put seems valuable if they have been fruitful in the development of human thought. In most cases the answers can only be the product of their period. As our knowledge of the facts is extended in the course of time, these answers are bound to lose their limited significance and must be modified.

As to Planck's discovery of the quantum of action, Heisenberg raised the question of its meaning to philosophy. In his view, it revived the problem that divided Plato and Democritus: For Plato, as discussed before, the elementary particles are not indivisible; the forms that Plato attributed to them are mathematically the simplest. Democritus' views were generally accepted as dogma by nineteenth-century scientists. Planck's various discoveries clearly suggested an unsteadiness in the natural occurrence of matter. His observations on thermal radiation were difficult to reconcile with the prevailing notions of the atomic structure of matter, thereby reviving Plato's notion in science, with strong emphasis on the belief that mathematical laws underlie the structure of matter. This new possibility was opened by Planck's discovery, thus relating it to a basic problem of philosophy, namely the structure of matter. Newtonian physics, dealing with macroscopic events, could not provide the answer. It deals with phenomena on a scale very large in comparison with the atom. Even Planck's quantum of action only demonstrated that the phenomena in the atomic field might display entirely different features.

Only a few years later, in Einstein's theory of relativity, a second constant, the velocity of light, underwent a drastic change in the interpretation of its role. The theory revealed relationships between time and space that seemed wholly independent, since our usual experience is concerned with processes that are slow compared with the velocity of light.

Thus two major new breakthroughs in our knowledge and understanding had occurred. These two theories brought about radical changes in our picture of the world, because they showed that our experience and perception are valid only in a restricted area and therefore do not represent an unshakable basis of science. But it took more than a quarter of a century of many additional dramatic breakthroughs for extending the emerging picture. Rutherford's and Bohr's atom models demolished the original atom notion and replaced it by a structure formed by a nucleus and electrons. From wave and quantum mechanics arose the notion of complementarity and the principle of indeterminacy. They required basic revisions of philosophical thinking. Even if the new developments are approached with some skepticism, one cannot deny that modern physics will be classified as being of the same magnitude and revolutionary character in the history of philosophical thinking as the ideas of Plato and the discoveries of Galileo, Kepler, and Newton.

Since nuclear physics and the problems of elementary particles are still at the center of physics, many people have raised the question of whether with the solution of these problems physics as a whole may come to an end. It may, therefore, be of interest to many readers to learn the views of Heisenberg (1974) on this question, which he discussed in a short article, "The End of Physics?" Since all matter consists of elementary particles, a complete knowledge of the laws governing their properties and behavior may permit, in a kind of "world formula," the establishment of the framework of all physical processes. In that case the questions of principle would all have been settled and fundamental research in physics would have come to an end.

As Heisenberg pointed out, such a thesis of possible completion of physics is contradicted by the experience of earlier periods in which it was also assumed that physics would soon reach the end of the road. Planck has recorded, as Heisenberg mentions in this article, that his teacher Jolly advised Planck against the study of physics, since after all it was essentially finished, so that for anyone who wanted to do active scientific research, it would scarcely be worthwhile to go into this field. That must have been in the 1870s, since Planck was born in 1858. One should therefore ask whether in the past history of physics there have been subareas in which final formulations of natural laws have ben obtained, and in which we can be confident that the phenomena will continue precisely in accordance with the same mathematically formulable laws.

Such "closed-off" subareas, as Heisenberg referred to them, undoubtedly exist. The laws of the lever, formulated by Archimedes, and Newtonian mechanics will always retain their validity. The moon travelers rely on these laws and act accordingly. But one may ask whether Newtonian mechanics have not been improved by, for example, the relativity and quantum theories. Has one not to take into account these refinements to obtain a higher degree of accuracy? If this were the case, even mechanics would still be far from finished. Comprehensive formulations of natural laws such as Newtonian mechanics are, however, concerned with idealization of reality and not with reality itself. The idealization is achieved because we approach reality with certain concepts that have proved themselves in the description of the phenomena. In mechanics we make use of such concepts as position, time, velocity, mass, and force, thereby restricting the picture of reality. If we remain aware of these restrictions, it is possible to claim that mechanical phenomena are brought to completion in Newton's theory; that is, it cannot be further improved. In that sense it is a completed theory. But we cannot claim that all physical phenomena can be described in terms of these concepts.

There are other areas of experience and other closed-off theories as well. In the nineteenth century the theory of heat, for example, took on a final form. The fundamental axioms of this theory define such concepts as temperature, energy, and entropy, which do not appear in Newtonian mechanics. Similarly, electromagnetism has been greatly developed in the last two centuries and there too earlier theories were inadequate. The theory of relativity emerged from the electrodynamics of moving bodies and has led to new insights into the structure of space and time. The quantum theory gives an account of processes in the interior of the atom, but it also incorporates Newtonian mechanics, as the limiting case in which we are able to objectify the events and can neglect the interaction between the object under investigation and the observer itself. These two theories Heisenberg considers also as closed-off theories, as an idealization of large areas of experience, valid everywhere and at all times, but again restricted to those areas of experience that can be understood by means of these concepts.

However, in modern atomic physics, in which large accelerators are used, new features of elementary particles have appeared. One could have assumed that even protons and neutrons could be further broken up by large forces using extraordinarily high energies. It turned out that elementary particles in these accelerators were transformed into new elementary particles, but they are *not* smaller than the original particles before the collision. Here transformation of energy into mass may have occurred. Although one may consider the great number of new particles, described in the last few decades, are really the smallest part of matter, here is certainly a field that cannot be considered as a closed-off theory. Let us however assume that one day a complete theory will be developed, accounting for all phenomena of elementary particles, which includes quantum theory and is expressed in mathematical terms. Could we then consider that a complete knowledge of the laws governing the behavior of these particles is equivalent to a complete knowledge of all physical laws, and claim that physics has come to an end? Heisenberg considered this conclusion inadmissible. Even a closed-off theory of elementary particles, although it may provide insight into a wide range of phenomena, will leave many areas unexplained. In any event, many physicists question that an end is in sight, in the field of elementary particles, and they believe that future research may reveal completely new and unknown regions.

A most conspicuous proof for the impossibility of explaining all aspects of science, even if a complete understanding of the properties and the behavior of elementary particles could be achieved, was for Heisenberg the field of biology. Although all biological objects consist of elementary particles, the concepts we use for the description of biological processes, for example, the concept of life itself, would by no means be elucidated. One could object that biology does not belong to physics, but the boundaries between physics and the neighboring sciences are so fluid that little would be gained by such limitations. This applies equally to other fields, such as mathematics or philosophy. The majority of physicists agree that, in view of these undefined boundaries, it would not be proper to speak of the end of physics. Heisenberg expressed his views on the limitation of understanding biological phenomena on several occasions. Some of them are quoted in the discussion on Meyerhof's philosophical views (see IV, D). In the light of the revolutionary achievements of modern physics in the first part of this century, and of biological sciences particularly— after World War II up to date—the problem of the limitations of science has become of paramount importance not only for scientists and philosophers, but for all intellectuals interested in the question of how far man will be able to penetrate into the mysteries of the universe and understand nature and humanity and life itself. It therefore appears to be pertinent, in the context of the effects of atomic physics on philosophy, to mention briefly the views of Shmuel Sambursky (1900–   ), a leading authority on the history and philosophy of science. He was a student of Planck's in Berlin and later was professor of physics at the Hebrew University of Jerusalem for many years; since 1958 he has been teaching history and philosophy of science at Jerusalem. A variety of topics pertinent to the problems discussed here appeared in a series of articles in *Naturerkenntnis und Weltbild* (Cognition of Nature and World Outlook) (Sambursky, 1977). The first article "Das physikalische Denken der Antike im Lichte der modernen Physik" (The Physical Thinking of Antiquity in the Light of Modern Physics), deals in its last section with the limitations of science in the cognition of nature. The views expressed are extremely pertinent and directly related to the ones just discussed.

Sambursky stresses that the notion of complementarity of Bohr and the principle of indeterminacy of Heisenberg signify the end of the naive-realistic concepts of phenomena that prevailed at the time of Newton. The phenomena of nature are extremely complex processes, in which the environment and the observer, the exterior and interior of man, are inseparably associated. In view of the increasingly deeper insight into reality that both physics and biology are providing, a return to classical theories of cognition appears exceedingly unlikely. At the beginning of this century many scientists considered the exclusion of all subjective elements from scientific knowledge as the desirable final aim. Now all indications favor the assumption that philosophical interpretations of science, of its basic principles and methods, will increasingly force scientists to abandon this view. The situation created by quantum theory, as many scientists recognize, means a limitation of the ideal of "objective" scientific research.

The question of the limitations of science has always been a touchy problem; many scientists refused to accept limitations. Nowadays when the seemingly unlimited power of science has become the myth of our time, such a question is almost a sacrilege. In 1872 Emil Du Bois-Reymond gave a lecture on the limitations of cognition of nature ("Über die Grenzen des Naturerkennens" (Du Bois-Reymond, 1881). It culminates in the famous words: *ignoramus, ignorabimus.* He treated two problems that in his opinion would never find a solution: the problem of matter and the question of consciousness. Today, after 60 years of quantum theory, we may consider his claim that we will never understand matter as having lost much of its validity. Before the discovery of elementary particles and the atomic structure one may have raised the objection that by a reduction of macroscopic phenomena of matter to the invisible ones one would find that the same laws apply to both; therefore, nothing would be gained from their understanding. Quantum theory has, however, shown that the macroscopic phenomena are dominated by much simpler principles and differ qualitatively from those on the atomic scale. When explanation has any meaning, we have made significant advances in the understanding of matter, even though we have behind us most likely only a very small part of the road.

But fundamentally different is the situation in respect to consciousness and its explanation by scientific analysis (that is, by methods of physics and chemistry). There Sambursky considers the *ignorabimus* still valid. Science is limited to reproducible phenomena, since its basic aim is the exploration of the laws of nature and the methods of this exploration consist in the observation of natural or artificial events that are reproducible. Consciousness in the notion of Du Bois-Reymond is the inner experience of an individual process that has the quality of a unique event and is always new, even when typical features may repeat themselves. Therefore, these processes can never be the object of scientific analysis and they escape treatment by the methods of science. In contrast, art uses the unique case as object of presentation, either in painting, music, or poetry, even if typical features may repeatedly occur. Art thus opens up realities that in their nature are not accessible to scientific analysis or theory. This limitation of science to the unspecific and reproducible event makes it apparent that the belief in the all-pervading power of science is unjustified.

These views and formulations of Sambursky are fully shared by the author. Consciousness, emotions, the unconscious, psychological phenomena, moral and ethical values, imagination and creative ideas—even if they are admittedly based on molecular events in our brains—are and will remain inaccessible to physicochemical analysis. This does not exclude the possibility that science will one day be able to analyze and understand some of the general principles and laws controlling the filtration processes and the ways of elaboration of our minds. Just as the laws of nature control all life processes, as extensively discussed by Eigen and Winkler (1975) for evolution, they will also control to a large extent processes such as learning, memory, and other events taking place in our brains, although chance may always play a certain if limited role in modulating and affecting the results. We must always keep in mind the extraordinary complexity of an organ such as the brain, which contains trillions of synapses; they offer a vast number of possibilities of interference with the regular filtration process and the workings of the brain. However, the uniqueness of the achievements of a brain in the field of creative art (music, poetry, painting, and the like) and the genius of a Newton, Einstein, or Planck, offer additional and in the view of the author unsurmountable obstacles to explaining the creative process in physicochemical terms. In this respect the author agrees with the *ignorabimus* as formulated by Sambursky. There are many scientists who share this view. Max Born (1965), for example, in an article on symbol and reality ("Symbol und Wirklichkeit"), in a collection of papers on the responsibility of scientists, considered scientific thinking not applicable to certain other domains, such as human and ethical values. He strongly emphasized there that in his judgment it is dangerous to apply scientific thinking to such areas as religion, art, literature, ethics, and all humanities. In these fields the methods are beyond the range of the validity of scientific thinking. This problem will be further discussed in the section on Meyerhof (see IV, D).*

It must be mentioned that there are scientists who disagree with the opinion expressed. In view of the spectacular achievements of biological science, particularly in the field of genetics, in our insight into evolution, the origin of life, they are convinced that one day we will be able to understand the manifestations of the human mind in terms of physics and chemistry.

A few comments should be made about Pauli's philosophical ideas. Although known to most physicists primarily as a brilliant colleague who decisively influenced the growth of modern atomic physics, a man equal in stature to Heisenberg and Bohr, he also had a deep and genuine interest

---

* It may be mentioned that the author, in his first discussion with Einstein, in Berlin in 1926, raised the mind–body problem and the dualistic concept of Descartes. He mentioned that he had attended a lecture, a few days earlier, about this topic. The lecturer had closed the lecture with the words *ignoramus, ignorabimus*. Einstein vigorously objected. A scientist should never say never. Who would have been able to foresee in 1890 the progress of atomic physics in the last two decades? The author wonders whether Einstein, with his more skeptical and cautious outlook in his later years on the limits of scientific cognition, and in view of his remarks on the primitive and childlike state of science mentioned on page 103, would have still objected so vigorously to the view.

in philosophy. He gave a number of lectures and wrote various articles about the theory and basis of cognition (see, e.g., Pauli, 1961). With extraordinary clarity he pitilessly dissected any theory or hypothesis whenever he found weaknesses. His devastating criticism was frequently combined with ironic remarks and a biting wit. When physicists came out with a new theory, for many of them one of the first questions was, What would Pauli say?, and they wondered whether he would discover any loopholes in the theory. Ehrenfest (1880–1933), a Jewish physicist from Vienna who had worked with Boltzmann and became the successor of H. A. Lorentz at the University of Leiden in 1912, once referred to Pauli as the "scourge of the Lord." Few were so much admired as well as feared as Pauli for his penetrating criticism. As Pauli stated in 1933 in his obituary for Ehrenfest, objective scientific criticism, no matter how sharp, is always stimulating and constructive (Pauli, 1961). One can understand Heisenberg's happiness when Pauli immediately endorsed the principle of indeterminacy. He was also one of the first to accept unreservedly Bohr's concept of complementarity. These two concepts were congenial to Pauli's philosophical outlook. On many occasions he not only vigorously defended them, but also stressed their far-reaching implications. Both Pauli and Heisenberg belonged to a generation brought up with the notion of the quantum of action and the theory of relativity. They were therefore much less reluctant to break with classical concepts of physics and philosophy than Planck, von Laue, Schrödinger, or even Einstein. Pauli expressed his regret that Einstein was unwilling to accept the concept of complementarity, which in his view presented a new way of thinking. It was a decisive turning point in the theories of cognition and opened many new aspects and possibilities for a better comprehension of the laws of nature. It offered the best hope of future developments in different disciplines that might lead eventually to a greater unitary concept of the universe, even though quantum theory was probably still far removed from its final form and left many problems unsolved.

Pauli fully shared Einstein's and Planck's views of the primary role of the concept, to be discussed subsequently. He too rejected the purely empiricist view whereby natural laws can be developed solely from experimental data. For him the creative spiritual process, for which no rational explanation exists, was the most powerful factor in the advance of the cognitive process though it must be based on experimental observations. He fully agreed with the views of those who emphasized the role of intuition in framing the concepts and ideas necessary for establishing a system of natural laws. There is no "unprejudiced" science based on simple observation of facts. Pauli was interested in the nature of a connecting link between sense perception on the one hand and concepts on the other hand. Pure logic is fundamentally incapable of constructing such a linkage.

Although Pauli was considered to be one of the most rationalistic thinkers, he was probably more interested than most physicists in the role of psychological problems, especially in their relationship to scientific notions. He was influenced by the ideas of the well-known Swiss psycho-analyst C. G. Jung. He published an article with him, "The Influence of Archetypal Ideas on Kepler's Construction of Scientific Theories" (Jung and Pauli, 1952). The role of the unconscious of the human mind has been hinted at by several philosophers, among them Kant and Hegel. Actually the psychological factor in scientific ideas had already been con-sidered by Plato and, 800 years later, by Neo-platonists, particularly by Proclus. However, it was Sigmund Freud who was the first to establish and to elaborate systematically the concept of the unconscious (*das Unbewusste*). It consists of a number of processes comprising many factors such as instinctive reactions (e.g., aggression and libido) and life experiences that occur in early childhood or later in life, but become repressed. Jung, a pupil of Freud, introduced new aspects to the idea of the unconscious, such as the notion of the collective unconscious and primeval images or archetypes (from Greek αρχέτυπος, "pattern, model, or prototype"). For Jung—as for Freud—the psyche, the mind, is the totality in which the conscious and the various elements of the uncon-scious are combined. It may be mentioned that Jung's theories are shared by relatively few psychiatrists, whereas Freud's concepts are much more widely accepted. Pauli accepted many of Jung's ideas, especially those of primeval images or archetypes. He considers them complementary to the scientific ideas arising from new knowledge. Some of these views are discussed in his article written on the occasion of Jung's eightieth birthday, "Naturwissenschaftliche und erkenntnistheoretische Aspekte der Ideen vom Unbewussten" (Pauli, 1961).

Pauli felt strongly that scientific knowledge was still limited. For him the question for the physicist was not whether the present theories would remain, but in what direction they would change. His attitude recalls the famous words of Blaise Pascal (1623–1662), who compared science to an ever-increasing ball: The larger it gets, the more contacts it offers with the unknown. Pauli's philosophical views were well analyzed, in 1959, in an article by his friend Heisenberg (1973).

Epistemology

Epistemology, the theory of knowledge (from Greek ἐπιστήμη, "knowl-edge"), is the philosophical discipline that deals with the aims, methods, and achievements of cognition, their validity, and so on. It played a con-siderable role in Greek philosophy, but with the beginning of modern science in the sixteenth and seventeenth centuries and the rapid advances of knowledge based on experiment, a new situation arose. Many new problems had to be considered, such as the question of how new theories emerged, the role of the experimental observation versus imagination and intuition, the value and limitation of interpretation, and a variety of other factors that influence cognition. Newton believed that all scientific progress is obtained by observation and experience; he considered our perception as the basis of all advances. Actually he did not quite realize how much his revolutionary contributions were not just the result of observation, but of his unique genius, his theories, imagination, and ideas, which led him to perform his experiments. This was already recognized by Leibnitz, who stressed the importance of theories in Newton's work. The views of Kant as to the limitations of observation as basis for scientific theory have been mentioned.

Few scientists have devoted so much time and effort to epistemology as the two great founders of modern physics, Einstein and Planck. Both produced by their contributions to epistemology a revolution in the theories of knowledge and cognition that is almost as profound in its impact on philosophical thinking as that which resulted in physics by the quantum of action and the theory of relativity. Many scientists perform their research without examining or paying special attention to the intellectual processes that induced them to perform their experiments. During the classical period in physics many scientists accepted Newton's view that the basic laws and concepts of physics are derived from experience by abstraction. The success of Newton's theory of gravitation was a main factor in preventing the realization of the constructive and imaginative nature of principles and theories. Einstein, in contrast, considered the general theory of relativity as showing in a convincing way the incorrectness of Newton's view that general principles and concepts are derived from experience. The theoretical system of physics is dependent upon and controlled by the world of sense perception. But there is no logical way whereby we can proceed from sensory perception to the principles that underlie the theoretical structure. An attempt to derive by logic the basic concepts and laws of mechanics from observations and experience is doomed to failure. There is no inductive method that can lead to fundamental notions or principles. It is the concept of the scientist that *precedes* the theory, although the theory must then be supported by experimental facts. The conceptual synthesis, which is the transcript of the empirical world, may be reduced to a few fundamental laws on which the whole synthesis is built.

Einstein clearly recognized that in scientific knowledge there are two distinctly different components. One is given empirically and is plainly indicated by observation, and the other is obtained by imagination and theory and has a different character than the empirically gained information. The contents of the theory on the one hand and the empirical data on the other hand present an intrinsic antithesis of the two inseparable components of our knowledge. In stressing the importance of empirical facts, Einstein is in agreement with Kantian philosophy. He was a great admirer of Euclid's geometry; it impressed him in his early youth as a miracle, as a logical system in which the deduced propositions could not be questioned. He considered Galileo as the father of modern science because he was the first to recognize that all empirical knowledge starts from experience. However, this empirical aspect is only one part of reality. Pure logical thinking and observation cannot provide the information required for the understanding of the empirical world. They can only be the starting points for the deductively formulated theoretical component of scientific knowledge, which requires of necessity the free invention of the human intellect. This alone can provide the axiomatic basis of theoretical physics (or of scientific theories in general). The way from the empirical data to deductively derived postulates or axioms is extremely difficult. It is here where the genius of the scientist manifests itself. Only completely free and intuitive imagination can lead to basic concepts and postulates of science. There is no other way to reach the goal. It follows from Einstein's views that a theory derived from observations and facts in nature is a creative process that is in its nature similar to that of the artist. It is striking to recall in this context Goethe's words: "In jedem Kunstwerk, gross oder klein, kommt es auf die Konzeption an" ("In each work of art, big or small, it is the concept that counts").

Einstein's views and attitude not only imply a complete break with positivism and with the purely empiricist philosophy of David Hume (1711–1776), but they also discard Kant's epistemological thesis that the deductively formulated systematic relatedness of scientific knowledge belongs to the a priori category.

It also follows from Einstein's thinking that theories and axioms of science developed by the imagination and the genius of the scientist on the basis of empirically obtained data have to be rejected and replaced by new ones when new facts emerge. Thus no theory can be established with certainty and for all time. Einstein for some time was convinced that there exists a right way and that eventually we will be able to find the key to the understanding of natural phenomena, and that pure thought can grasp reality. However, this optimistic and affirmative attitude seems to have gradually shifted when he grew older, to a more cautious and skeptical view. Such a change is reflected in the statement quoted by Hoffmann (1972): "One thing I have learned in my long life: that all our science, measured against reality, is primitive and childlike—and yet it is the most precious thing we have."

It may be of interest in this context to mention remarks that Einstein once made to Heisenberg. In the spring of 1926 Heisenberg, at that time 24 years old, was invited to give a lecture at a colloquium of Nernst. This was a great honor: These colloquia were usually attended by leading physicists, among them Einstein, Planck, von Laue, and Schrödinger. Nernst had asked Heisenberg to present the newly developing quantum theory. Einstein was apparently so fascinated that he invited Heisenberg to walk home with him so they might discuss the new ideas at greater length (Heisenberg, 1971). He seemed particularly interested in the philosophical background of the new theory and inquired what had led Heisenberg to his assumptions. In the course of this revealing and interesting discussion, Einstein, although stressing the importance of observable magnitudes, made the following statement:

> On principle, it is quite wrong to try founding a theory on observable magnitudes alone. In reality the very opposite happens. It is the theory which decides what we can observe. . . only theory, that is knowledge of natural laws, enables us to deduce the underlying phenomena from our sense impressions. When we claim that we can observe something new, we ought really to be saying that, although we are about to formulate new natural laws that do not agree with the old ones, we nevertheless assume that the existing laws—covering the whole path from the phenomenon to our consciousness—function in such a way that we can rely on them and hence speak of observations.

In his answer, Heisenberg stressed that it ought to be possible to think up many experiments whose results should be predicted by the theory. And if the actual experiment should bear out the predictions, there is little doubt that the theory reflects nature accurately in this particular realm. With this view Einstein was in agreement. "Control by experiment is, of course, an essential prerequisite of the validity of any theory."

Einstein and Planck, despite their mutual admiration and affection, disagreed scientifically in several respects. But their philosophical views and their attitude toward epistemology were very close, one may even say fundamentally similar. Planck too devoted much thought to these problems and presented his ideas frequently in his writings and lectures. Especially in his later years, in the 1920s and 1930s, many public lectures were devoted to the problems of cognition and the limitations of science, and to critical evaluations of scientific methods and theories.

Planck was strongly influenced and inspired by Plato and Neoplatonism. For him, just as for Einstein, each theory was the product of the completely free speculation of the human mind. The scientist was for him like the artist who is capable of handling ideas and whose imagination cannot be bold enough. Physical theories are not built on measurements. The observation becomes meaningful only by the interpretation derived from theory. Even in observations using the highest degrees of precision, conclusions can be drawn only on the basis of theory. There is hardly a claim that has created more damage than the assumption that science is free of preconceived assumptions. The firm basis of science is unquestionably the material available from experience. But it is equally true that the material alone or its logical elaboration does not form the essence of science. The material is always full of gaps, even if it is formed by a number of individual facts, sometimes a great number of facts. Therefore the material must always be supplemented and the gaps closed. This process is possible only by ideas connecting the gaps. These connections cannot be achieved by reason and logic, but only by the creative imagination of the scientist. One may call it working hypothesis, theory, or belief; the essential point is that the content of the theory far surpasses the facts obtained by observation. It is in this sense that Planck once made the statement, "Auch in der Physik gilt der Satz, dass man nicht selig werden kann ohne den Glauben" ("The thesis applies even to physics, that one cannot be blessed without faith") (Planck, 1922). This view is supported by the way progress in science has been achieved in the last few centuries.

There is, of course, a danger that the available data may be wrongly interpreted and that the contradicting data may be simply ignored. In that case science becomes pseudoscience, an empty construction that will inevitably collapse. There is only one protection against this danger: the respect for facts. The more imagination and ideas a scientist has, the more urgent it is to keep in mind that the detailed facts form the fundamental basis without which science cannot exist. A careful and conscientious examination must be applied to evaluate all the available facts. Particularly disastrous is the attitude of a scientist, possibly misguided by the success of his work, who omits new data presented by other scientists, ignores them and closes his eyes, or tries to minimize them. He will sooner or later pay a high price for the insight he rejects. But if one keeps these dangers in mind, it still remains true that real science cannot be built on empirical material alone, but requires the creative idea by which a satisfactory interpretation, a new concept, is achieved.

Despite the great strides made by physics in this century, Planck realized how far we still are from a real understanding of the physical world, how many areas remain obscure. Whether man will ever be able to explain the universe is a question we are unable to answer at present There may be areas that remain closed to scientific rational analysis. However, he stressed on several occasions the incredible advances that were made possible by the developments of physics. This fact alone shows that our knowledge continuously increases and how greatly mankind has benefited from the advancement of knowledge. To theories as to new discoveries applies, as the most important criterion, what Goethe expressed with the words, "Was fruchtbar ist, allein ist wahr" ("For what creative is, is true"). The fact that Planck frequently emphasized this view shows how much importance he attached to this criterion in the evaluation of new facts or ideas.

F. THE EFFECTS ON CHEMISTRY, BIOCHEMISTRY, AND OTHER FIELDS

Chemistry is the science of substances, their structure and properties. Structure has to do with matter; that is, with atoms and the small particles that make up all matter, or the combination of atoms with other atoms to form molecules. Substances are subdivided into classes: elementary substances (or elements) formed by atoms of only one kind, and compounds formed by two or more different kinds of atoms. The understanding of chemistry has been revolutionized in this century by atomic physics. To illustrate this effect a few of the outstanding features and principles based on the understanding of the atom should be at least briefly mentioned. An extended presentation would not be pertinent to the subject of this book, since most of the decisive contributions were achieved outside Germany, particularly in the United States even before Hitler, and still greatly extended afterward. The discussion of the impact of atomic physics on this field is therefore limited to a few pages.

In the nineteenth century the English chemist and physicist John Dalton (1766–1844) introduced, in 1805, the notion of the atom in the Greek sense as the smallest unit of all matter in the rapidly developing field of chemistry. This notion continued to be used until it was drastically changed by the models of Rutherford and Bohr; they demonstrated, as described before, that atoms are made up of a nucleus circled by electrons moving with a speed approximately 1% of that of light. The nucleus is formed by protons and neutrons (see p. 34), except for the hydrogen atom, which in its usual form has only one proton and no neutron. However, there is a form of the hydrogen atom that does have both neutron and proton. This form, called *deuterium*, was discovered by Harold Urey (1893–   ) in 1932; only 1 in 5000 hydrogen atoms occurs as deuterium. In Bohr's atom model, described earlier, the distance of the electron orbits is quantized. The electron orbits, in the simplest atom, hydrogen, are strictly limited as to their distance from the nucleus (0.53 angstroms, 4 × 0.53 angstroms, and so on). The energy associated with the electron varies in a stepwise manner from one orbit to the next. The greatest energy is in the most distant orbit; the smallest energy, in the closest orbit. When an electron falls from a high to a low level of energy by changing its orbit distance it emits energy in the form of electromagnetic radiation. Normally atoms have quite stable structures. Electron jumps from orbit to orbit do not normally occur. Only when an atom is "excited" does it emit radiataion. Such an excitation takes place if atoms are heated and appear in vapor form. One example is the incandescent lamp formed by sodium vapors; the radiation emitted evidences characteristic wavelengths for each atom and is responsible for the spectra with its lines, as discussed before. Bohr's discovery was a tremendous stimulus for the whole field of spectroscopy for the analysis of atoms and molecules.

The number of electrons in an atom determines its *atomic number*. For the hydrogen atom with 1 electron, the atomic number is 1; for helium, with 2 electrons, 2; for lithium, with 3 electrons, 3; for nitrogen, 7, and for oxygen, 8. Uranium, known to every layman as the substance from which nuclear energy may be derived, has 92 electrons and thus its atomic number is 92. But the *weight* of the atom depends on its nucleus. Protons are 1840 times heavier than electrons. Neutrons have about the same weight as protons. The number of protons and neutrons determines the *atomic weight*.

Whereas the number of the positively charged protons is the same in the different elements or atoms, the number of the uncharged neutrons may vary. Atoms of the same element with different numbers of neutrons are called *isotopes*. Oxygen, for instance, has 8 protons and 8 neutrons; its atomic weight is therefore 16. However, some oxygen atoms have 9 or 10 neutrons and they are called isotopes of ordinary oxygen. They have atomic weights of 17 and 18, respectively. The common form of uranium has the atomic weight of 238; it has 92 protons and 146 neutrons. There are several isotopes of uranium, but the most important one has 143 neutrons and an atomic weight of 235. It is this particular isotope of uranium that is suitable for fission and therefore required for the production of nuclear energy, including the production of the atom bomb, as will be discussed later (see II, I).

Atoms (elements) have a definite combining power or valence (from *valencia*, "capacity"), which determines the number of other atoms with which an atom can combine and form molecules, either with the same kind of atom or with a different kind. The atoms of some elements, for instance hydrogen (symbol H), combine only with one atom of hydrogen or of another element; they are *univalent*. Other elements have the ability to combine with two, three, four, or more atoms; they are bivalent, trivalent, quadrivalent, and so on. Oxygen (symbol O) is bivalent, nitrogen (symbol N) is trivalent or quadrivalent, and carbon (symbol C) is quadrivalent. Valence is represented by dashes connecting the atoms. In some molecules atoms are connected by a double or even triple bond, expressed by two or three dashes. Thus, water is formed by one O and two H, ammonia by one N and three H, methane by one C and four H; carbon dioxide has one C connected by two double bonds with the two O's; they are written as follows:

<pre>
    H   H   H
    | | |
 O—H   H—N—H   H—C—H   O=C=O
                                  |
                                  H

 water   ammonia   methane   carbon dioxide
</pre>

When atomic physics elucidated the atomic structure of the elements, it became possible to explain the nature of valence and the forces binding the atoms together. Gilbert N. Lewis (1875–1946), at the University of California, recognized in 1916, a few years after Rutherford and Bohr proposed their models, that atoms are bound together by sharing a pair of electrons (Lewis, 1916). Irving Langmuir (1881–1957) introduced for this electron-pair-sharing bond the name *covalent bond* (Langmuir, 1919). After the development of quantum mechanics, in 1927, a detailed quantitative theory of the covalent bond was elaborated. The dashes between atoms, representing for a century the valence bond formulas, could now be replaced by the shared pair electron symbol. Instead of representing the hydrogen molecule by H–H, it could be written H:H, the two dots signifying the two shared electrons. Similarly water and methane could be written

$$
\begin{array}{cc}
\text{H} & \text{H} \\
\ddot{\text{:O:}}\text{H} & \text{H:}\ddot{\text{C}}\text{:H} \\
 & \text{H}
\end{array}
$$

water   methane

although chemists usually still use the dashes since they are for several reasons more convenient.

The electrons of the atom orbit the nucleus not in a simple circle, but in very fast movements filling the space. A more realistic picture of the hydrogen atom and the hydrogen molecule is shown in Figure 1. In the space closest to the proton the density of the circling electron is highest, and it becomes smaller as the distance to the nucleus increases. In the molecule the electrons orbit the two protons not separately but combined. The impossibility of determining the precise site of the electron and its velocity simultaneously was, as may be recalled, the beginning of quantum theory and the notions of indeterminacy and complementarity. In the

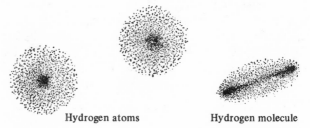

Hydrogen atoms   Hydrogen molecule

Figure 1. Drawings of hydrogen atoms and a hydrogen molecule, showing time-average distribution of electrons. It is conventional to represent atoms by drawings of spheres including the volume in which the electron-distribution function is concentrated.

decade following Bohr's description of the hydrogen atom structure, the electronic structure of almost every element was elucidated by the study of physical phenomena. Although it soon became clear that there was something wrong with the original theory, as discussed before, rapid progress was made; the predictions turned out to be almost, but not exactly, right. The most important tool was spectroscopy; the emission of spectral lines from the vapors of strongly heated atoms provided a large amount of information. This new knowledge was instrumental for the understanding of molecular structure of atoms and many of its behavioral features; it initiated a new era in chemistry.

The heavier the atom, the greater the number of electrons. Some electrons are close to the nucleus, whereas others are at increasingly larger distances. They form concentric "shells" around the nucleus. Some of the heavier atoms may have as many as six shells (called K, L, M, N, O, and P shells). They are also represented by the numbers 1, 2, 3, 4, 5, and 6, respectively, which are the values of the quantum numbers that enter into the treatment of the motion of the electrons of the quantum theory. The various shells are occupied by a certain number of electrons. Starting with the helium atom, the smallest atom to have 2 electrons, the K shell always has 2 electrons and the L shell has 8 electrons; the greatest number of electrons is found in the N shell, which has in some atoms 32 electrons. The shells are divided into subshells, which also have a given number of electrons.

The Pauli exclusion principle (see p. 40) requires that no more than two electrons with two opposite spins occupy a single orbit. The most stable orbit in every atom is the K shell, with its two electrons. The normal hydrogen atom has only one electron, but in the helium atom its two electrons have opposite spins. In order to form a stable bond in the hydrogen molecule, its two electrons must have opposite spins—like the helium atom's—and these molecules are also stable. Electrons with opposite spins are referred to as *paired*, whether they occupy the same orbit in one atom or are involved in the formation of a bond.

When electrons are associated with the molecule as a whole, as in the hydrogen molecule, they are called π electrons and the bond they form is called a π bond. They are the kind of bonds formed when atoms are united by a double covalent bond. Such bonds are particularly frequent in carbon compounds. One of the most important features of carbon is its ability to combine with other carbon atoms by single or double bonds. This ability can be explained on the basis of carbon's electronic arrangement. No other atom has this power to the same extent. It is probably because of this feature that carbon atoms play a decisive role in the substances in the living cell.

A simple example of a carbon compound with a double bond is ethylene. It has the structure

$$H_2C=CH_2 \text{ (structural: } \underset{H}{\overset{H}{>}}C=C\underset{H}{\overset{H}{<}})$$

Here again the electrons between the two carbons are π electrons. Another important example of the role of π bonds is the benzene ring conceived by Kekule (see p. 152):

In this molecule, formed by six carbon atoms, three of the carbon atoms form double bonds; but this does not properly represent the electronic distribution. The π electrons are not restricted to the position indicated by the double bond, but are spread over the whole ring system.

In benzene and many other compounds the double bonds may readily change their position. This shifting of the double bonds is called *resonance*. The concept of resonance was introduced into quantum mechanics by Heisenberg in connection with the quantum states of the helium atom (Heisenberg, 1926). It was greatly elaborated for use in interpreting certain features of chemical compounds, particularly by the pioneer work of Linus Pauling (1901–   ) (Pauling, 1948). It was frequently difficult to assign to a molecule a satisfactory single electronic structure in terms of the valence bond type; two or more electronic structures seemed to fit equally well. The concept of resonance was able to explain the electronic structure of the molecule. There exists only one kind of molecule, but because of the electron shifts between the carbon atoms there are several "resonating" structures. As a consequence of this resonance the molecule is stabilized by a certain amount of energy, called the *resonance energy*. Benzene provides an excellent illustration of a resonating molecule; the shifting of the double bond due to the shift of electrons results in the most stable bond structure. Because of this complete resonance, the benzene ring is a surprisingly unreactive and stable structure.

The great variations of electronic structure, depending on the atoms, double or triple bonds, and many other factors are too extensive to discuss here. Figure 1 may be adequate to illustrate the complexity of electronic distribution even in the simplest molecule, formed by two hydrogens; it indicates that even using dots to represent shared electrons adds little or no information about the complexity of electronic distribution.

There are certain atoms that are chemically inert. They do not combine with other atoms under ordinary conditions. They occur in gas form and are called noble gases. Helium, with its two electrons, is such a gas and has the striking property of existing only in its free state. Its atoms do not even combine with each other; they remain as separate atoms in the gas. Six atoms belong to this group: they are helium, neon, argon, krypton, xenon, and radon; they have in their electron shells 2, 10, 18, 36, 54, and 86 electrons respectively. It is the electronic structure that accounts for the stability of these gases. The energy required for removing an electron from these atoms is very high, much higher than from other atoms. Similarly, the electronic affinity—the energy liberated by the addition of an electron—is zero for the noble gases; they do not have the power to add an electron.

Just the opposite is true in the case of another group of atoms. Their special property is that of easily adding another electron or losing one. Atoms that accept an additional electron become negatively charged, since this extra electron is not neutralized by the protons. Atoms that lose an electron become positively charged. Such positively or negatively charged atoms are called *ions*. The negatively charged ions are *anions*; the positively charged ions, *cations*. The forces acting between anions and cations are called *ionic* (or *electrostatic*) *bonds*, in contrast to covalent bonds. A hydrogen atom, for example, can easily lose its electron and become a positively charged proton (symbol H^+; the plus sign stands for positive charge). Among well-known ions are the halogen ions (*halogen* means "salt producing"), to which belong the chloride (Cl^-) and bromide (Br^-) ions (the minus sign stands for negative charge); they are the ions of the chlorine and bromine atoms. They combine readily with a group called *alkali metals*, to which belong, for example, sodium (Na) and potassium (K) and their ions (Na^+ and K^+). The compounds formed by a combination of halogens and alkali metals contain the same number of electrons as one of the inert noble gases; their stability and their lack of reactivity may thus be attributed to the similarity of the electronic configurations.

In the nineteenth century it was discovered that the elements can be classified in a certain number of groups characterized by several similar properties, for instance the halogens or alkali metals mentioned before. As the number of known elements increased, these group characteristics became increasingly apparent. Lavoisier (1743–1794) knew of 23 elements; by the early nineteenth century more than 50 elements were known, and by the turn of this century their number reached 92. The most important step in classifying the elements into distinct groups was the development of the *periodic table* of the elements, in 1869, by the Russian chemist Dmitri I. Mendeleev (1834–1907). He first proposed a table containing 17 columns, but soon revised it to 7 columns, based on the relation between atomic weights of the elements and their various physical properties, with special attention to their valences. At about the same time, working independently, the German chemist Lothar Meyer (1830–1895) proposed, in 1871, a similar table of elements. After the discovery of the inert noble gases by Lord Rayleigh (1842–1919) and Sir William Ramsay (1852–1916) in 1894 and the following years a "zero" group was added to the periodic table, which thus in its final form consists of eight groups.

The theories of Rutherford and Bohr of the atomic structure and the determination of the exact values of the atomic numbers by X-ray spectra enabled Niels Bohr to interpret the periodic table in terms of electronic structure. The understanding of the meaning of valence, of covalent and ionic bonds, of resonance, and so on, permitted an explanation of some of the most basic properties that the elements of the eight groups have in common. By these explanations of the properties of the elements, based on their atomic structure, the notion of periodicity, and the various properties of the atoms, chemistry increasingly became a part of physics. The apparent multiplicity of the properties of the elements could be attributed to a small number of fundamental characteristics of electronic structure. The distinction between physics and chemistry as two entirely different fields of science lost its meaning.

These few examples, described in a very perfunctory way, may be adequate to illustrate how chemistry was revolutionized by modern atomic physics. Many chemical reactions were previously based on experience, on a trial-and-error basis, helped by intuition and imagination. But now many processes could be explained and predicted on the basis of electronic structure. The field referred to as *physical organic chemistry* was greatly developed in the following decades; it led to an understanding of the precise mechanisms of chemical reactions and thereby permitted a rapid expansion of almost unbelievable dimensions. The fantastic development of chemistry during the twentieth century is thus directly linked to the understanding of atomic structure—and some other aspects of modern physics—developed in the first three decades of this century.

Since organic chemistry forms one of the pillars on which biochemistry is based, it was almost unavoidable that, after a certain period of transition, the new insight into the forces directing chemical reactions would increasingly influence biochemistry and what is today called *molecular biology*. Physical organic chemistry became one of the decisive factors in the spectacular growth of biochemistry in the last decades, starting in the 1940s. A milestone that influenced the thinking of many biochemists was a book by Linus Pauling (1948) entitled *The Nature of the Chemical Bond*.

It may be mentioned that in the 1920s—that is, during the period when rapid advances were being made in biochemistry at the Kaiser Wilhelm Institutes (see IV)—and even in the 1930s few biochemists were familiar with the rapid and dramatic advances and new concepts in chemistry that were brought about by the understanding of electronic structure. The situation changed only in the 1940s when many modern trained chemists entered the field of biochemistry.*

Widespread Impact of Atomic Physics on Pure and Applied Science and Industry

The effects of atomic physics on chemistry, permitting the understanding of the electronic structure of atoms and molecules and the forces of interaction, have been briefly described. In the last four decades, since about 1940, a vast variety of highly diversified fields have been revolutionized by the knowledge provided by atomic physics. Although nowadays even many laymen are aware of them, it may be appropriate to recall just a few examples.

The development of highly refined instruments and methods made possible by advancements in physics greatly contributed to biochemistry—and biological sciences in general. X-ray diffraction, for example, has become an indispensible tool for understanding structure. It provides knowledge of the precise configuration of molecules. This information had a spectacular effect when applied to macromolecules, polymers, and biopolymers, such as proteins, nucleic acids, and other cell constituents. The knowledge obtained of the tridimensional structure, in combination with many other methods, has revealed how proteins, including enzymes, nucleic acids, and polysaccharides work. Another example is the electron microscope. Its ability to magnify several hundred thousand times—and this may not be the limit—provides insight into the intracellular organization, the various substructures and organelles of the cell, which could not be seen with the light microscope. Cell membranes, for example, postulated since the turn of the century but never seen until the 1950s, have been found to be the site of some of the most essential cellular functions, such as power supply (the oxidation–reduction system), vision, never-impulse conduction, and active transport across the membranes. Used in combination with many other biochemical and biophysical methods, electron microscopy raised the exploration of all these cellular functions to an entirely new level. The new and exciting field of genetics in the postwar period would have been impossible without these new tools. A whole series of powerful optical methods became available, new types of spectrophotometers, optical rotary dispersion, circular dichroism, and so on. A variety of ingenious methods were developed for measuring extremely fast reactions taking place in a billionth of a second or less. They have

---

* The author vividly recalls the revelation it was for him when Leonor Michaelis introduced him, in many discussions in Woods Hole in the 1940s, into the completely new era of chemical concepts. Subsequently, Pauling's book deeply influenced his thinking and his research work.

opened new fields of investigations in physics, chemistry, and biochemistry. The use of stable and radioactive isotopes has been tremendously extended and they have become necessary routine methods. As every layman knows, they are widely used in medicine. Isotope techniques that permit accurate dating have completely transformed anthropology, archaeology, and geophysics. It would be impossible—and it seems unnecessary—to enumerate all the great changes that are due to the use of these new methods in pure and applied science, in technology, and in industry, changes that greatly affect our daily lives. But it must be added that we are at the beginning of the road and that new and highly sophisticated methods are continually being developed, all based on atomic physics. Their effects cannot even be foreseen.

However, one type of instrument must be briefly mentioned, since its revolutionary impact may equal or even surpass the others discussed: the masers and lasers (acronyms for *m*icrowave *a*mplification by *s*timulated *e*mission of *r*adiation, and *l*ight *a*mplification by *s*timulated *e*mission of *r*adiation). They are instruments of radiation of specific wavelengths by stimulation instead of spontaneous changes of energy levels of atoms or molecules. Masers produce, as the name indicates, radiation in the range of microwaves—that is, between the length of light waves and those of very low frequency. Lasers produce radiation of waves in the range of visible light. The effects are based on the following principle: Electrons circling the nucleus of the atom have a certain energy level. When they absorb photons of electromagnetic waves, they jump to a higher energy level. When they return to their initial state they emit radiation of specific wave frequencies. It is possible to direct waves of one specific frequency in any desired and well-defined direction.

The intensity of light can be tremendously increased. Lasers are able to transmit much more information than radar. Because of their intensity they have an extremely high power. They can cut concrete and steel, and vaporize diamonds. Lasers, in particular, have been developed since the 1950s for use in a great number of diversified fields. They are probably one of the most versatile inventions to emerge from atomic physics. They may revolutionize communication, being capable of transmission with extremely high speed contained in a very small volume using very thin filaments. They have already provided powerful tools for science, medicine (e.g., eye surgery), and industry and they have unfortunately been used for an arsenal of terrifying and appallingly effective military weapons, already used in several wars. They can destroy earth-orbiting satellites and missiles. They may take photographs 300 miles in outer space with incredibly high resolution.

One of the most important potentials of lasers may be their ability to produce thermonuclear fusion, thus opening an inexhaustible source of energy for mankind. It may take 50 years or more until all the serious technical difficulties for large scale production will be overcome; but the theoretical progress achieved thus far is most promising. According to the views of some competent physicists, through the use of lasers, fusion energy may become producible on a small scale in laboratories within a decade. Moreover, the use of magnetic fields, electron beams and other similar approaches seem to offer equally great promises. The blessing this will provide for mankind, at present in the throes of a serious energy crisis, does not have to be explained. In contrast to fission, fusion would produce little or no radioactive wastes, so its production would be less hazardous. The fact that the United States had a budget of half a billion dollars for 1978 for research on lasers (the Soviet Union may have had an even higher budget) indicates how important these developments are considered to be. Few other results of atomic physics may thus influence the fate of mankind—for good or evil—as much as the laser. Masers and lasers were first developed in the 1950s by Charles H. Townes (1915–), in collaboration with A. L. Schawlow, at Columbia University. Townes received the Nobel Prize in 1964.

Most of these developments took place after World War II. The biochemists described in Chapter IV belong to the first part of this century, and they were trained in classical chemistry. The far-reaching effects of the methods just mentioned are not included in this work, since they took place after the Nazis came to power in Germany; only some of them may be occasionally mentioned in the discussion of the postwar period. The postwar generation of scientists, familiar with the new possibilities, have opened a new era in the biological sciences. But one should not minimize the paramount role of the pioneer work of the great chemists and biochemists described in Chapter IV. Their work provided the conceptual basis and laid the foundation for many of the breathtaking discoveries of the present generation, just as Newtonian physics, electromagnetic waves, thermodynamics, and the many other achievements of the nineteenth century were not only the forerunners, but the prerequisites for the great revolution in physics in this century.

G. THE LATE NINETEEN-TWENTIES; RISE OF THE NAZI MOVEMENT

On October 1, 1927, between the physics congress in Como and the Solvay Conference in Brussels, Heisenberg was appointed *Ordinarius* of theoretical physics at the University of Leipzig, at the age of 25. His extraordinary brilliance, his youthful enthusiasm, his dynamic personality, and his international reputation soon attracted many outstanding physicists from all over the world to his institute. In a short time it became one of the great centers of atomic physics. In 1932, at the age of 31 he received the Nobel Prize. Among the brilliant scientists, many of them future

Nobel laureates, who joined Heisenberg in Leipzig were Felix Bloch (1905–   ), Nobel laureate in 1952; Edward Teller (1908–   ); Hans Bethe (1906–   ), Nobel laureate in 1967; and Lev Davidovich Landau (1908–1968), Nobel laureate in 1962. The new challenges offered by quantum theory for a deeper understanding of atomic properties were eagerly taken up. Throughout this time Heisenberg maintained very close personal and scientific contacts with Bohr and Pauli. A yearly visit to Copenhagen for at least two weeks was obligatory. No important results were published without consultations with Bohr.

With Einstein, Planck, von Laue, and Schrödinger in Berlin, Otto Stern with a large and flourishing school in Hamburg, Franck and Born in Göttingen, Sommerfeld in Munich, and now Heisenberg in Leipzig, Germany had become one of the greatest centers, probably the greatest, in atomic physics. After the years 1924–1927, frequently referred to as the "heroic" time of this field, all these centers were actively engaged in research associated with the great breakthrough initiated by wave and quantum mechanics. Physicists from many countries spent at least some time in one or several of these centers.

In the United States atomic physics was still poorly developed. It is thus not surprising to read a statement made by Edward U. Condon (1902–1974), that for American physicists interested in this field it was almost obligatory to spend some time in Germany (see Morse, 1976). Condon himself worked first with Born in Göttingen and later with Sommerfeld in Munich. Among the American physicists who went to Germany was Isidor Isaac Rabi (1898–   ). Born in Austria, he had come to the United States as a young child. He was destined to become one of the most illustrious scientists of this country. He not only influenced physics, but was one of the chief promoters of science in the United States in general, when he served for many years as a member and chairman of the President's Scientific Advisory Committee at the time of Eisenhower. During this period he played an instrumental role in building up the United States as the greatest scientific center of the world (Rabi, 1960). Among his many activities, which earned him the fame of being a statesman–scientist, was the organization of the movement of atoms for peace that resulted in the building up of the great European center for high-energy physics known as CERN (*Centre Européen Recherches Nucleaires*) of which Bohr was the first president. He has repeatedly visited Israel and is a member of the board of governors of the Weizmann Institute.

In Germany Rabi first joined Otto Stern in Hamburg in 1927, where he was attracted to the use of atomic and molecular beams for the analysis of atomic behavior. During the year spent in Hamburg he utilized a novel method of deflecting the beams and was able to obtain highly accurate measurements of atomic magnetic moments. He then spent another year in several laboratories, particularly in those of Bohr, Pauli, and Heisenberg, before returning in 1929 to the United States. He became full professor at Columbia University in 1937.

On his return to the United States Rabi applied a new technique for studying the structure of atomic energy levels. In 1933 Otto Stern and his co-workers had shown that the magnetic moment of the proton was about $2\frac{1}{2}$ times larger than expected from a theory proposed by Paul Dirac, one of the most brilliant British mathematicians and physicists, who shared the Nobel Prize with Schrödinger in 1933. Stern obtained his results with molecular hydrogen. Using atomic hydrogen and deuterium, Rabi and his co-workers not only confirmed these results but developed an independent way of measuring the magnetic moment of the proton and deuteron. In the course of these investigations Rabi invented new methods for the utilization of resonance phenomena. Rabi's many outstanding achievements are far too many to outline here, but it may be mentioned that his methods were later generalized and had many important implications, for instance in the development of nuclear magnetic resonance, the maser and laser, and other new methods. During World War II he was involved in the development of radar and in the Manhattan Project. Rabi received the Nobel Prize in 1944, at the same time as Otto Stern, who had been awarded the Nobel Prize in physics for 1943. Stern, after having left Germany in 1933, came to the United States as a research professor of physics at the Carnegie Institute of Technology in Pittsburgh. The few comments on Rabi's work are another illustration of the worldwide repercussions of atomic physics at its acme in Germany in the 1920s.

Another famous American physicist who obtained his training in Germany was Robert Oppenheimer (1904–1967). He received his Ph.D. in Göttingen with Max Born. During the war, as is well known, he was the director of the atomic energy project in Los Alamos, which succeeded in building the first atom bomb.

In Brussels, at the 1930 Solvay Conference, the quantum theory was exposed to a new test. Three years had passed since the confrontation between Einstein and Bohr. Einstein, still vigorously opposed to the theory, had worked out a new "ideal" experiment, this time involving observations based on the theory of relativity. Einstein's new proposal, which in essence tried to undermine the notions of complementarity and indeterminacy, raised serious problems. Bohr and his colleagues spent the whole night trying to find out where there might be an error, because they did not doubt that Einstein was wrong. The next morning they came to the conference with a complete answer. Every aspect, including mathematics, was worked out. Einstein's own previous observations were used against him. Einstein could not disagree with the explanation offered. The arguments were impeccable. Once again quantum mechanics had triumphed. Einstein had lost, although he still remained unreconciled. He never accepted, up to his death in 1955, the new theory. All attempts of Born over many years to convince Einstein in an exchange of letters (Born, 1969) were unsuccessful.

It should be stressed that the personal relationship between Einstein and Bohr was not affected in the slightest by the quantum theory disagreement. The two titans were closely associated not only scientifically, but by their many other aims and regard for human values. Einstein was always on Bohr's mind, and his death was for Bohr a terrible shock. Bohr described Einstein as one of the greatest and most creative minds mankind had ever produced. He praised his deep sense of responsibility, his humility and modesty, his hope that a better understanding of the physical world would eventually help the progress of civilization. Actually, Einstein had the same opinion of Bohr. He considered him to be one of the greatest thinkers of the age, combining a rare blend of boldness and caution, a man whose intuitive grasp made him one of the greatest discoverers mankind has produced.

Between the two Solvay conferences of 1927 and 1930 dramatic events had taken place that were rapidly changing the whole world situation. In the mid-1920s a euphoric outlook prevailed in the Western world. Prosperity seemed to grow; even in Germany the economic outlook appeared to brighten, thanks to American generosity in helping the country, and in spite of the burden of reparations. The continuous growth of science raised many expectations for further progress. The efforts of Stresemann, Briand, and Chamberlain were considered as a hopeful sign of movement toward a peaceful world. Stresemann died early in October 1929. The crash of the New York Stock Exchange in 1929 and the following depression had disastrous effects on the world economy. The depression took on unprecedented dimensions. In the United States there were 10 million unemployed; in Great Britain, in 1932, 3 million.

But the greatest catastrophe took place in Germany. There were 8 million unemployed in a country with a population of less than half that of the United States. The middle class had lost all its financial reserves during the runaway inflation in the early 1920s. The economic and social turmoil, the repercussions following a decade of continuous humiliation due to the Versailles treaty, and the general gloom and despair led to the violent reactions that helped the growth of National Socialism and brought Hitler to power. In 1930 the National Socialists had 103 seats in the Reichstag; in 1932, 230 seats. About 40% of the population voted for Hitler. On January 30, 1933, Hitler became chancellor. After the Reichstag fire the Communists were outlawed and the Nazis, who previously had 40% of the members of the Reichstag, had the absolute majority. The powers given to Hitler by special laws, in April 1933, made him the dictator of Germany with absolute and unrestricted powers. The people of Germany did not fully realize the extent of what had taken place, and neither did the people of other countries and their governments. It took years for them to recognize the true nature of the Nazi regime, too late to take the appropriate measures to prevent the war that led to the devastation of Germany as well as Europe.

As mentioned before, the racist civil service laws were introduced in April 1933. The devastating effects on the University of Göttingen have been described. Jewish scientists everywhere were forced to give up their positions; many resigned voluntarily. In the years 1933–1935 about 1150 Jewish or partly Jewish scientists left Germany. Among them were many illustrious leaders who had established large schools; many of them were Nobel laureates or future Nobel laureates. The Kaiser Wilhelm Institutes in Berlin suffered a terrible blow, similar to that of Göttingen, although it was softened by the fact that they were not as completely dependent on the government as the universities since they received considerable financial support from private sources. The events there will be described in Chapter IV. German science suffered a frightful setback, not only by the loss of many highly qualified scientists, roughly estimated to be about 25%, but because science in Hitler Germany was relegated to an insignificant role. Scientific excellence was much less important than party loyalty, endorsement of Nazi ideology, loyalty to the new regime, and other values considered to be more essential for the future of Germany. Several outstanding scientists who were only partly Jewish or not Jewish at all voluntarily left the country because they were unable to live in the atmosphere created by the Nazi regime; among these were Erwin Schrödinger, Max Delbrück, and Hans Gaffron.

Atomic physics was particularly hard hit. The loss of Einstein, James Franck, Max Born, Edwin Schrödinger, Otto Stern, and many of their associates was accompanied by the rise to power of two strongly anti-Semitic physicists, Johannes Stark (1874–1957) and Philipp Lenard (1862–1947). They considered the whole field of atomic physics as a Jewish fraud. Quantum theory and relativity were for them "Jewish" physics, in contrast to "German" (or "Aryan" or "Nordic") physics. German physicists who accepted and taught Jewish physics were called "white Jews."

Both Stark and Lenard had been first-rate physicists early in their careers. They had made excellent contributions and were Nobel laureates. But even before World War I they became increasingly detached from the mainstream of physics. They were in violent opposition to the German Physical Society, because it was too much dominated by Jews. After the war they identified themselves with the Nazi party. On Lenard's office door in Heidelberg was a note: "Entrance to Jews and members of the German Physical Society not permitted." Stark had been ostracized by the universities since 1922; he was unable to obtain a position and worked in industry. Stark's and Lenard's hatred and accumulated resentments exploded in 1933 and they became the great leaders of physics and advisers of the Nazis. Their personal history, their actions during the Nazi period, their fights with German physicists, many of whom were not willing to accept their leadership, the detrimental effects on physics in Germany do not actually belong here. The Readers interested in these aspects are referred to the excellent, detailed, and well-documented book of Beyerchen (1977), who describes the politics and the physics community in the Third Reich from 1933 to 1945.

On the other hand, it is obviously pertinent to describe the attitude of at least some of the most prominent atomic physicists who remained in Germany during the Nazi period. In view of the many close personal friendships between Jews, half-Jews, and non-Jews, it is essential to know how those most deeply involved reacted to the expulsion or voluntary departure of their friends. Without this information the picture of the era of collaboration would be incomplete; it would be difficult to evaluate the genuine character of the friendships.

The attitude of Planck, the dean of German physicists, has been described before (see II, C). A few words may be added to understand his behavior and his actions or lack of them. As most German scientists, he was politically naive. He distinguished between service to the government on the one hand and service to the state and the nation on the other. This attitude prevailed among many of his colleagues and in fact was widespread among other sections of society, especially during the period of the Weimar Republic. He was deeply devoted to the nation and felt a strong responsibility to help strengthen it for the future, and to fight for the survival of its ethical, spiritual, and cultural values. Governments were thought to be temporary phenomena. In the beginning, perhaps influenced by his son Erwin, Planck thought that the Nazi regime could not last long. He went for his vacation to Sicily. Only on the way back in May did he learn from Max Born, when they met in Tyrol, about the shocking events in Göttingen, and that Haber and many other scientists had left the Kaiser Wilhelm Institutes. Planck was in 1933 the president of the Kaiser Wilhelm Society. He had held, and still held, important official positions in the academic world. Coming from a family of civil servants for generations, the tendency of loyalty to the "legal" government was strong and it was difficult to break with this tradition. Even though he hated and despised the Nazis, how should he proceed in his exposed position most effectively? He faced frightful problems, a dilemma with which he was unable to cope. For him the main problem was to save German science from complete destruction in a rapidly deteriorating situation. He felt he had to be cautious and avoid too great provocations, which might lead to still greater disaster.

One of the most courageous personalities among the atomic physicists in rejecting any cooperation with the Nazis was von Laue. He was as devoted to his country as Planck and had an equally strong sense of duty. In the first few months he too thought that the regime would last only for a very short time, but after a few months he began to see the reality and became very pessimistic. Von Laue became the symbol of courage, the "champion of freedom" and of the active fight against the Nazis; he remained so during the whole Nazi period. This was well known abroad and he was greatly admired for the audacity of his actions. Einstein and Born were deeply impressed. During the Newton celebrations of the Royal Society in 1946 he happened to be in London and one of the members took him to the festivities. He was warmly welcomed by his colleagues.

A few examples may be given. At the annual meeting of the German physicists in the fall of 1933, von Laue spoke about the horror of the Inquisition and the attitude toward Galileo. In his closing remarks he quoted Galileo's famous words "And it still moves." The audience was fully aware that his speech was symbolic and that he referred to the attitude of the Nazis toward Einstein. His speech was followed by strong applause. The German Physical Society was one of the few organizations that dared to refuse alignment with the Nazis. Von Laue was one of the main promoters who worked to keep the society independent and to maintain the election of officers without outside interference. To the great consternation of the Nazis, the society allowed Jews to be members.

After Haber's death, in 1934, von Laue published two articles praising Haber's great contributions to Germany and deploring his tragic fate (see IV, A). He sent his son, in 1937, to the United States to remove him from the influence of the Nazis. He kept up his friendship with those of his Jewish friends still remaining in Germany. Among them was Arnold Berliner, the editor of *Naturwissenschaften*, one of the most important German journals. Berliner was forced to resign the editorship in 1935. When the situation became intolerable, Berliner committed suicide, in 1942. Von Laue attended the funeral at the Jewish cemetery.

In view of the degree to which von Laue dared to defy the Nazis, it appeared surprising that nothing happened to him and that he was not silenced by the Nazis. One can only speculate about the reasons. One factor may have been his fame and high international standing; he was Planck's most distinguished pupil, a Nobel laureate. Another factor may have been that he came from an old and prominent Junker family, with many connections to the top of the *Reichswehr*.

Perhaps the most difficult situation was faced by Werner Heisenberg. His attitude and behavior during the Nazi period were frequently questioned and criticized by some colleagues in the United States. In view of his leading role in the development of quantum theory, the worldwide recognition of his genius, his many close friendships with Jews, it is of obvious interest to describe briefly his actions and reactions during the Nazi era. From his first student years Pauli had been his best friend, and the friendship lasted until Pauli's death in 1958. Born was his admired teacher, Niels Bohr his idol, Einstein in his eyes one of the greatest geniuses of mankind. In Copenhagen Heisenberg was deeply impressed by the international atmosphere, an experience that left a deep and lasting imprint. When he became *Ordinarius* in Leipzig, in 1927, he dreamed of close and intimate collaboration with Pauli in Zurich and with Bohr's institute. His frequent and regular trips to Bohr were for him an integral part of his scientific activity for critical discussions, exchange of views and information, and so on. Most of his pupils in Leipzig were Jews, and Felix Bloch was his chief assistant; they were attracted by the charm of his personality, his simplicity, his youthful enthusiasm, his brilliant mind. His life was completely absorbed by physics, music, and his love of nature. Since his student years, he loved to hike with friends, particularly in southern Germany; he adored the forests, the lakes, the quaint old cities, and above all the Bavarian mountains. He frequently went skiing. He was deeply attached to Germany. Music and hiking were his recreation, but his obsession was physics.

Of special interest in regard to his attitude toward the Nazis is, however, his political attitude during the Weimar Republic. As mentioned, he was internationally minded owing to his frequent visits to Bohr's institute and his many international friendships there. He was extremely far from narrow nationalistic sentiments. In spite of his love for Germany, he felt himself to be European. His general views were outspokenly progressive and liberal. He taught in the *Volkshochschulen* established by the Social Democrats for the working class in order to increase their standard of education. These institutions were oriented toward the views and the liberal ideas of the Weimar Republic. (They were immediately abolished when the Nazis came to power.) Only people with a strongly positive attitude to the republic would associate themselves with these institutes and teach there. Many elements of the middle and upper class, especially in the universities, were skeptical, unfriendly, or even hostile toward the new regime. Quite a few of the older generation, even among Jews, looked back with nostalgia to the glorious days of the *Kaiserreich*. The difficulties for the young democracy were compounded by the continuous humiliations and hardships imposed by France and England, acts that shocked even Einstein and Born, and greatly strengthened the reactionary and anti-Semitic elements. To someone intimately familiar with the political events in Germany of that era (as the author was), his teaching at the *Volkshochschulen* is a most illuminating factor, a strong indication of his genuinely liberal views and his closeness to the Social Democratic aims, although he did not belong to any political organization. Many of Heisenberg's views were so progressive that some of his friends considered him a *sozi* ("socialist"). The Social Democrats of that era were the most vigorous opponents of the Nazis, rejecting any idea of totalitarian or authoritarian regime. The Social Democrats and the Nazis were extreme opposites, although some liberal democrats and left-wing members of the *Zentrumspartei* (the Catholic party) were very close to the aims and views of the Social Democrats.

The rise of the Nazis to power hit Heisenberg like a thunderbolt out of

a blue sky. When he spent some days with Bohr in March he was optimistic that the storm would blow over in a short time. But the illusion
did not last long. The events of the spring of 1933 appeared inconceivable. He was greatly shocked and became deeply concerned. Not only
did German physics lose its great leaders, he lost his most brilliant associates, among them Bloch and Bethe. He made vigorous efforts to keep
at least some of the leading Jewish physicists in their positions, and to help
his Jewish students find places to continue their work. He and some colleagues of the faculty, especially Friedrich Hund and Karl-Friedrich
Bonhoeffer, were infuriated and in despair. How should they react? Was
there anything they could do to improve the situation? They seriously
considered resigning. They realized this would mean emigration, although
they were deeply attached to Germany. But even these drastic steps would
have no effect. Moreover, did they have the right to abandon their colleagues and students? This was a question particularly important for
Heisenberg; as the torchbearer of the new quantum mechanics he was in
the best position to train a new generation in the field of atomic physics
and the theory of relativity.

In this dilemma he turned to Planck for advice. Both Planck and von
Laue strongly urged him to stay on and continue his research and teaching. It was to them apparent that a catastrophe was, sooner or later, unavoidable. They urged all colleagues, including those Jews who for some
reason were able to continue to work, to stay on in Germany, to continue
to work and train a new generation. This would help to rebuild a new
and decent Germany after the collapse of an insane and criminal regime.
Heisenberg in particular was a great hope; he was young and vigorous,
destined to become one of the foremost leaders after the collapse of the
regime. Leaving the country would have no effect, staying on would be
an encouragement for all opponents of the regime. Even his admired
teacher and friend Niels Bohr advised him to stay in Germany. In retrospect, it seems fair to admit that the advice of Planck, von Laue, and
Niels Bohr, and the resulting decision of Heisenberg and others proved to
be important for the reconstruction of a new Germany in the post-Hitler
period; the destruction of physics and science would have been still worse
and no leaders of stature would have been there trying to rebuild a decent
Germany and to overcome the political, spiritual, intellectual, and moral
devastation. Nobody, of course, could have foreseen the extent of the
final destruction—not only of science, but of the whole country; the
criminal activities, the Holocaust. Although Heisenberg realized that staying on in Germany meant a difficult time, the accepting of disgraceful
compromises and humiliations—in essence, paying a high price—he accepted Planck's advice. He continued his research and teaching in the
same spirit as before. His institute became an "island of freedom."

When he met Bohr at the Solvay Conference in November 1933, he
learned that he would receive the Nobel Prize of 1932, together with

Schrödinger and Dirac. Heisenberg was fully aware how much he owned to his teachers Pauli, Born, Franck, and above all Bohr. He expressed his feelings to Bohr in a moving letter, recalling how decisive all he had learned in Copenhagen had been for the work for which he received the award. He also wrote a letter to Born that he was ashamed that the prize was not shared in view of Born's decisive contributions to quantum mechanics (see p. 81). About the time it became known that Heisenberg would receive the Nobel Prize, a great number of outstanding German scholars, among them the surgeon Sauerbruch, the philosopher Heidegger, and the art historian Pindar, met in Leipzig to prepare a document expressing their attachment to Hitler. When they asked Heisenberg to sign the declaration, he flatly refused.

Heisenberg began to feel increasingly lonely and depressed. He repeatedly expressed his feelings to Bohr. His only consolation was a still greater intensification of his work and his music. But the worst period was still to come. The groups around Stark and Lenard became increasingly influential and aggressive. Heisenberg and Sommerfeld were among their chief targets and came under heavy attack. Sommerfeld had probably trained the greatest number of students in Germany in quantum theory and relativity. He continued his teachings without change in the Hitler period. He was one of the greatest admirers of Einstein, and in the exchange of letters between him and Einstein (Hermann, 1968), he assured Einstein during the Hitler period that his students would still hear in the lecture halls about him and his work. Even after his retirement he continued to teach "Jewish" physics, until a successor was found, who turned out to be a disaster. When Heisenberg was proposed as Sommerfeld's successor, the "Aryan" physicists started a vigorous fight, full of personal attacks, slander, and vituperation. The *Völkische Beobachter* and other leading Nazi newspapers tried to prevent his appointment, picturing him as the man who represented the spirit of Einstein (*Geist vom Geiste Einsteins*), the greatest enemy of Germany. Heisenberg was deeply depressed; it was a trying time. He tried to fight these attacks by a memorandum, signed by 75 professors, defending the importance of his teaching in physics. Obviously, it had no effect. It was a waste of time and effort. The fight continued with the worst possible insults; he was called a Jew in character, in spirit and attitude. He started to consider emigration seriously again. His only happy days were the trips to Copenhagen, where he met many of his cherished friends.

In 1937 he married Elisabeth Schumacher, daughter of Professor Hermann Schumacher, *Ordinarius* of *Nationalökonomie* at the University of Berlin. Pauli and Bohr congratulated him warmly. They were quite unhappy and worried about his predicament, his frustration and anguish. They hoped the marriage would be a strong help and moral support, and this was indeed the case. The couple spent two weeks in the spring of 1939 in Badenweiler. It was evident that the war was coming. Heisenberg was in a deep depression. He had turned down offers that would have permitted him to leave the country. He faced a frightful dilemma. He might have had the right to accept for himself the risks of war, out of feelings of responsibility toward the future of his country, even if it was then dominated by a detested and criminal regime. But he was married and had two children. Did he have the right to expose his family to the horrors of a war when he could have easily escaped?

The author has had the privilege of receiving from Mrs. Elisabeth Heisenberg a preprint of her memoirs about her husband and permission to mention them here. As has been repeatedly stressed, he was one of the greatest geniuses in atomic physics of the twentieth century, and much has been written about him (see, e.g., Hermann, 1977). But nowhere does one find such a fascinating and complete picture of his personality as in Mrs. Heisenberg's memoirs. With brilliance and great psychological insight, Mrs. Heisenberg has described the agony, the mental torment, the painful dilemmas her husband suffered through during the Nazi period. Hers is the first analysis to explain vividly and with masterly clarity the reasons for many of his actions and decisions. It is extremely difficult for anyone who was not there to understand what went on in Germany during those years: the organized terror, the assassinations, the executions, and the concentration camps that threatened all who dared to criticize the regime. At the time Heisenberg could not convince his friends abroad of his real motives because they simply were unable to conceive of what was going on in Germany during that time, and also because the risk involved in explaining was so great, especially to his family. He was frequently in despair, knowing that his decision to stay was endangering their lives. Consequently his behavior during the Nazi period was poorly understood and even misinterpreted by some colleagues. Mrs. Heisenberg's memoirs fill this important gap. She tells how Heisenberg was torn between his hatred of the Nazis and his love for his native country. He felt that it was his moral obligation to continue teaching modern physics to the new generation and to be in a position to help the reconstruction of Germany after the inevitable collapse of the Nazi regime. Mrs. Heisenberg's memoirs reveal the psychological details underlying his motives and decisions, details only she as his wife could give. The memoirs are of paramount importance for all those who want to understand Heisenberg's personality and why he acted as he did during the Nazi period. In addition, they are an invaluable historic document portraying one of the chief targets of Nazi attacks and describing with rare clarity the torment, helplessness, and despair of the many German intellectuals who realized that their country was dominated by a criminal group.

The author is familiar with the attitudes and psychology of leading German scientists of that era; he has personally known some of them. He was among those who were extremely sensitive to anti-Semitic or pro-

Nazi sympathies. He is compelled to express his deep regret and his vigorous objection to the misinterpretations of the great sacrifice made by Heisenberg and some of his young colleagues. Such misjudgment shows a complete ignorance of their true motives, as Mrs. Heisenberg shows in her memoirs. They knew the heavy price they would have to pay; some lost their lives. They stayed on in spite of their hatred of the Nazis and against their personal interests. They felt the obligation to stay on in order to help one day in rebuilding a new, free, and democratic Germany. It is not just their own country but mankind that owes them a great deal of gratitude and respect for their courage and their idealism.

These few examples of the attitudes of those physicists who remained in Germany illustrate their dilemmas and motives, the accompanying frustrations, the anguish and tragedy. The view is frequently expressed that if these leaders had resisted the Nazis more strongly and publicly, even at the risk of their lives, they would have prevented the worst excesses. Those who hold this view do not realize the deliberately created atmosphere of terror, organized to perfection, the ruthlessness and sadistic brutality, with which the Nazis suppressed every resistance. Opponents were sent to concentration camps, were assassinated, or simply disappeared. Public notice of death were not permitted, so that these protests were totally ineffective. Under these conditions one must admire rather than criticize the extent to which these opponents of the Nazi regime dared to defy the Nazis. One must fully sympathize with von Laue, who was quite upset and expressed these feelings in his letters to his son in the United States after the war about the extremely critical views concerning the attitude of German scientists; he was hurt by the generalization blaming all his colleagues for not being sufficiently forceful in their protests. The comments here are limited specifically to the attitude of the atomic physicists in view of the close friendships and collaboration in the pre-Hitler era; many of them were taken up again in the postwar period.

Another factor, frequently not appreciated, was the apparent lack of any practical value of atomic physics. It was an esoteric science, of purely theoretical interest, more a kind of philosophy and an intellectual exercise, in contrast to many fields of experimental physics of potential value for industry, agriculture, and the war effort. Thus the voice of the most eminent leaders in atomic physics carried little weight, much less than that of many less prominent scientists whose work may have turned out to be useful. Even leaders of these groups, powerful industrial leaders like Karl Bosch and Albert Vögler, had no influence upon a regime that despised science and was violently anti-intellectual.

It may be useful to remind the reader in this context of a related aspect in the history of the Vietnam war. What was the effect of the strong appeals of many scientists to stop the war? Many leading universities—Harvard, Yale, Columbia, Cornell, and others—protested against a senseless and cruel war that produced terrible destruction; the lives of many young Americans were sacrificed and hundreds of billions of dollars were wasted. The United States is a country with deep democratic roots and with complete freedom of expression. President Johnson was so infuriated about the attitude of the academic world that funds for research and the support of the universities were drastically cut, an action that had —and is still having—a serious effect on the development of science. How then could one expect the few atomic physicists in Germany to have the slightest influence, even if they were to risk their lives? They would have committed suicide; the only result would have been a still greater deterioration of scientific research.

Finally, whereas the governments during the Weimar Republic, even that of Gustav Stresemann, supported by Aristide Briand and Houston Chamberlain, were unable to obtain any concessions from England and France to soften the humiliations resulting from the Versailles treaty, Hitler succeeded in obtaining one concession after the other, especially from the English government, and was even repeatedly greatly encouraged. This situation obviously weakened the strength of the opposition, but this aspect cannot be discussed here.

Since the problem of the strength of the resistance of the atomic physicists is relevant to the question of how real and genuine the friendships were between Jews and non-Jews, it was necessary to recall these few facts. The Bible states that one should not judge one's friend until one is in the same situation. When Hannah Arendt, in her well-known book about the Eichmann trial, sharply criticized the Jews for their lack of courage and spirit of resistance, leading eventually to the Holocaust, many of her Jewish friends—and the author was one of them—were deeply hurt and upset. Much has been written about the incorrect statements and incomplete descriptions in her book, how poorly she was informed, how unfamiliar with the huge amount of existing material, although mostly unpublished or written in Hebrew. Although the problem does not belong in this book, the parallelism is obvious. The most vociferous criticism of the attitude, the behavior, and the motives of the German atomic physicists has been voiced by people poorly informed about the facts and the reality of the tragic situation in Germany during the Nazi regime, which killed thousands of the intellectual leaders of the old German families.

The attitude of scientists in general will not be discussed, except for those who will be mentioned later. Many scientists endorsed the Nazi regime, but quite a few put up a strong resistance. In general, the public knows very little about this whole problem. Very few know about the strength of the resistance movement in Nazi Germany. Only recently has an informative and well-documented account of the extent of the resistance been published (Hoffmann, 1977).

Many of the Jewish (or half-Jewish) scientists who left Germany were welcomed by their colleagues in various countries, especially in Great Britain and France, despite the economic depression. Great efforts were made to permit them to continue their work. Niels Bohr accepted a great number of physicists in his institute until they could find a suitable position elsewhere. He was extremely active in placing Jewish refugee scientists. But by far the greatest number eventually went to the United States. As was written in an English journal in regard to the persecution of scientists, Germany's loss was the world's gain.

H. NUCLEAR FISSION

The disaster that hit German atomic physics did not stop its progress. There were other important centers in Europe and in the United States where great advances were made in the 1930s. One particular aspect must be discussed in the context of this book.

Increasingly the nucleus of the atom was becoming the center of interest. In 1919 Rutherford had bombarded nitrogen nuclei with α particles, and hydrogen nuclei (protons) came out. In following up these experiments with James Chadwick, it was established that the α particles had not just removed a proton from the nitrogen, but they had actually entered the atom. The nitrogen nucleus, by absorbing an α particle and ejecting a proton, was transformed into an oxygen nucleus! Rutherford had achieved the transformation of one element into another.

At that time Rutherford and other physicists suggested that another subatomic particle might exist in atoms comparable to the mass of the proton but without an electric charge. Experimental tests to substantiate this idea were unsuccessful. In 1930 the German physicist Walther Bothe (1891–1957), with the assistance of Herbert Becker, bombarded beryllium with α particles. Beryllium is a light atom with the atomic number of 4; it has one more electron and proton than lithium and its atomic weight is 9. The beryllium gave off an unusually high penetrating radiation. Bothe was a student of Planck. He had worked at the *Physikalisch-Technische Reichsanstalt* in Berlin. In 1934 he became the director of the Institute of Physics in the Kaiser Wilhelm Institute for Medical Research in Heidelberg, the same institute to which Otto Meyerhof had moved three years before (see IV, D). For his many achievements in atomic physics, which need not be discussed here, Bothe received the Nobel Prize in 1954, together with Max Born.

In 1932 Frédéric and Irène Joliot-Curie repeated Bothe's experiments, putting a paraffin screen behind the beryllium. They found that the penetrating radiation caused protons to be ejected from the paraffin at high speed. In view of the energy of these protons the observations were difficult to reconcile with the assumption that the radiation consisted of γ rays.*

The correct interpretation of these experiments was provided by James Chadwick. His observations showed that the ejection of protons could be explained by assuming that the radiations consisted of uncharged particles: the long-predicted neutrons. His observations were amply confirmed. Neutrons have about the same mass as protons. A new chapter was initiated in the exploration of the nucleus of the atom. Chadwick received for this and his many other achievements the Nobel Prize in 1935. Heisenberg soon suggested a scheme according to which the nucleus of the atom consists of the two particles, protons and neutrons, both of which were called *nucleons.*

In the same year, John D. Cockcroft (1897–1967) and Ernest T. S. Walton (1903–   ), working in Rutherford's laboratory, bombarded lithium (atomic number 3) with a stream of protons. They found that this element, formed by three protons and four neutrons, absorbed a proton and subsequently split into two helium nuclei (α particles), each formed by two protons and two neutrons. The mass of the two α particles is less than the combined mass of a lithium nucleus and a proton. According to the Einstein equation $E = mc^2$ this loss of mass should be transformed into energy and exhibited by the energy with which the α particles were ejected from the target. Cockcroft and Walton measured the energy of the α particles and the data agreed with the Einstein equation. They shared the Nobel Prize in physics in 1951.

When Enrico Fermi in Rome learned about the discovery of artificially produced radioactivity by the Curies, he decided to test whether the use of neutrons instead of α particles might also produce radioactivity. Since α particles and protons have electric charges, it should be more difficult for them to overcome the barrier posed by the electric charge of the nucleus than for the electrically neutral neutrons. Thus neutrons might more easily penetrate atoms and disrupt their structure. Experiments with elements of small atomic number were not satisfactory. But when uranium was bombarded with neutrons, it became very radioactive. Some accidentally performed experiments, not directly connected with the uranium bombardment, suggested that if the neutrons passed through paraffin or water, the radioactivity they produced would be markedly increased. Fermi interpreted this effect as a slowdown of the neutrons

---

* The Curies later bombarded aluminum with α particles and found that a new substance was formed, phosphorus, but the compound vanished in 30 minutes. It gave off γ rays and turned to silicon. For this discovery of artificial radioactivity they received the Nobel Prize in 1935. But they did not pursue their beryllium experiments and they missed the discovery of the neutron.

during the passage through water or paraffin by collision with the similar-sized hydrogen nuclei. He assumed that the slow-moving neutrons were much more likely to be captured and to remain in the nucleus than high-speed particles, which would simply shoot through the atom. Bohr, the Curies, their associates, and other physicists, were greatly inspired by Fermi's observations and quite an amount of work was devoted to the effects of neutrons; several theories were offered about the processes taking place inside the atom. Most scientists assumed that uranium captured the neutrons and that "transuranium" elements might be formed, with atomic numbers 93 or 94; Fermi, who had thought of this possibility, actually remained skeptical.

Some physicists speculated that the new element might be protactinium, which has the atomic number 91, an element that Otto Hahn and Lise Meitner had discovered in 1917. It was, therefore, quite natural that both became interested in the effects produced by neutrons. At that time, in the mid-1930s, their main interest in radiochemistry was turning increasingly to nuclear chemistry. However, before continuing the discussion of the scientific developments, it is appropriate to describe the history of the collaboration of Hahn and Meitner, in view of their pertinence to the subject of this book.

Otto Hahn (1879–1968) was born in Frankfort on the Main. His father was descended from Rhenish peasant stock but became a glazier and artisan and later a businessman. Hahn studied chemistry and received his Ph.D. at the University of Marburg. After spending two more years with his teacher, Professor Theodor Zincke, he actually wanted to go into industry. One company accepted him but suggested that he first spend half a year abroad. In 1904 he joined Sir William Ramsay in London. Ramsay was famous for his discovery of the noble gases (see p. 112), and while working with him Hahn discovered radiothorium. Ramsay was impressed by Hahn's abilities and encouraged him to stay in research. Although Hahn could have joined Emil Fischer at the University of Berlin, he had become so interested in radiochemistry that he decided to spend some time with Rutherford, the leading figure in this field, at McGill University in Montreal. Rutherford was an inspiring and enthusiastic teacher and greatly influenced Hahn (Hahn, 1966). He spent about a year there and discovered another new substance, radioactinium. Many years later these "elements" were recognized as isotopes.

When Hahn returned to Germany, he joined Emil Fischer. Radiochemistry was not yet a regular part of the activities of the institute. There were no facilities for radioactive measurements. Fischer offered him a place in the basement. When he had to perform chemical experiments, he was able to use the regular laboratories in the evening hours. His work appeared strange to many of his colleagues, but Fischer recognized the great potential of the new field and Hahn became *Privatdozent* in 1907. Three years later he became *Extraordinarius*. He frequently attended seminars in the department of physics, where Professor Rubens was chairman. His interest in physics brought him together with many outstanding physicists, and he became a friend of Rubens, Walther Nernst, Emil Warburg, Max von Laue, James Franck, and Gustav Hertz.

While attending the seminars in the physics department, he met Lise Meitner (1878–1968). She was the third of eight children of a lawyer in Vienna. She was Jewish, although baptized. She had studied physics in Vienna and received her Ph.D. there in 1906. She then went to Berlin to continue her studies with Max Planck. She had worked in Vienna in Boltzmann's institute on the absorption of α and β rays and became interested in theoretical atomic physics. She joined Hahn on a part-time basis in 1907. Hahn realized the importance that a collaboration with an atomic physicist might have for his work. Meitner had originally planned to spend only two years in Berlin. Instead, the collaboration and friendship between Hahn and Meitner lasted 30 years, disrupted for only a few years by the Nazi regime.

Fischer did not usually accept women in his institute. Although extremely liberal in general, he was in this respect very conservative.* Meitner worked only in the basement. She was not permitted to work in the other laboratories.

When the Kaiser Wilhelm Institute of Chemistry was opened at the end of 1912 (see chapter IV), Fischer, who had a great influence on and played an important role in the creation and promotion of the Kaiser Wilhelm Society, suggested that Hahn and Meitner should get a special department of radiochemistry in the Institute of Chemistry. They moved there in 1912. Meitner had in that year become an assistant of Max Planck. In 1917 she became director of an independent physics department in the chemistry institute. Hahn was a superb experimentor, but theoretical physics was not his strength; thus the collaboration of Hahn and Meitner proved invaluable for both partners. During World War I Hahn joined the army and worked with Fritz Haber (see IV, A). During a short period that Hahn spent in his own institute in Dahlem, in 1917,

---

* He disliked the Prussian militarism and sent his son Hermann O. L. Fischer, after his *Abiturientenexamen*, for a year to a college in Cambridge for a liberal education, hoping that Hermann would become immunized against the Prussian spirit. The result was that Fischer's son left Germany immediately when the Nazis took over. He went first to Bâle, but since this place was for him uncomfortably close to Germany he left for Canada. He finally accepted an appointment at the University of California at Berkeley, where he was very happy. He was a very close friend of Otto Meyerhof (see IV, D) and the author met him frequently together with Meyerhof. Emil Fischer had a deep affection and admiration for his teacher Adolf von Baeyer. His strong rejection of anti-Semitic feelings was well known. When a colleague once asked him whether he had never experienced any anti-Semitic feelings, he answered that he came from the Rhineland and had no reason to have inferiority complexes.

he and Meitner isolated protactinium from pitchblende. After the war the close collaboration continued.

In 1933, when most Jewish scientists left the Kaiser Wilhelm Institutes within a short time, Hahn, like Planck and von Laue, was deeply shocked and desperate. For a few years he tried hard to fight the Nazi regime and showed great courage. He resigned his position at the University of Berlin in order to become more independent than a university professor could be. He read the speech prepared by Bonhoeffer at the Haber Memorial Service (see IV, A). After a few years it became extremely difficult and almost senseless to resist the Nazis openly and he gave up fighting them actively. But after the war he became the first president of the successor society, the Max Planck Society; in this position he visited, as will be described later, the Weizmann Institute and was instrumental in establishing links between the two institutions (see V, E).

Lise Meitner was for a while, after 1933, able to continue her work since she was an Austrian citizen. After Hitler marched into Austria in 1938, and the country was annexed, she became, after the *Anschluss*, a German citizen and was forced to resign. A 30-year-long close and intimate collaboration, invaluable for both partners, which had produced many brilliant results, was suddenly disrupted. There are few examples of such a lasting creative and close scientific German-Jewish collaboration than that between Hahn and Meitner. Meitner was not permitted to leave Germany, but with the help of German and Dutch scientist friends she was able to escape to Holland. In the fall of 1938 she received an invitation to continue her work at the Nobel Institute in Stockholm.

To return to the research of neutron effects on uranium, Hahn and Meitner's interest was enhanced, as mentioned before, by the suggestion that protactinium might be produced. Hahn was by then one of the world's foremost authorities on radiochemistry. The chemical identification of the elements produced by neutron bombardment of uranium was obviously a great challenge to his ingenuity and skill and promised to clarify the actual effect on the uranium structure. The work eventually became a turning point in science and one of the most fateful experiments ever performed, with effects nobody had foreseen. As we will see, they were destined to change the future of mankind for better or worse; they may lead to devastating effects on our planet and wipe out whole civilizations, or they may open a new era of inestimable benefit for mankind.

For several years Hahn and Meitner, later joined by Fritz Strassmann, devoted their efforts to identify chemically the "transuranium" products. Hahn soon excluded the possibility of protactinium formation. It was in the middle of these activities when Meitner was forced to leave. Hahn and Strassmann continued the work. Soon they obtained some unexpected results. It became obvious, after many checks and rechecks, that barium (atomic number 56) appeared in the products obtained by the effects of neutrons on uranium. This was extremely puzzling. There was nothing in the knowledge of nuclear physics at that time to explain how barium could be produced from uranium. All attempts to find an explanation failed. In view of Hahn's competence nobody could question the experimental data. Barium could only be produced when the atom was split and that was impossible, as everybody knew.

In this impasse Hahn reported the results in a letter to Meitner in December 1938. The letter arrived just as she was leaving Stockholm to spend Christmas with friends near Göteborg. She had invited her nephew Otto R. Frisch (1904–   ) to join her for Christmas vacation. Frisch had worked from 1930 to 1933 in Otto Stern's laboratory. After leaving Germany he spent some time in Blackett's laboratory in London and then joined Bohr in Copenhagen. Thus he had just to cross the Kattegat to come to Göteborg. It was a memorable and joyous reunion. They had spent many Christmas vacations together before the disruption by the Nazis. After all the turmoil and difficult times they were able to spend their vacation in a relatively relaxed atmosphere. But the meeting soon turned out to have momentous consequences.

Meitner was excited about the letter of Hahn and showed it at once to Frisch. He was skeptical; the appearance of barium as a result of a split of the atom was unthinkable. Everybody then believed that the atom could not be split. Frisch suggested the possibility of an error in the chemical analysis. This was immediately excluded by Meitner. Hahn was too competent, too meticulous and careful to make such a mistake. Was it not after all possible that uranium was split by the impact of the neutron? They went for a walk, discussing all possible interpretations of this extraordinary finding. Meitner made several calculations based on the Einstein equation. If one-fifth of the mass of the atom was split, 200,000,000 electron volts would be set free, an amount of energy large enough to split the atom apart. They both became increasingly excited. If the process was a splitting of the atom, or as Frisch later referred to it, a *fission*, it was an event of incalculable consequences, a breakthrough of tremendous dimensions.

Both were extremely anxious to find out what Bohr's opinion would be. They wrote down some notes for his information. When Frisch returned to Copenhagen, Bohr was just about to leave Copenhagen for the United States with his son Eric and his associate Leon Rosenfeld, to spend a few months at the Institute for Advanced Studies in Princeton. Frisch informed him of Hahn's discovery and showed him the notes with his and Meitner's interpretation. Bohr instantly grasped the idea. This was indeed exactly what must happen. He was terribly excited, realizing its tremendous importance. His great teacher and friend Lord Rutherford had died in 1937. Shortly before Rutherford's death, they had discussed the possibility of splitting the atom and using the energy released; Rutherford was extremely skeptical and expressed the view that man would never be able to use nuclear energy by splitting the atom. Now the impossible had happened: the atom has been split!

Frisch immediately tried to demonstrate experimentally that the uranium atom had indeed been split by the action of neutrons. He succeeded. Two papers were mailed to England on January 16, 1939, the first on the interpretation of the barium appearance as atom splitting by Meitner and Frisch, the second on the experimental confirmation by Frisch. The first paper appeared on February 11, the second on February 28.

Unfortunately the claim of Meitner and Frisch to have been the first to recognize and to confirm experimentally one of the most momentous events in the history of science—and mankind—remained unknown and in doubt for a long time. It was many years before they received credit for their brilliant contribution. It was a monumental documentation of the ingenuity and vision of Meitner that she found the right interpretation of experimental facts that seemed incompatible with prevailing concepts. If one accepts Einstein's and Planck's views on the decisive importance of intuition and imagination in recognizing the significance of experimental facts, here was a striking and almost classic confirmation of their ideas.

Bohr had promised not to talk about the data and ideas presented to him. But he immediately informed Rosenfeld. He was too startled to keep this terribly exciting news to himself. During the whole trip from Copenhagen to New York they discussed all aspects of the problem and made numerous calculations on the basis of the notes Frisch had given to Bohr. By the time they arrived in New York, Bohr was fully convinced of the correctness of Meitner's and Frisch's views: The atom had been split. John A. Wheeler from Princeton and Enrico and Laura Fermi were at the pier to welcome Bohr. The Fermis had seen Bohr only a few weeks earlier. They had left Italy to go to Stockholm for the Nobel Prize award. Mrs. Fermi was Jewish and they had left Italy with the firm intention not to return. Fermi worked at Columbia University. Bohr informed Wheeler that the atom had been split, but that nothing could and should be said about it.

That night, January 16, 1939, Wheeler invited Rosenfield to a meeting of a group of leading physicists in Princeton. Bohr had completely forgotten to inform Rosenfeld that nothing should be said publicly until the various publications of Hahn and Strassmann and Meitner and Frisch about the experimental findings and their interpretations had appeared. Rosenfeld therefore presented the findings and the interpretation to the group. On can easily imagine the great excitement of the audience.

The next morning, Rosenfeld informed Bohr that he had discussed the findings of Hahn and Strassman and the interpretation of Meitner and Frisch of the splitting of the atom. Bohr was terribly upset. He then, for the first time, informed Rosenfeld that it had been agreed to keep the news secret until the publications of this phenomenal breakthrough had appeared. He immediately foresaw a race to test the splitting of the atom, since these experiments were not difficult to perform. He was afraid that Meitner and Frisch would not get proper credit for their extraordinary contribution because of his oversight. He cabled Copenhagen and urged Frisch to proceed rapidly with the experiments and the publication. In his anxiety he prepared within two days a note to *Nature* describing the whole story and giving proper credit to all involved. But all his efforts were too late. On January 26, 1939, a conference on theoretical physics was held at George Washington University in Washington. Leading American physicists were present. Fermi told Bohr that he had heard about the exciting explanation of Meitner and Frisch and that he was doing experiments to test the interpretation. Bohr did not feel free to say anything. But at the meeting a copy of the issue of *Naturwissenschaften* in which Hahn and Strassmann reported their results was available. As Hahn explained later, he was reluctant as a chemist to announce a revolutionary discovery in physics. But he hinted that there might be a possibility of a "bursting" of the atom. At this moment Bohr felt that he was free to speak out. The Hahn–Strassmann experiments were now published. The more widely known the findings were, the greater the possibility that Meitner and Frisch would after all get the proper recognition. Bohr's announcement produced a veritable explosion. Physicists from several universities from all over the country called their colleagues asking them to start experiments. The story of a tremendous release of energy by the splitting of an atom reached the newspapers, although in most reports the revolutionary impact of the discovery was not realized. Physicists started within hours to test the hypothesis. In the Carnegie Institute Merle Tuve invited Bohr, Rosenfeld, and Edward Teller to watch the test that night. They saw the splitting of the atom. Bohr's anxiety grew. He was not sure whether Frisch had performed the experiment and had mailed it to *Nature*. The next morning, January 27, it was revealed that the experiment had also been performed in Berkeley, Baltimore, and New York. Although Bohr had called attention to the tremendous potential energy of the atom in 1930, its release seemed unreal. Only a few writers, among them H. G. Wells, speculated about the doomsday that might come to pass if scientists succeeded in splitting the atom.

Bohr continued to worry that the story should be written for history in the proper way. If the publications of other physicists came out before those of Meitner and Frisch, the latter would not get the credit they deserved and he would be responsible. Despite Bohr's frantic efforts to get newspapers to mention the story properly, Meitner's and Frisch's names were not mentioned. In New York papers the large conversion of mass into energy was attributed to Columbia University. Bohr sent a note to *Physical Reviews* early in February, describing the sequence of events and giving proper credit to all concerned in a precise way. Bohr's struggle for the truth was for many years not really successful. In 1944 the Nobel Prize was awarded to Hahn and Strassmann. Hahn was among the scientists who were brought to England by Allied troops in April 1944 and interned near Cambridge. It was there that he learned of the award. Because Meitner had worked for many years with Hahn on the effect of neutrons on uranium, the paper would have been published with the proper interpretation and with Meitner as coauthor if she had not been forced to leave a few months before. Thus partly because of the Nazi regime, partly because of Bohr's revealing the story to Rosenfeld without proper warning, she was left out in the recognition of the crowning result of the 30-year collaboration. But history slowly straightened out the truth. In 1960 the Fermi Award, the American equivalent of the Nobel Prize in physics, was given to Hahn, Meitner, and Strassmann. The injustice was repaired.

Meitner and Frisch, working in Bohr's laboratory and using a technique first employed by Meitner decades earlier, soon demonstrated that uranium bombardment by neutrons produced krypton (atomic number 36) in addition to barium (atomic number 56). Further experiments in the following years by Hahn and by American, British, and Canadian scientists soon found that a huge number of fission products appeared, including about three dozen elements. But there were also real "transuranium" elements—as was shown by Glenn T. Seaborg (1912–   ) in the United States—such as neptunium (93) and plutonium (94).

Two more events may be briefly mentioned. George Placzek (1905–1955), a Jewish scientist from Czechoslovakia, working in 1939 in Bohr's laboratory, asked him while they both were at Princeton why only slow-moving neutrons produced fission, whereas the fast-moving ones had no effect. Bohr, after some thinking and discussions with Rosenfeld, came out with the correct answer. Most of the uranium was the isotope 238; these atoms would capture the neutron but not undergo fission. Only the uranium isotope 235, only about 0.7% of naturally occurring uranium, would be split. Uranium-235 has an uneven number of neutrons and protons; one more neutron and uranium-236 would be formed. A nucleus with an even number of particles would be more tightly bound together than one with an odd number. The perturbation by the arrival of the neutron in the former atom with an even number would be strong enough to offset the forces holding the nucleus together and cause the nucleus to split with a large release of energy. Uranium-238 would become 239 with the addition of a neutron. As an odd-numbered nucleus it would be less likely to be split; the neutron would simply be captured and retained. Bohr's extraordinary vision and intuition again proved to be correct: a year later it was verified that only uranium-235 is responsible for the splitting. If one would succeed in separating the isotope 235 from the rest, huge amounts of energy would be released by slow neutrons.

A second important factor was the discovery that during the splitting of uranium neutrons are released. If they led to the fission of other uranium atoms, a chain reaction would result. The idea of a chain reaction produced by neutron bombardment had occurred to Leo Szilard (1898–1964) in 1934. Szilard, born in Hungary, had studied in Germany and received his Ph.D. degree in physics at the University of Berlin in 1922. He was *Privatdozent* there in 1933 and, being a Jew, left Germany in March 1933, seeing clearly what was coming. While in England in 1934, he wanted to test the ideas of neutron release with beryllium, but it was an expensive experiment; it would cost £2000. He was unable to get both the support of other physicsts and the money. But he had actually applied for a patent describing the laws governing such a reaction (Szilard, 1969). In 1939, after the splitting of uranium, Szilard was in New York and tried to get the help of Fermi at Columbia University. Fermi was at first skeptical, but after some encouragement from I. I. Rabi, Fermi agreed to test the idea. They soon found that neutron bombardment and splitting were associated with the release of neutrons. For political reasons they discussed the advisability of publishing their data. Just at that time, in March 1939, a paper by Halban, Joliot-Curie, and Kowarski appeared in *Nature*, reporting that uranium splitting was associated with the release of neutrons (Halban *et al.*, 1939). The results of Szilard and Fermi were now immediately submitted for publication. The realization that uranium-235 was the isotope readily split, and the possibility of its isolation and of a chain reaction opened the way to the release and use of large amounts of nuclear energy.

I. NUCLEAR ENERGY AND THE ATOM BOMB

The developments leading to the production of the atom bomb obviously do not belong in this book. Moreover, there is a vast literature about the atom bomb; many books are well written for the layman (see, e.g., Smyth, 1945; Goudsmit, 1947; Hecht, 1947; Laurence, 1947; Fermi, 1954; Strauss, 1962; Davis, 1968; Szilard, 1969; Blumberg and Owens, 1976). However, a few comments appear appropriate in the context of this book and of some general aspects of science and society. The construction of the atom bomb would have been impossible without the phenomenal growth of atomic physics in the first 30 years of this century; the role of German-Jewish scientific collaboration during this period has been outlined in this chapter. In the 1930s atomic physics began to develop rapidly in the United States, owing to a young and vigorous group of brilliant physicists. I. I. Rabi, Robert Oppenheimer, and Edward U. Condon were already mentioned; but there were many other outstanding physicists, several of them Nobel laureates, such as Glenn T. Seaborg;

Ernest O. Lawrence (1901–1958); John A. Wheeler (1911–   ), a pupil
of Bohr; and Arthur Compton (1892–1962), all of whom became leaders
in the field. Nevertheless, it can hardly be questioned that the development of physics in the United States was greatly enhanced by the influx
of many brilliant European physicists, not only Jews from Germany, but
also Jews and non-Jews from other European countries who escaped from
authoritarian regimes (Weiner, 1969). Many of them were either trained
or had spent some time in the great German centers. An important factor
was the warmth and cordiality with which they were received by their
American colleagues; they were accepted with open arms and generosity
in providing them with the facilities to continue their work, in spite of
the difficult economic situation in a country trying to recover from a
deep depression.

German-Jewish physicists and other Jews trained in Germany, particularly Hungarian Jews, greatly contributed to the construction of the
atom bomb. Without going into any details, at least a few of the famous
names may be mentioned: James Franck, Hans A. Bethe, Felix Bloch,
Leo Szilard, Eugene P. Wigner, Edward Teller, Eugene Rabinowitch,
and John von Neumann; additional physicists came from England (see
below). Moreover, Enrico Fermi, shortly after his arrival from Italy, and
Niels Bohr, after his escape from Nazi-occupied Denmark, also contributed. Many people ask whether physicists should be unhappy to have
developed a weapon capable of destroying mankind. In the discussion of
this question some pertinent facts may be recalled to provide an answer.

When the possibility of building an atom bomb became apparent, in
1939, war with Hitler Germany was imminent. Many physicists feared
that Germany would be able to build an atom bomb. Even though many
illustrious atomic physicists had left Germany, there were still enough brilliant and competent physicists left who might be capable of producing
the atom bomb. Few realized the tremendous technical difficulties involved. The United States had almost unlimited facilities, funds, and
brainpower, yet it took about four years of intense effort to develop the
bomb. It is interesting that Niels Bohr, when asked in 1939 about the
feasibility of building an atom bomb, replied that it would probably be
possible, but it would be necessary to transform the United States into a
huge factory. Szilard, who had witnessed with horror the growth of the
fanatic, criminal, and ruthless Nazi movement, convinced Einstein to write
the famous letter to Roosevelt. Both were ardent pacifists, as were many
of their colleagues. But the nightmare of Nazi Germany producing the
bomb convinced them of the need to work on it, in case Nazi Germany
succeeded in building the bomb. Both were later deeply depressed about
their initiative and had terrible guilt feelings after the disaster of Hiroshima
and Nagasaki. The true story is that Einstein's letter had virtually no
effect. Only after Pearl Harbor at the end of 1941, when the United
States became involved in the war with Germany and Japan, did the gov-

ernment begin to support the famous Manhattan Project on the tremendous scale required. Moreover, Great Britain's Defense Ministry, under the leadership of such brilliant physicists as James Chadwick and Sir John Cockcroft, had already started work on developing nuclear energy. One has to remember that in the summer of 1940 the Nazis had overrun France, Denmark, Norway, and other countries, and the British were in a desperate position. They had already devoted great efforts to producing nuclear energy, but they had neither the facilities nor the funds required to build an atom bomb. The widespread destruction in England by the German bombings, especially during the first years of the war, had reduced the British industrial potential, and the country suffered from the lack of many vital supplies due to the submarine war. To build the vast industrial plants required for a successful production of the atom bomb, under the continuous threat of further bombings, was hopeless. Many scientists, however, came to the United States to work on the atom bomb, among them Franz Simon, Otto A. Frisch, Rudolf Peierls. For many physicists it was a horrible thought to help produce such a monstrous weapon, and they faced an awful dilemma between the moral problem and the defense of the free world against a ruthless enemy. The shock and depression among scientists in general about the disaster produced in Japan was very deep. The Franck Report has been mentioned. Many were seriously disturbed and distressed about this "monstrous perversion of science," as it was called by Churchill.

It is, however, naive to believe that if the atom bomb had not been built at that time by the United States it would not exist today, and that mankind would live in peace and security without the nightmare of a nuclear war. Once the possibility of building such a weapon became apparent, it was unavoidable that it would be built sooner or later, somewhere. The USSR had many brilliant atomic physicists, and the atom bomb was built there soon after the war. In fact, the Russians built not only the atom bomb using atom fission, but also the hydrogen bomb using atom fusion.

Only the defeat of Germany and its occupation permitted the Allies to find out whether German nuclear physicists had developed the use of nuclear energy and whether they had tried to produce an atom bomb. In 1939 great changes had begun to take place in Germany. After the discovery of the fission of the uranium atom and the demonstration of chain reactions, it became apparent that the use of nuclear energy was an obtainable goal, and this source of energy presented a potential of unprecedented dimensions, including the possibility of developing explosives far more powerful than anything existing before. "Jewish physics," suddenly emerged as a reality with a huge potential. This was, at the end of 1939, not just a severe blow to Aryan physics and its leaders; it was a total defeat. Atomic physicists regained, at least in the scientific community, the place of respect and eminence they had enjoyed before Hitler.

But the impact of the preceding years was serious. In addition to the loss of many eminent leaders, poor appointments of professorships and greatly reduced numbers of students had created a critical situation in the field.

About the time the war broke out, the potential use of nuclear energy came to the attention of the Ministry of Defense. In September 1939 an important meeting took place there. Heisenberg, widely recognized as one of the most eminent leaders in the field, was asked to attend. The discussions centered on the question of whether the use of nuclear energy was a real possibility. Heisenberg was asked to prepare a memorandum about the possibility of the construction of what is today known as a nuclear reactor. Actually it was not his particular field, since he was not an experimental, but a theoretical physicist. Nevertheless, within a few months he succeeded in working out a memorandum with the conclusion that the building of a nuclear reactor was feasible and would provide an important source of useful energy. Many of the basic principles discussed were so fundamental that they became important guidelines for the construction of nuclear reactors in the postwar period, and his ideas were greatly admired by experts in the field.

The Ministry of Defense then took over the Kaiser Wilhelm Institute of Physics in Berlin–Dahlem and began in the following year to assemble various scientists working in the field of nuclear physics. The laboratories used for uranium research were referred to as the "virus house" to conceal the real purpose. After the outbreak of the war, Peter Debye, who had been the director of the institute, was asked to give up his Dutch citizenship and accept German citizenship. He requested a leave of absence and went, in 1940, to Cornell University in Ithaca in the United States; he never returned to Germany. Dr. Kurt Diebner was appointed as deputy director. One of the prominent physicists in the institute was Karl Wirtz. He had been associated in Leipzig with Karl-Friedrich Bonhoeffer, a close friend of Heisenberg, and was a strong anti-Nazi. Debye had invited Wirtz to join him when he moved to Berlin. When the army began to assemble nuclear physicists in the Physics Institute, Wirtz and some of his close friends recognized the dangers. First, there was the possibility that Nazi functionaries would take over the institute as directors. This had happened in many institutes when their leaders had been chased away by the Nazis. But an even greater worry was that one of the physicists who belonged to the group around Stark and Lenard and were fanatic Nazi supporters, would become the director of the institute. In that case all scientific freedom would be eliminated and serious research would become impossible. The most ominous threat was, however, the possibility that such individuals might try to build explosive weapons, possibly an atom bomb, with the use of nuclear energy, despite their inadequate abilities and poor judgment. The Nazi government would give a great deal of financial support to such a project, probably at the expense of many other important research projects.

In this situation Wirtz and his friends tried hard to involve Heisenberg in a consulting capacity, in lectures at the institute and at the university, in order to prepare the ground for his election as leader and chairman of the institute which by now had become the most eminent and distinguished position in the field of nuclear research in Germany. Heisenberg gave much thought to the problems involved and the advisability of accepting the chairmanship, and discussed the matter at length with close friends. They were fully aware that the construction of an atom bomb had become a possibility and that this fact raised a vast number of political and ethical problems and involved heavy responsibilities and great danger. Heisenberg realized that the existence of atom bombs would create problems that would outlast the war and have serious implications for the future of mankind.

In view of the conflicts and dilemmas facing him, Heisenberg decided to discuss them with his admired teacher and friend Niels Bohr. The unfortunate outcome of this visit, which took place in 1941, has been discussed by several authors (see, e.g., Moore, 1966; Hermann, 1977; E. Heisenberg, in preparation). The entirely different psychological situations in which the two old friends found themselves, owing to the political and military events, created barriers that were difficult to bridge. In particular, Heisenberg was forced to keep silent when Bohr inquired about his present activities; he knew that he risked his life no matter what kind of answer he gave, since his statement would unavoidably become widely known and would be considered in Germany as treason. One may recall the extraordinary precautionary measures taken even in the United States in respect to the secrecy of the Manhattan Project. Executions in Germany at that time were frequent and rapidly carried out for even harmless remarks displeasing the Nazis. Bohr unfortunately completely misinterpreted Heisenberg's silence. Both men were most unhappy about this meeting.

Despite his many worries, doubts, and anxieties, and the serious problems that faced him, Heisenberg decided to accept the chairmanship and took it over on April 24, 1942. He was most uncomfortable and torn by conflicting sentiments. But he felt that, as the recognized leader in the field, he had the moral obligation to accept the responsibility, whatever unpleasant personal consequences might result. Only a few weeks later the developments reached a climax and fully justified this step. On June 4, 1942, the decisive meeting took place in which the building of an atom bomb was discussed. It was held in the guest house of the Kaiser Wilhelm Society in Dahlem in the presence of Albert Speer, the powerful minister in charge of the mobilization of all resources for the war effort. A general asked whether it was possible to build an atom bomb. Ernst Telschow, the secretary general of the society, stated this was the first time he heard the words *atom bomb*. Heisenberg replied that the building of an atom bomb was indeed feasible. Answering the inquiries of Speer about what the project would require, he explained that it would necessitate the building of many gigantic industrial plants and would take many years of immense efforts of a huge number of physicists and engineers; the cost would amount to many milliards (billions) of marks. Heisenberg's evaluation was realistic. But he also was sure that such a project could and would not be considered. The concept of a *Blitzkrieg* (i.e., a relatively short war) still prevailed. At that time (June 1942) the United States and the Soviet Union were already at war with Germany, and in view of the heavy and devastating bombing attacks, building huge industrial plants was, just as in England, totally unrealistic. In any event, with his statements Heisenberg had virtually eliminated the idea of building an atom bomb. Speer abandoned the plan and the army returned the Physics Institute to the Kaiser Wilhelm Society. The correctness of Heisenberg's evaluation, and similar evaluations expressed by Niels Bohr in 1939, has been borne out by the story of the Manhattan Project: the huge industrial plants that had to be built, the great number of physicists and engineers who participated, and the funds required were in full agreement with Heisenberg's estimates. It also took four years, until July 1945, for the first two bombs to be built.

However, Heisenberg convinced the Ministry of Defense of the usefulness of the construction of a nuclear reaction as a source of energy. He received relatively modest funds for his research; the priority was not higher than that for almost all work performed by physicists. As von Laue explained in a letter to his son in the United States after the war, the label "decisive for the war effort" was a trick used to keep as many physicists as possible away from the army. Nevertheless, 2000 out of 6000 physicists were killed. Projects completely unconnected with the war, such as Heisenberg's work on cosmic rays, were labeled the same way.

Heisenberg's acceptance of the highest position in Germany in the field of nuclear research during the war has been misinterpreted as indirect evidence for his intention to build the atom bomb; it has been said that the construction of the nuclear reactor was planned as the first step. Since this view raises questions extremely pertinent for judging Heisenberg's attitude toward the atom bomb and the Nazis in general, as well as his human qualities and his personality, it appears imperative to analyze his motivations in more detail.*

It may be recalled that Heisenberg had gained international admiration and fame at a very young age. He had become *Ordinarius* at the age of 25, Nobel laureate at the age of 32. His close friendships with Bohr, Pauli, and other great scientists, as well as the impression his lecture in 1926

---

* I am deeply grateful to Mrs. Elisabeth Heisenberg for providing me with invaluable information. The reader interested in a more comprehensive presentation of Heisenberg's personality is referred to Mrs. Heisenberg's memoirs (E. Heisenberg, in preparation).

made on Einstein, have been outlined. The rise of the Nazis to power initiated many painful years of frustration, anguish, and humiliation. He was exposed to the most vicious attacks and slander. He was denied the famous chair of Sommerfeld. Now after the collapse of Aryan physics he was offered the foremost position with the highest prestige in nuclear physics in Germany.

The many conflicting sentiments, the many problems facing him if he accepted this position, have just been described, as well as the moral obligation he felt to accept the responsibility in the decision of whether or not to try to build the atom bomb. But building a nuclear reactor presented a great challenge to his ingenuity and had, therefore, a great appeal to him. From his early youth he had an irresistible drive for excellence and for achievement; he was a perfectionist whether science, music, sport, or play was involved. He was ambitious not in the personal sense of longing for honors or fame, but in the sense of being able to meet a challenge. Scientists know that the deepest and most gratifying satisfaction of research is that of unraveling the secrets of nature, of providing solutions and answers to difficult problems. As previously discussed in the section on epistemology, for both Einstein and Planck great discoveries and achievements in science were creative works of the human mind comparable to those of great artists. Heisenberg's genius had manifested itself in one of the most glorious chapters in the history of science. In accepting the position offered he hoped to be able to build the reactor, which not only required great ingenuity but also promised to be of great practical value long after the collapse of the Nazi regime. He had the self-assurance to be able to build the reactor. Once he had convinced himself that building an atom bomb was out of the question, he felt, although with great reluctance, that the factors in favor of accepting the chairmanship were stronger than those of refusing it.

A vivid and fascinating account of Heisenberg's feelings and thoughts during that period is given by Mrs. Heisenberg (1979) in her memoirs. The family lived at that time in a house in Urfeld at the Walchensee. There Heisenberg could speak freely without any danger. During a weekend visit shortly after the decisive meeting in June 1942, Heisenberg informed his wife that it was now certain that the building of an atom bomb was possible. Mrs. Heisenberg was deeply disturbed and asked what he and his friends would do if they were forced to produce the bomb. He calmed her down and assured her that she did not have to worry. The efforts required would be so immense, so costly and time-consuming that Germany would not able to supply them. Thus they were in the fortunate situation of not having to make a real decision; the reality of the situation saved them from this dilemma and they must be grateful for it. He also added that even if Hitler did want to force them to build the bomb, such force would be useless. The project would require imagination and originality, which could not be obtained through force. Creative work requires complete freedom and can only be achieved when scientists are willing to do it.

These few comments may suffice to show that the suspicions that Heisenberg accepted the chairmanship to build not only the reactor but also the bomb were groundless. The truth is just the opposite: he wanted to prevent such a development, and he succeeded.

The situation in Germany deteriorated rapidly. The bombing of Berlin had increasingly devastating effects and forced the removal of the scientific institutes to safer places. The Physics Institute was moved to Hechingen, a small village in Swabia; about 10 miles from there was a small place, Haigerloch, where large caves provided protection against bombing attacks. There Heisenberg continued to work on the reactor. The prevailing conditions were most unfavorable and the work proceeded very slowly. Moreover, Bothe, one of the foremost experimental physicists in Germany, was convinced that graphite was unsuitable as a moderator for the chain reactions in the reactor, and following his advice, Heisenberg used heavy water. When at the end of the war Haigerloch was occupied, the work on the reactor was not yet completed.

Von Laue, in discussing the problem of the atom bomb, wrote in one of his letters to his son in 1946 that Germany just did not have the means that the United States and England had to build such a bomb. He too stressed the fact that "none of us wanted to lay such a weapon into the hands of Hitler." Nobody familiar with the personalities of Planck, von Laue, Hahn, and Heisenberg, with their actions and reactions during the Nazi period, will question this statement. Whether there were physicists with a different attitude is not a matter for discussion here. Irving (1967) concludes in his book that the German atomic physicists devoted all their efforts (in the field of nuclear energy) exclusively to the construction of a nuclear reactor. Beyerchen (1977) arrives at the same conclusion. In 1947 Heisenberg was invited to lecture in Cambridge. He was warmly received by his colleagues and got ovations from the students. In England, after all the suffering caused by the war, strong feelings of hatred still prevailed. The English scientists were well informed. The early invitation of Heisenberg clearly indicates that his anti-Nazi attitude was well known and fully appreciated.

It may be stressed in this context that in the new Federal Republic of Germany Heisenberg became one of the key figures in rebuilding German science, in the Max Planck Institutes, the universities, education, and elsewhere. He had many long discussions with Konrad Adenauer, who attached great weight to his advice. He had many friends among leading Social Democrats, quite in line with his lifelong record of progressive views, his humanistic ideals, so superbly expressed in his many philosophical writings. His service was, as Planck had foreseen, invaluable and a great asset for the rebuilding of a new, democratic free Germany.

A few remarks about fusion may be appropriate. In 1938 Hans A.

Bethe (1906–) studying the possible sources of the tremendous heat prevailing in the sun (about 20 million °C) and emitted continuously from the sun, proposed a "carbon cycle" formed by a series of reactions in which carbon acted as a catalyst: the cycle terminates with the re-formation of carbon. One of the reactions, suggested as a source of solar energy and heat production, is the fusion of deuterons to produce helium.* This fusion releases tremendous amounts of energy. Although only 1 of 5000 hydrogen molecules in water exists in the form of deuterium, deuterium can easily be obtained in pure form, much more easily than uranium-235 from 238. Thus many physicists immediately thought of the possibility of using this fusion process for the construction of an atom bomb. But the extremely high temperatures required for the reaction appeared insurmountable.

Bethe was born in Strasbourg, France, at that time Germany. His father was the distinguished physiologist Albrecht Bethe (1872–1954), who was for many years chairman of the department of physiology at the University of Frankfort on the Main. His mother was Jewish. He received his Ph.D. in physics, in 1928, in Munich with Sommerfeld. After spending short periods at other universities, he joined Heisenberg in Leipzig. He left Germany in 1933. After a short stay in England, he joined the faculty of Cornell University in Ithaca, New York, in 1935, as professor of physics. He is one of the most illustrious physicists of this generation. Although best known for his studies on solar energy, the breadth of his contributions to physics is almost unique. He received the Nobel Prize in 1967. He married Rose Ewald, the daughter of Peter P. Ewald, who had been a teacher of Bethe in Stuttgart. Ewald was not Jewish, but his wife was; they were anxious to emigrate and finally managed to come to the United States before the war. Hans Bethe met Rose in Washington through the Tellers; he knew Teller from Leipzig, when they both worked in Heisenberg's institute.

After World War II the political situation changed drastically. The Soviet Union and the United States became strongly antagonistic. The Soviet Union produced the fission bomb much earlier than most experts in the United States had anticipated. In view of the cold war that had emerged between the two superpowers, some physicists raised the problem of trying to build the hydrogen bomb; the high temperatures produced by fission might be used for initiating the reaction. In this case the opinions betwen physicists were sharply divided, on the basis of moral, political, technical, and other factors. One of the strongest proponents of going ahead with the building of the hydrogen bomb was Edward Teller (1908–). The unfortunate confrontation between Teller and Oppenheimer is well known. There is some difference of opinion whether the

---

* A deuteron is the nucleus of deuterium, the hydrogen molecule that contains a neutron in addition to a proton.

USSR or the United States was the first to have the hydrogen bomb, but in any event around 1955 both superpowers had the two types of bombs.

Teller was born in Hungary. Both parents were Jewish. His father Max was a lawyer; his mother Ilona, *née* Deutsch, was the daughter of a successful banker. Both families were fully assimilated and leaned strongly toward German cultural tradition. In Ilona's family the language spoken was German. After the *Abiturientenexamen* in a humanistic gymnasium in Budapest, in 1926, Teller left Hungary, at that time under the strongly anti-Semitic and fascist regime of Horthy. Teller first studied chemistry at the *Technische Hochschule* in Karlsruhe. Among his teachers, those who had the greatest influence upon him were Herman F. Mark and Peter P. Ewald, whose daughter, as mentioned, later married Hans Bethe. But actually Teller's main interest since his school years had been in physics and mathematics, and he followed the exciting developments with great interest. In 1928 he went to Munich to study with Sommerfeld. There he had a severe accident in which he lost one foot. After his recovery he joined Werner Heisenberg in Leipzig and received his Ph.D. in 1930. At that time it was one of the most exciting centers and he met many of the most brilliant physicists of his generation. After his Ph.D. he spent some time in Göttingen. In 1933 he left Germany and went first to Copenhagen and England, and in 1935 he came to the United States as professor at George Washington University in Washington. He married Augusta Maria (known as Mici), *née* Harkanyi. Both her parents were Jewish, but baptized. Part of his family lives in Israel and he has close ties to Tel Aviv University.

In view of the great number of outstanding Hungarian physicists who played an important role in atomic physics, first in Germany, and later in the United States, a few remarks about the historical background seem appropriate. Jews in Hungary had enjoyed relative freedom and prosperity from the middle of the nineteenth century to the outbreak of World War I. In 1914 there were about 1 million Jews in Hungary, 5% of the population of about 20 million. A disproportionately large number of Jews were professionals. After World War I, Hungary went through terrible times, first under a ruthless communist regime, then under the fascist regime of Horthy. In the 1920s many scientists left for Germany and later for the United States.

The most important features of the development of nuclear energy are, however, not the aspects discussed so far. For almost a quarter of a century nuclear energy was almost exclusively identified by most people with the atom bomb and its inherent threat to the survival of society. This was an important factor in the widespread hostility against science in general among a large fraction of the population, particularly in the United States. Many people all over the world are haunted by the nightmare of a total destruction of mankind. Nobody can say with certainty whether or not this threat will ever become a reality. In the last two decades, however, scientists, industrialists, economists, and specialists in the field of energy became increasingly aware of another aspect of nuclear energy. Many years ago it became apparent that the conventional sources of energy available for the urgent needs of society—coal, oil, and natural gas—were limited. The rate of energy use is continuously increasing, owing to the population explosion, the vast expansion of industry and modern forms of agriculture, present means of transportation, among other factors. The amount of energy used per year throughout the world has increased since the turn of the century more than fiftyfold and is still rising. Scientists and many experts vigorously warned of the serious crisis threatening our civilization by the rapid exhaustion of the available sources of conventional energy. Little or no attention was paid by governments or the general population.

The "energy crisis" after the Arab oil embargo in 1973 and the sudden sharp increase in the cost of oil opened the eyes of governments and broad sections of the population and forced them to realize that the supply of conventional energy was rapidly dwindling. They finally recognized the necessity of looking for new sources of vast amounts of energy. Conservation is able to extend the period of grace for the transition for a few years and for finding a solution; it cannot solve the problem. A statement signed by 75 leading physicists and biologists of the United States stressed emphatically and convincingly that the only serious hope for meeting the ever-increasing vital needs of mankind and ensuring its survival is the proper development and competent use of nuclear energy.

It may be useful to recall a few figures for illustrating the amounts of energy that may be obtained by nuclear fusion or fission. Every reader knows the powerful explosive forces of the uranium bomb; 1 kilogram of uranium-235 can yield by fission an explosive force 20 million times more powerful than that of 1 kilogram of TNT. In common units the fission of 1 kilogram of uranium may provide energy equivalent to 23 million kilowatt hours. This is about 5 million times as much as 1 kilogram of coal. The amounts of energy derived from fusion are even more impressive. Helium has less weight than its parts. The loss of weight when one-half gram of protons and the same amount of neutrons form 1 gram of helium is 7.5 milligrams. According to Einstein's formula $E = mc^2$, 7.5 milligrams times 300,000 kilometers squared (the square of the speed of light) is equivalent to about 200,000 kilowatt hours. One kilogram of helium formed by fusion provides about 19 million kilowatt hours. These few figures illustrate the gigantic amounts that nuclear energy is able to provide.

Unfortunately, there are at present many technical and scientific problems to be solved before society can make full use of nuclear energy for its needs. There are many books about these problems. Only a few aspects will be mentioned here. Fission energy is already widely used. In the Western world there are at present about 80 reactors, most of them in highly industrialized countries. Many more are under construction. One serious problem is that the estimates of amounts of uranium available vary widely; no one knows for sure how much is available and how long it will last. Breeder reactors increase the energy yield fivefold, but the uranium-235 now used in these reactors produces plutonium, which is used for atom bombs; thus the reactors increase the danger of atom bomb production. More efficient breeder reactors are being tested with uranium-233, which does not produce plutonium, as a starting material. Finally, fission produces dangerous radioactive material. Many competent scientists consider the present reactors not safe enough to prevent disastrous leakages; they insist that much work must be done to improve safety. Disposing of the radioactive material formed is yet another problem. Until recently the efforts to solve the problem were most unsatisfactory. Now some promising results are being achieved, especially in Canada and Germany. In any event, with sufficient effort by governments and adequate funding, great improvements can and certainly will be achieved.

Entirely different problems are offered by the use of fusion for obtaining energy. Here the sources are unlimited. The oceans provide billions of liters of deuterons and could supply mankind with adequate amounts of energy for millions of years. There is no danger of radioactive waste products. But although we have succeeded in using fusion to build the hydrogen bomb, we are still far from achieving the methods for using fusion as a source of energy. Many competent scientists are optimistic, and believe that the problem will eventually be solved. Rutherford did not dream that, less than a decade after his death, nuclear energy would become available, although unfortunately first for destruction.

In any event, nuclear energy has opened a new age in the history of mankind. Science is neither good nor evil. How knowledge of nature and the exploration of its secrets are used depends on man, on society and its leaders. This problem will be discussed at the end. Atomic physics has opened new territories. Heisenberg once remarked that the greatest feature of Christopher Columbus's discovery of America was his decision to leave the known regions of the world and to sail westward far beyond the point from which his provisions could have gotten him back home again. "In science, too," he wrote, "it is impossible to open up new territories unless one is prepared to leave the safe anchorage of established doctrine and run the risk of a hazardous leap forward" (Heisenberg, 1971). Columbus did not foresee that his discovery would lead to the destruction of the native population by the conquerors and to the great new possibilities America has provided for hundreds of millions of people. Planck and Einstein, Rutherford and Bohr did not foresee what the new territories opened to mankind by their genius would mean for the future of mankind. It is up to society whether it will commit suicide and destroy the world or provide undreamed-of benefits for coming generations, whose future is now seriously threatened by the rapid exhaustion of the conventional sources of energy. Scientists can greatly help in shaping a better future; they do not have the power to decide the course of history.

In summary, the first 30 years of this century are generally considered one of the most brilliant, revolutionary, and exciting chapters in the history of physics and of science in general. The new era started with Planck's introduction of the quantum of action, followed by Einstein's theory of relativity and culminating in the wave and quantum mechanics of de Broglie and Schrödinger, and Niels Bohr, Heisenberg, and Born respectively. Let us consider this era in the specific perspective of this book, the collaboration between German and Jewish scientists. We find, just among the most eminent figures, the following ethnic distribution: Max Planck, German; Albert Einstein, Jew; Arnold Sommerfeld, German; James Franck, Jew; Gustav Hertz, half-Jew; Max Born, Jew; Niels Bohr, half-Jew; Max von Laue, German; Otto Stern, Jew; Erwin Schrödinger, German; Wolfgang Pauli, half-Jew; Otto Hahn, German; Lise Meitner, Jew. This is a list of some of the most prominent personalities and is far from complete. In particular, it does not include the many illustrious scientists outside Germany, who during this early period made decisive contributions to atomic physics, such as Lord Rutherford, James Chadwick, Paul Dirac, Louis de Broglie, Enrico Fermi, and Frédéric and Irène Joliot-Curie, nor many other Germans or German-trained Hungarian Jews, who made their most outstanding contributions after having left Germany. This chapter does not consider the development of atomic physics in general but looks at only a few main aspects of German and Jewish collaboration before the destructive effect on German science of Hitler. To most scientists this ethnic affiliation may appear irrelevant, and stressing this aspect may seem ludicrous. But if one tries to document for posterity the achievements and contributions that resulted from the pioneer work of German and Jewish scientists, the chapter of atomic physics in Germany during the first three decades of this century is one of the most striking and dramatic demonstrations of the close association and the mutual stimulation between the two ethnic groups. It is this special feature that makes this enumeration of interest. In no other field was the interaction between the two ethnic groups such a genuine reality as in the field of atomic physics. One may perhaps question whether there was a genuine dialogue and mutual stimulation between the two ethnic groups in many other fields. A variety of views, from one extreme to the other, have been presented in the Year Books of the Leo Baeck Institute. In the field of atomic physics the remarkable strength of personal ties and friendships, the passion and intensity of discussions that were essential elements of this magnificent chapter, cannot be questioned.

As stressed repeatedly, science is by nature international. The problems of atomic structure, of the molecules of living cells, of the forces driving the universe, are the same everywhere. International scientific collaboration is increasing at a rapid rate, greatly facilitated by the improved communication facilities. Nevertheless, the mutual stimulation between German and Jewish scientists had a very special character. Common language, similarity of cultural and educational background, the impact of physical and spiritual environment, the sharing of many common experiences created an unusual and uniquely favorable atmosphere. The problem will be further discussed at the end of Chapter IV.

III

Developments in Chemistry and Physiology in the Nineteenth Century

A. CHEMISTRY

Although in the early nineteenth century France was the leading country in the field of chemistry, Germany soon reached a level equivalent to that of France and became, in the second part of the century, the most prominent center. Two chemists played an essential role in this development: Justus von Liebig (1803–1873) and Friedrich Wöhler (1800–1882). Both are considered the founders of organic chemistry. Perhaps the most brilliant and influential chemist promoting the rise of German chemistry was Liebig. He became interested in chemistry at an early age. After he received his Ph.D. he went to Paris and joined the outstanding chemist and physicist Joseph Louis Gay-Lussac (1778–1850). He followed Gay-Lussac's lectures with great enthusiasm and took part in some of his experimental work.

On his return to Germany, Liebig became professor at Giessen. He was only 21 years old. In the following year he became *Ordinarius*. By the vitality, the brilliance, and the enthusiasm of his personality he attracted and trained a great number of excellent students, and in the following decades these students became the leaders in the rapid growth of German chemistry, both at the universities and in industry. Although Liebig made a number of outstanding experimental contributions, his great stature derived primarily from his concepts, his inspiration, and his many brilliant ideas, even though some of them were later made obsolete by the remarkably fast growth of chemistry. Especially after he moved to the University of Munich, in 1852, his efforts were devoted much more to the promotion of his ideas and concepts than to laboratory work. He expressed in many writings and lectures his great visions of chemistry's potential to provide essential information for physiology and agriculture. As was stated by Sir Frederick Gowland Hopkins (1861–1947) in his address, "The Influence of Chemical Thought on Biology," given at the Harvard Tercentenary Conference of Arts and Sciences in 1936: "It was the genius of Liebig that started modern organic chemistry on a triumphant career, and Liebig's great desire and one which directed his own efforts was to see chemistry render full service to animal physiology and agriculture." It is interesting that in this address Hopkins, comparing Liebig and Pasteur, the other great genius of that era, believed that Liebig lacked biological training and thus a biologist's instinct; he remained too much the chemist. This impeded to some extent his influence on biologists. Pasteur, with his great successes in the biological field, became too much a biologist and thereby delayed to some extent the confluence of chemistry and biology.

Wöhler was a pupil of the famous Swedish chemist Jöns Jakob Berzelius (1779–1848). The two became close friends and maintained this intimate relationship until the death of Berzelius. Wöhler's most famous contribution was the synthesis of urea from ammonium cyanate. This synthesis demonstrated that it was possible to produce an organic compound in a test tube starting with inorganic substances. The experiments, performed in 1828, represent a landmark in the development of chemistry. Urea is one of the end products of nitrogen-containing metabolites excreted in the urine. Wöhler's observation led to the breakdown of the assumption that there was a fundamental difference between organic and inorganic chemistry. Until then it was widely assumed that the formation of organic compounds required a vital force (vis vitalis). This view was still strongly emphasized in 1827 by Wöhler's teacher and friend Berzelius, who maintained this belief even after Wöhler's success. Although many chemists soon abandoned the distinction, it was many years before the view was finally and universally accepted. Another important work was the isolation of a soluble enzyme obtained from bitter almonds, subtilisin, performed by Wöhler in collaboration with his friend Liebig. The significance of this finding will be discussed later in the proper context. Wöhler became Ordinarius in Göttingen in 1836. There he created one of the most important schools of chemistry in Germany. From his school, just as from that of Liebig, a large number of leading chemists emerged. These two schools must be considered as essential catalysts in the remarkable growth of chemistry in Germany during the second part of the nineteenth century.

One of Liebig's most outstanding pupils was August Kekule von Stradonitz (1829–1896). His most famous contribution was his concept of the ringlike structure of benzene. Molecular structures were at that time the center of interest for chemists, and Kekule's concept was a milestone in the elucidation of aromatic compounds such as naphthalene, and anthracene. The concept was greatly elaborated by other chemists, especially Jacobus Hendricus van't Hoff, although the final understanding was only achieved in this century, after atomic physics permitted electronic structure to be understood (see II, F). Kekule's lifework became an essential factor in the systematic industrial synthesis of dyes and some drugs, two areas in which Germany soon became dominant.

After he received his Ph.D. Kekule spent some time in France and England. He became *Privatdozent* in Heidelberg in 1856 and *Ordinarius* in Bonn in 1865. He was an inspiring teacher and greatly influenced the thinking of his many pupils and associates.

Another outstanding chemist who was important for the growth of the field was Robert Wilhelm Bunsen (1811–1899). He made many important contributions to chemistry, but his most outstanding one was the development of spectrum analysis.* He collaborated closely with the physicist Gustav Robert Kirchhoff, who was responsible for the construction of the spectrometer and for many improvements. He became *Ordinarius* in Heidelberg in 1852 and kept the chair almost until his death at the age of 78. He attracted a great number of pupils by his impressive personality, and he was chiefly responsible for the rapid spread of spectral analysis.

The development of chemistry at the University of Berlin was slow. In 1825 Eilhart Mitscherlich (1794–1863) became the first chairman. He was a pupil of Berzelius and an excellent chemist. In the 1830s he showed, independently and at the same time as Charles Cagniard de La Tour (1777–1859) in France and Theodor Schwann (1810–1882) in Germany, that alcoholic fermentation requires the presence of yeast cells; that is, living organisms. This discovery was not only of practical importance because of its role in alcohol production. The nature of "fermentation," a term used for many reactions in the living cell, was a hotly discussed topic. Vigorous controversies about the interpretation of this process involved Liebig, Berzelius, Pasteur, Berthelot, and many other illustrious scientists. It is unnecessary to go into the details of these historical developments, since they are not essential to the aim of this book. Moreover, the real explanation came only in the twentieth century and will be dicussed later (see IV, E).

After the death of Mitscherlich, August Wilhelm von Hofmann (1818–1892) became, in 1865, his successor. He had been a student of Liebig's and was a brilliant chemist. He had obtained the chemical compound aniline from coal tar, an achievement essential to the development of the

---

* A spectrum is formed when a source of light is dispersed by a prism and falls on a screen. The spectra are characteristic for different sources of light and may therefore be used to compare these sources. Spectrometry has been developed for all kinds of radiant energy and has been continuously refined. Today it is an important tool in many scientific and industrial fields.

powerful German dyestuff industry in the second part of the nineteenth century. Queen Victoria and Prince Albert, who had met Hofmann in 1845 during a visit to Bonn, were so impressed by his personality that they offered him the chairmanship of the Royal College in London, where he established a large school. When he accepted the offer to join the University of Berlin, an institute of chemistry was finally built and he attracted many good students; Berlin soon emerged as one of the important centers of chemistry, comparable to those in Heidelberg, Göttingen, and Bonn. Nevertheless, Berlin could not yet quite match the glory attached to the chemistry department in Munich. When Liebig died in 1873, after having held the chair at Munich for two decades, J. F. W. Adolf von Baeyer became his successor. Since Liebig was considered the greatest genius in chemistry of his era, his chair was the most coveted position in the country. Baeyer not only maintained the glamour of this center, but increased it by the brilliance of his achievements and his personality. For several reasons it seems appropriate to give a little more detail about Baeyer; one reason is his family background, which is of special interest in the context of this book.

J. F. W. Adolf von Baeyer (1835–1917) was born in Berlin. His father, J. J. Baeyer, was attached to the German General Staff, but he was more a scholar than a military man, performing geodetic measurements first in East Prussia and later elsewhere. He became president of the Geodetic Institute in Berlin. He was an honorary member of the Prussian Academy of Sciences. His bust was put in the Hall of Honors in the National Gallery in Berlin. Baeyer's mother, Eugene, *née* Hitzig, was of Jewish descent. Baeyer's grandfather, Julius Eduard Hitzig, was a highly respected Kammergerichtsrat (a member of the supreme court in Prussia) in Berlin. In view of the grandfather's great literary interest, his house was a center of Berlin literary life. Among his friends were the poets E. T. A. Hoffman and Adelbert von Chamisso. Several members of the family received great honors rarely accorded Jews at that time, such as the award of two *Pour le mérite's* (the highest decoration for service to the country), one in the field of arts and sciences, and one for heroism in battle. Baeyer's wife was Lydia, *née* Bendemann, of a prominent Jewish family with many distinguished members. Her father was *Wirklicher Geheimer Bergrat im Ministerium* (a high ranking ministerial position). A cousin received the *Pour le mérite*. A brother of Mrs. Baeyer's was a Prussian general and a cousin of hers was an admiral in the navy.

Despite the atmosphere in his home, created by his grandfather Hitzig's great interest in the humanities, the young Adolf had no interest in literature. Apparently his main interest from childhood on was chemistry. Only for a few years in the gymnasium in Berlin did he become attracted to physics and mathematics, owing to an inspiring teacher. But when he entered the University of Heidelberg, his interest returned to chemistry and he studied with Bunsen. Since the University of Berlin at that period, in the 1850s, had no adequate facilities, students went to other universities. Heidelberg attracted many because of a number of brilliant scientists in Bunsen's laboratory, among them Kekule. Baeyer was deeply influenced by his ideas. Many of Baeyer's achievements are due to the insight this extraordinary teacher gave him into the structural properties of chemical compounds.

Baeyer returned to Berlin in 1860. He became *Privatdozent* at the university, with a thesis on uric acid. But the lack of facilities forced him to look for a better laboratory. He joined another institute, which was later relocated in Charlottenburg and became part of the *Technische Hochschule*. There he began the work he was to devote most of his life to: the development of dyes. He developed a new method by which he achieved the formation of indigo blue from indole. Indole is one of the principal nitrogen containing compounds extracted from coal tar by distillation. It is a simple structure formed by one six carbon ring combined with a ring of four carbons and one nitrogen. Indigo blue is a rather complex structure, in which two carbons of two indoles are linked by a double bond; by some additional modifications it becomes an extensively conjugated structure with a strong blue color. Baeyer spent many years on this project which required many new steps and great ingenuity. The interesting properties of this compound were still analyzed by many chemists in this century. The synthesis of many other artificial dyes soon followed, and Baeyer is generally considered as one of the founders and main promoters of the German dye industry. In Berlin, inspired by Baeyer and using his new method, two of his pupils, Karl Graebe (1841–1927) and Karl Liebermann (1842–1914), worked out procedures for producing alizarin, a dye of great industrial value that could now be easily prepared from coal tar. Before this work, this widely used dye could only be obtained by extraction from certain plant roots. It may be mentioned that Karl Liebermann belonged to a well-known Jewish family and was related to the famous painter Max Liebermann.

Continuing his work on uric acid, Baeyer developed a number of related compounds, among them barbituric acid. It was from this compound that his famous pupil Emil Fischer (1852–1919) later developed the barbiturates, still widely used as hypnotics. In one of his lectures Baeyer mentioned that he chose the name barbituric acid because he was in love with a Miss Barbara. Through his work on uric acid, which is an important metabolite in the human body, Baeyer became interested in some problems of physiological chemistry. He made some contributions to photosynthesis, a problem of extraordinary importance for life on our planet, as will be discussed in the next chapter (IV, C), since the real mechanism of this process found an explanation only in the twentieth century.

In 1870 Baeyer was offered the chair of the newly founded University of Strasbourg. There he was joined by a group of excellent collaborators, among them the cousins Emil and Otto Fischer. But in 1873 Liebig died, in Munich. His chair was, for the reasons discussed, a symbol of great prominence. In 1875 Baeyer was elected as Liebig's successor and he went to Munich. There he remained as chairman until 1916, a year before his death. In this period he succeeded in synthesizing an amazingly large number of compounds. The brilliance of Baeyer's many achievements is well documented in a special volume of *Naturwissenschaften* which appeared in 1915 on his 80th birthday (Baeyer, 1915). Among the many outstanding contributors were R. Willstätter, H. Wieland, P. Karrer, O. Dimroth and W. Schlenck. As his pupil Richard Willstätter wrote in his autobiography, Baeyer's main work was achieved in a period that fitted his great talent and his preference. It was a period when the simple test tube was the main tool. Moreover, he did not like to use complex materials, preferring simple substances that were easy to buy or to prepare. Their transformation into new compounds was his aim. Most of the work was essentially qualitative, except for the analysis of the elements involved. Simple exploration of chemical reactions was his great art. In that he combined extraordinary imagination and intuition, skill, patience, devotion, and energy. He was not fond of proposing theories but if he did he had no hesitation about withdrawing it when the facts contradicted it. He also disliked working with complicated equipment such as that which became increasingly necessary in organic chemistry after the turn of the century. In addition to a great number of other honors and distinctions, he received the Nobel Prize in chemistry in 1905 for his outstanding achievements in the synthesis of dyes.

However, Baeyer's importance for the development of chemistry rests on more than his scientific contributions; he inspired and trained an extraordinary group of brilliant pupils. Students from all parts of Germany and many different countries joined his laboratory. There is no question that it was not only the quality of his work, but also his extraordinary and forceful personality and his wisdom that attracted this large number of outstanding students. He apparently had a great ability to explore problems without prejudice and succeeded in conveying his approach and way of thinking to his students. His integrity, his excellent judgment of people, which enabled him to select only students of high quality, the way he tried to promote the independence of his young collaborators, his devotion to his work and his enthusiasm, all these characteristics explain his success in building up one of the greatest schools of the nineteenth century. From his school emerged not only a number of highly qualified organic chemists, but also several leaders who were instrumental in laying the foundation for the spectacular growth of biochemistry in the twentieth century. Many of his pupils were associated with a chapter of paramount importance in the many areas that marked a new era in biochemistry.*

When A. W. von Hofmann died, Emil Fischer (1852–1919), who had left Baeyer's laboratory in 1882 for Erlangen and then for Würzburg, became in 1892 *Ordinarius* of chemistry in Berlin. Emil Fischer is widely recognized as one of the greatest chemists, one of the real giants of the nineteenth century. He probably contributed more than anybody else to the elucidation of the structure of the most important components of living cells. His classic work on the structure of proteins and sugars, carried out in the last decades of the last and at the beginning of this century, represents one of the decisive breakthroughs in our knowledge of the essential cell components and the methods for their determination. These achievements opened the way for many rapid and crucial advances in biochemistry.

When Emil Fischer moved to Berlin, the size of Hofmann's institute was no longer adequate and a new institute was built for him. It permitted him to accept and train many more students and to introduce modern equipment and instruments, which became available at an ever-increasing rate. Among his pupils who became leaders in biochemistry may be mentioned Otto Warburg, Carl Neuberg, and Max Bergmann, who will be described later (see IV, C, E and V, D). One of his students was Hans T. Clarke, who became, in 1929, head of biochemistry at Columbia University in New York. When a few years later the Nazis seized power and many German-Jewish biochemists were forced to leave Germany, Clarke received several of them in his laboratory and supported their work in a most generous way. His department became one of the world's great centers of biochemistry (see V, C).

Fischer was extremely liberal and internationally minded. His family came from the Rhineland and he disliked the Prussian military spirit. He sent his son, Hermann Otto Laurentz Fischer (1888–1960) after the *Abiturientenexamen* to Cambridge for a year to imbue him with a truly democratic spirit. As a result, his son left Germany in 1933, disgusted by the Nazis, although he was 100% "Aryan." He went first to Basel, moved later to Toronto, and finally moved to Berkeley, where he was received with great warmth and was greatly honored. He contributed greatly to the development of chemistry and biochemistry.

Fischer and Max Planck persuaded the Academy of Sciences to offer a position to Jacobus Henricus van't Hoff (1852–1911), who had revolutionized chemistry by his contributions to the understanding of the struc-

---

* In 1933 the two sons of Baeyer—Otto, professor of physics in Berlin, and Hans, professor of medicine in Heidelberg—were immediately discharged; their mother was Jewish, their father a half-Jew. According to Nazi classification they were three-quarters Jews.

tural arrangement of substances in space (i.e., their structural configuration). He was one of the founders of physical chemistry and his ingenious work, after the usual vigorous objections in the beginning, and his inspiring personality were generally admired. He was the first Nobel laureate in chemistry, in 1901. He greatly influenced Svante Arrhenius (1859–1927) and Walther Nernst. Van't Hoff moved to Berlin in 1896. When Nernst joined the University of Berlin in 1906, another giant was added to the galaxy of stars at the university, which became one of the world's leading centers in physics, physicochemistry, and chemistry. Fischer was also one of the chief advisers to Adolf von Harnack in the planning of the Kaiser Wilhelm Institutes in Berlin–Dahlem, as will be described in Chapter IV.

B. PHYSIOLOGY

There were a few isolated forerunners in the seventeenth and eighteenth centuries, but systematic and widespread efforts to explore and to understand phenomena of life, the working of the human body, began only in the early nineteenth century. After the founding of chemistry by Lavoisier, "animal" chemistry began to play an important role. France, where chemistry was most advanced at that period, was the leading center in the newly developing field, especially in the early phase. But even later, the greatest biologist of the nineteenth century was Louis Pasteur (1822–1895). In fact, his genius and his work still had a great influence on the thinking of the leaders in modern biological science, which started only in the twentieth century. Another outstanding leader was Claude Bernard (1813–1878), who greatly influenced physiological thinking in the nineteenth century. He made many important contributions to physiology and physiological chemistry (see, e.g., Holmes, 1974). But beyond that, his general concepts and notions about the approach to experimental medicine present exciting reading even today.

However, with the centers of chemistry and especially of organic chemistry shifting to Germany around the middle of the century, animal physiology at German universities developed rapidly and soon reached high levels. Johannes Müller (1801–1858) in Berlin trained a group of outstanding pupils. Among them were Hermann von Helmholtz, whose brilliant contributions to sensory physiology, before he turned to thermodynamics and physics, were mentioned before (II, A); Emil Du Bois-Reymond (1818–1896), the founder of electrophysiology, who established that nerve impulses in nerve and muscle fibers are propagated by electric currents; and Rudolf Virchow, one of the primary promoters of the central role of the cell in the body (*Omnis cellula e cellula*) and one of the most outstanding pathologists of the century. Another pupil was Theodor

Schwann (1810–1882), who was particularly interested in physiological chemistry. His contribution to the discovery of the role of yeast cells in alcoholic fermentation has been mentioned before. He also identified the enzyme pepsin as the active agent in gastric juice. A great center of physiology was established by Carl Ludwig (1816–1895) in Leipzig; he was one of the most outstanding physiologists of the nineteenth century. His inspiring personality and his interest in different approaches to problems attracted a great number of students. In Bonn was Eduard F. W. Pflüger (1829–1910), who played an eminent role in the development of physiology in Germany. He strongly opposed the division of physiology into what we would today refer to as biophysics and biochemistry, since he considered the two indivisible. Although this view is basically correct, the result of this attitude was harmful for an independent and intensive development of biochemistry, or physiological chemistry, which are not really different fields. It may be mentioned that in Königsberg the chairman of physiology was the distinguished Jewish physiologist Ludimar Hermann (1838–1914).

It was fortunate that, despite much opposition, Felix Hoppe-Seyler (1825–1895) established the first independent institute of physiological chemistry in 1861 at Tübingen University. In 1872 he moved to Strasbourg, after the annexation of Alsace–Lorraine. Both places became great training centers of physiological chemistry. Among the pupils of Hoppe-Seyler was Friedrich Miescher (1844–1895), who was the first chemist to isolate nuclein. This was an achievement of great importance, since it was the beginning of the development of research on nucleic acids, still today in the center of interest in biochemistry. Nucleic acids were recognized in the 1940s as the genetic material that transmits hereditary properties and controls cell growth. Another important center of this field was Heidelberg, under the direction of Friedrich Wilhelm Kühne (1837–1900). The discovery of two important cell components is associated with his name: *myosin*, a protein essential in muscular contraction, and the light-sensitive *rhodopsin*, essential in the elementary process of vision.

One remarkable scientist who greatly influenced biological research must be mentioned at least briefly: Paul Ehrlich (1854–1915). Although the field on which he had the greatest impact was medicine, many of his almost prophetic visions and experimental contributions greatly stimulated several border fields of biomedical research, among them physiology and biochemistry. A full portrayal of his personality and his lifework belongs in a book on the history of medicine. A project describing the collaboration between Germans and Jews in this field is in fact in preparation under the sponsorship of the Leo Baeck Institute. However, in view of Ehrlich's contributions to basic biological problems a short description appears appropriate here.

Among Ehrlich's important achievements is the extraordinary refinement and extension of the use of staining methods and their application to the study of physiological problems. For his work in another field, immunology, he received the Nobel Prize in 1908. But the most striking manifestation of his genius is probably the development of "specific" chemotherapy. He is the uncontested founder of modern chemotherapy. Although his fame is frequently associated with the development of Salvarsan, his real greatness was the lucidity and clarity with which he established some of the basic notions and principles of chemotherapy. Some of the theoretical assumptions had to be later modified, because they were proposed when the cell interior and its structure and organization were obscure and unexplored. That the basic principles turned out to be correct makes his vision all the more admirable.

Ehrlich was born in Strehlen, Silesia, into a prosperous and highly cultured family. He attended the gymnasium in Breslau. In his high-school years he spent much time in the house of a cousin on his mother's side, Karl Weigert (1845–1904), a pathologist at the University of Breslau and later professor at Frankfort on the Main. Weigert was an expert in staining techniques. He frequently used aniline dyes. The problem of staining with aniline dyes was for the young Ehrlich an exciting topic. Even in his school years he gave much thought to the mechanism by which the dyes act on tissues. When he had to choose a profession he decided to study medicine. His real passion was actually chemistry. He never thought of practicing medicine; his aim from the beginning was to go into research.

He first studied at the University of Strasbourg. At that time Adolf von Baeyer was professor of chemistry there and he greatly stimulated the enthusiasm of Ehrlich for chemistry. From his first student years on, his aim was to understand cell nutrition and the action of toxic compounds and drugs by chemical and probably similar basic mechanisms. He believed that staining was achieved by chemical reaction and not by purely physical attachment: *Corpora non agunt nisi fixata.* This fundamental notion guided his work throughout his life. He always stressed the necessity of theory, although only the experiment decides whether the theory is correct. Even a wrong theory seemed to him to be more productive than a purely empirical approach, which records facts without attempting an explanation. After his stay in Strasbourg he returned to Breslau. He continued his studies on staining in the physiological institutes of Rudolf Haidenhain and Julius Cohnheim.

After having received his M.D. in Leipzig, in 1878, Ehrlich joined the clinic of Th. von Frerichs at the University of Berlin, where he spent nine years. He used staining methods to obtain an insight into several physiological problems. Referring to the great success of the use of aniline dyes in the textile industry, he stressed that the great potential offered by microscopic analysis of staining with dyes was not yet being adequately used as a research tool. One of his earliest great achievements with staining methods was the creation of the basis of modern hematology. He emphasized the significance of vital staining; that is, staining when the fundamental cell activity is still going on. Another important finding was the selectivity of dyes in staining cells. By using many different dyes he found that certain dyes stained specifically certain organs or types of cells. For example, methylene blue stained nervous tissue but not other cells. Not only were these findings of obvious practical value, but they raised intriguing questions for which there were yet no answers. But they influenced Ehrlich's thinking in a direction that became most pertinent for his later ideas, on which he based the principles of chemotherapy. They led to his theory of side chains attached to the proteins of the protoplasm: Just like the dyes, biologically active substances, toxins, antibodies, and drugs, must react with specific cell constituents attached to side chains. In pursuing these ideas he later introduced, in 1898, the notion of specific receptors for these compounds, a brilliant concept of the greatest importance, widely used today in biochemistry and molecular biology.

A series of other observations of that period must be mentioned. Certain dyes were reversibly oxidized or reduced in the cells. After the dyes alizarin blue or indophenol were injected into the circulation of animals, some organs were colored blue, whereas others reduced the dye to a colorless product. From these experiments Ehrlich deduced the relative ability of tissues to take up oxygen and thus obtained an indication of oxygen consumption. Ehrlich (1885) summarized these observations in a monograph: *Das Sauerstoff-Bedürfnis des Organismus. Eine farbenanalytische Studie* [The Oxygen Requirement of the Organism. A Study Based on the Analysis with Dyes]. His experimental observations and his discussions described in the monograph, including many interesting comments on views and findings of Liebig, Berthelot, Hoppe-Seyler, Pflüger, and others, reveal an extraordinary imagination and an intuitive gift of analyzing data and obtaining insight into the complex problem of oxygen requirements of cells. Leonor Michaelis, a pupil of Ehrlich's and an authority in the field of redox potentials, stressed the importance of Ehrlich's observations and conclusions; many completely new concepts were proposed, far ahead of the knowledge of the period. Ehrlich's ideas are all the more remarkable since little was known about intracellular processes. Emil Fischer had just started his pioneer research on the structure of carbohydrates, but decades passed before a biochemical approach was applied to the problems of oxidation.

By his friendship with two outstanding scientists, Robert Koch (1843–1910) and Emil von Behring (1854–1917), Ehrlich became attracted to immunology and turned increasingly to the study of its problems. Ehrlich had suffered from tuberculosis for two years. In 1888 he went to Egypt, recovered, and returned in 1889. He joined Koch in 1890 and worked with him on tuberculin. But Koch soon abandoned his efforts. Ehrlich joined Behring, who tried to develop a serum against diphtheria. Here again Ehrlich was particularly interested in the mechanism of the reaction between antigens and antibodies. He recognized the chemical nature of the problem, trying to demonstrate that the interaction between toxin and antitoxin must be a chemical reaction and not due to physical forces as proposed by some investigators: The toxins (antigens) combined with specific receptors in the cell just as other active compounds did. Since different sera are all specific, it was necessary to assume a specific configuration of the active agent. This chemical specificity was for him an essential feature of immunological reactions.

Ehrlich's ideas and his active collaboration with Behring were essential factors in the preparation of a highly efficient and successful serum against diphtheria. A very close personal friendship developed between these two men. Ehrlich continued his immunological studies for several years. It was for his contributions in this field that he received the Nobel Prize in 1908, together with Ilya Mechnikov (1845–1916). In his Nobel Prize lecture he pointed out that the era of the microscope was nearing its end and that only research on cell chemistry would provide a real understanding of the mechanisms essential for the progress of medicine.

By the time Ehrlich received the Nobel Prize his main interests had shifted to the development of chemotherapy. His concept of specific receptors, its application to evolve the treatment of various infections on the basis of specific agents, and his achievements in this area are probably his most outstanding contribution to medicine and to biological science in general. He began this work with studies of infections caused by trypanosomes. Trypanosomiasis is the cause of sleeping sickness, a serious problem in many parts of Africa. He found that a dye, trypan red, had curative properties when applied to mice experimentally infected by a certain type of trypanosome. It was the first successful treatment of an experimentally produced disease by a synthetic organic substance of a known chemical structure; but its usefulness was limited. It was effective against one but not all kinds of trypanosomes. Soon arsenical drugs were found to be much more promising. Koch had found Atoxyl, an arsenobenzene derivative, to be effective in the treatment of sleeping sickness. It had, however, toxic side effects, among them blindness. Ehrlich recognized the great potential of Atoxyl, since it offered the possibility of the synthesis of hundreds of modifications by the substitution of the amino group attached to the benzene. He was firmly convinced that some homologues would have the desired specificity of destroying the microorganism without any damage to the host cell, that is, the organism. This was the basic idea that led to the development of Salvarsan, the famous compound 606, which proved to be effective in destroying the microorganism *Spirochaeta pallida*, which produces syphilis, thus curing the dreaded disease. Soon a still more effective modification was found, Neosalvarsan. It was a great triumph of the human mind, the greatest achievement of chemotherapy before the discovery of sulfonamides, followed soon afterward by the development of penicillin and other antibiotics (see V, D).

Ehrlich became one of the most celebrated scientists of his time. In 1883 he had married Hedwig Pincus, and when he returned from Egypt, his father-in-law generously provided funds to permit him to devote his time to pure research free of any administrative or other obligations, and to devote his time to his real interests. In 1899 he became the director of the Institute of Experimental Therapy in Frankfort. Impressed by his brilliance, several people promoted his research, in particular Arthur von Weinberg, a great industrialist, and Friedrich Althoff, permanent director of the Prussian Ministry of Education (see p. 25). In 1905 the widow of Georg von Speyer gave Ehrlich the funds for a new institute for the development of chemotherapy. He had pupils from all over the world; his fame had spread everywhere. He frequently visited England, where one of his many famous pupils was Sir Henry Dale, a leading pharmacologist. On his visit to the United States in 1904 he was enthusiastically welcomed and everywhere he encountered triumphant receptions. In New York he gave the first Christian Herter lecture, established for outstanding contributions in biochemistry. In addition to the Nobel Prize he was honored by many awards and elected to membership in leading societies. At the meeting of the *Gesellschaft für Naturwissenschaftler und Ärzte* (society of scientists and physicians) in Königsberg in 1910 he received the greatest ovation ever given there. On his sixtieth birthday he was honored by his pupils and colleagues from all corners of the world. He died at the age of 61. At his funeral at the Jewish cemetery in Frankfort, Rabbi Lazarus conducted the services. Professor Alexander Ellinger, a distinguished colleague and friend, referred to him as "a prince of science, who left us rich in honors, still richer by what he gave to the world." Most moving were the words of Emil von Behring. He called Ehrlich a king in the science he established. "You created a school as hardly anybody before you and you became a *Magister Mundi* in the science of medicine." There are many biographies of Ehrlich; an excellent one is that of Greiling (1954).

In 1976 the Weizmann Institute of Science in Rehovot, Israel, opened a special wing of a building devoted to the biological sciences, dedicated to the memory of Ehrlich. At the ceremony Professor Hans Herken, chairman of the Institute of Pharmacology of the Free University of Berlin, a pioneer in the application of biochemistry to the analysis of pharmacological and pathological mechanisms, gave a superb memorial lecture (Herken, 1976). In his evaluation of the great achievements and the personality of Ehrlich, Herken emphasized not only his originality and ingenuity but also the high moral and ethical principles that guided him, his caution in testing and applying new drugs to humans, the feeling of responsibility for his patients. This attitude is of particular importance at present, when new drugs are being developed all the time, and sometimes tested without enough attention being paid to possible side effects.

Let us return again to the development of physiological chemistry in general. Although its importance for biomedical research was increasingly recognized, it still advanced rather slowly. For instance, it was not until 1928 that an independent institute of physiological chemistry was established at the University of Berlin. Most biologists interested in chemistry worked in laboratories in clinical hospitals or pathological institutes. In France and England the development was also very slow. In England, the first professor of biochemistry was Sir Frederick Gowland Hopkins (1861–1947), appointed in 1914 in Cambridge. In the United States, on the other hand, physiological chemistry found strong support in the last years of the nineteenth century and early in this century, especially with the opening of the Rockefeller Institute for Medical Research in 1901. One of the most influential and most brilliant promoters of this field was Jacques Loeb (1859–1924). He insisted on the necessity of applying physical chemistry and biochemistry to the study of living organisms. Loeb was a German Jew who had emigrated to the United States. He became professor at the Rockefeller Institute. But there were a number of others who established biochemical schools in the United States, notably Russell H. Chittenden (1856–1943) at Yale University, who had studied with Kühne in Heidelberg. His successor at Yale was Lafayette B. Mendel (1872–1935).

However, it must be recognized that the slow development of biochemistry was not due to any lack of recognition of its importance or lack of adequate efforts. In 1847 Justus von Liebig wrote, in his introduction to *Thierchemie*, that no manifestation of life is conceivable without molecular (i.e., chemical) reactions, and he emphasized time and again that we will not be able to understand life processes without knowing the underlying chemical reactions. Many leading physiologists accepted this view. The attitude of Ehrlich has just been discussed. But the crucial factor preventing rapid progress in biochemistry was that the foundations were missing. The development of two fields changed the situation and opened the way for the rapid development of biochemistry. First, the advances of organic chemistry, in which the brilliant chemists who had emerged from Baeyer's school, as mentioned before, played an outstanding role (Emil and Otto Fischer, Richard Willstätter, Heinrich Wieland, Eduard Buchner). The chemical structures of a great number of important cell constituents were elucidated and methods for their determination were elaborated. This knowledge was a prerequisite for systematic studies of the intermediary metabolism. Second, the great advances made in physical chemistry by many outstanding scientists in Germany and elsewhere, such as Wilhelm Ostwald, Jacobus Henricus van't Hoff, Svante Arrhenius, Walther Nernst, Fritz Haber, and Gilbert N. Lewis, permitted the analysis of many factors necessary for the understanding of essential features of the reactions in living cells, such as thermodynamic, kinetics, electrochemistry, electrolyte behavior, and so on. Both fields were of paramount importance for the great achievements of what is referred to as dynamic biochemistry in the first four decades of this century, especially for enzyme chemistry and the exploration of metabolic pathways. A pioneer role in these developments was played by the Kaiser Wilhelm Institutes, to be described in Chapter IV.

Atomic physics, resulting in the knowledge of the electronic structures of atoms and molecules and subsequently in a huge number of highly refined physical instruments and methods—such as electron microscopy, the use of stable and radioactive isotopes, nuclear magnetic resonance, new types of optical methods, lasers, the oscilloscope, X-ray analysis, and methods for measuring very fast reactions—added new dimensions to biochemistry, especially after World War II. These developments catalyzed the spectacular rise of biochemistry in this century. Although biology in the last century was essentially descriptive phenomenology, it has emerged in this century as a science becoming increasingly capable of analyzing processes in living cells on a molecular level.

IV

Chemistry and Biochemistry in the Early 20th Century. The Kaiser Wilhelm Institutes in Berlin—Dahlem

The spectacular rise of science in German universities and *Technischen Hochschulen* in the latter part of the nineteenth century laid the foundation for the world leadership of German industry and thus formed the basis of the economic affluence of the *Kaiserreich*. A favorable atmosphere was created for the expansion of research in general and the establishment of institutes devoted exclusively to research. Such institutes already existed in France, England, and the United States (e.g., the Pasteur Institute, the Lister Institute, and the Rockefeller Institute). Farsighted leaders in science, industry, commerce, and government in Germany recognized the necessity of creating pure research institutes in which brilliant and outstanding scientists would have complete freedom and all the necessary facilities to perform their research work under the most favorable and efficient conditions.

Friedrich Althoff had already begun, in the last two years of his life, to outline plans for this type of institute. He envisaged a great center that would include physics, chemistry, physical chemistry, electrochemistry, physiological chemistry, brain research, biology, anthropology, and others. After his death Kaiser Wilhelm became seriously interested in carrying out these plans, through several meetings with Nernst. He had been greatly impressed by Nernst's brilliance and his outspoken way of presenting his views. The kaiser recognized the great importance of such institutes for maintaining and increasing the leadership of Germany in science and industry, the basis of German power. He was eager for Althoff's plans to be vigorously pursued. Friedrich Schmidt-Ott, at that time, in 1908, *Ministerialdirektor* and later *Kultusminister* (head of the ministry of cultural affairs) in Prussia, was asked to continue the work on these plans. He asked Professor Adolf von Harnack to submit a detailed memorandum. Harnack was a brilliant scholar and an extraordinary personality. Although his strength was in the humanities—he was professor of theology—he had a great passion and admiration for science. He was a grandson of Justus von Liebig. With the help of leading scientists —among them Emil Fischer, Walther Nernst, and August von Wasserman —Harnack produced, at the end of 1909, a classic document that must be considered the basis of the concept on which the Kaiser Wilhelm Society was founded. It not only presented convincingly the necessity for the society, but also outlined the basic principles of its organizational and financial structure. Harnack particularly stressed the need for complete independence and freedom, and the cooperation of industry, science, and government. Only in this way would a completely free development of science be ensured. He lucidly pointed out the many dangers that might arise if only the state, or only industry, were in charge of financial support. The brilliant insight that was so remarkably well expressed in this document has been confirmed by many later developments of science in this century, perhaps most drastically in the United States.

At the centenary of the University of Berlin in 1910, Kaiser Wilhelm presented the plans developed by Harnack and asked for their strong support. He gave instructions for adequate space to be reserved for these institutes in Dahlem, a suburb of Berlin that originally belonged largely to the domain owned by the Hohenzollern family. The constituent assembly of the Kaiser Wilhelm Society took place in 1911. Leading industrialists and bankers, among them many Jews, contributed most generously to the financial support of the proposed institutes. The total annual income amounted to about 10 million marks, an almost astronomical figure considering the value of the mark at that time and the relative smallness of the equipment required for research compared to that needed in present-day science. Harnack was elected first president of the society. His superb qualities of leadership greatly contributed to the magnificent development of the institutes. He kept his office until his death in 1930. His successor was Max Planck. Harnack belonged to a large family, prominent in universities, industry, and government. Because of its well-known liberal and progressive attitude many members of the family were executed by the Nazis.

The first institute completed was that of chemistry, with four independent departments. Ernst Beckmann was the director of inorganic chemistry; Richard Willstätter, of organic chemistry; Otto Hahn, of the chemistry of radioactive compounds; and Lise Meitner, a physicist and pupil of Max Planck, who had worked with Otto Hahn since 1908 in Emil Fischer's institute, of the section for the atomic physics of these compounds. The second institute was that of physical chemistry, of which Fritz Haber became the director. The money for this institute was provided by the Jewish banker Leopold Koppel, who had given it on the condition that Haber would be appointed director. The opening ceremonies of these two institutes took place late in 1912 with great pomp, in the presence of the kaiser, members of the government, leaders of industry, bankers, and leading scientists of the University of Berlin. The third institute was that for experimental therapy, also opened in the presence of the kaiser. Its director was August von Wasserman (1866–1925), a pupil of Paul Ehrlich's. The building of the fourth institute, for biology, was started in 1913 and finished in 1915, in spite of the difficulties caused by the outbreak of the war. An institute for physics was decided upon in March 1914. Einstein was supposed to be the director. He had come to Berlin in 1913. War broke out, so the plans for a physics institute were suspended and taken up only in the 1930s, when the institute was built with the financial support of the Rockefeller Foundation. When it was finished, Einstein had already left. It is noteworthy, in the context of this book, that in the first two institutes three directors were Jews, and that two of the other institutes were assigned to Wasserman and Einstein, also Jews. This reflects on the attitude of the leadership of the Kaiser Wilhelm Society and will be discussed later.

The Kaiser Wilhelm Institutes in Berlin–Dahlem (there were other institutes elsewhere) were destined to become one of the greatest and most brilliant centers of that era. It may be recalled that there were many brilliant scientists at the University of Berlin and several other important institutions. After the turmoil of the first postwar years, an extraordinary renaissance also took place in literature, art, music, theaters, operas, and other areas, so it is not surprising that Berlin, in the 1920s, impressed visitors from abroad as the scientific and intellectual capital of the world, in spite of the lost war.

A. FRITZ HABER (1868–1934)

This century has produced many giants in science and many in technology. However, hardly anyone can match the extraordinary genius of Haber in contributing to a truly amazing degree to both fields. By his incredibly wide range of knowledge, by his undisputed competence in both areas, by his vision and broad perspectives, he became one of the chief architects in promoting the relations between science, industry, and agriculture, the relations between science and government, and the use of scientific knowledge in the many transformations of society in this century. The only other example of such an extraordinary combination of ingenuity in both aspects—what is usually referred to as basic and applied science—may be Louis Pasteur in biology in the nineteenth century. Pasteur was a genius capable of using his great achievements in fundamental science to solve urgent problems of health, industry, and agriculture. Haber's fame is usually associated with the fixation of nitrogen from the air by its combination with hydrogen to form ammonia, which could be used to produce nitrates essential as fertilizers. By this spectacular achievement he saved mankind from the threat of starvation and it is for this work that he received the Nobel Prize in 1918. The Swedish Academy congratulated him on this "triumph in the service of his country and of all mankind." But actually this was only one of many extraordinary contributions. He was strongly imbued with Fichte's idea that the immediate purpose of science must be its own development, *rerum cognoscere causas*, but that the final aim should be to influence and improve the life of the society. Just as for Pasteur, basic and applied science were for him two sides of the same coin. The responsibility of scientists toward mankind is today a frequently discussed topic; nobody actually questions this aspect seriously. But few have this rare combination of deep basic knowledge and the ability to recognize the practical implications of science for the benefit of man.

Family Background and Career

Haber was born in Breslau as the eldest son of Siegfried Haber, a prosperous chemical and dye merchant, a city councilor of the town in which his family had been prominent for many decades. He was a man of the highest integrity, honest to the extreme. Siegfried's father Jacob had been a successful wool merchant; he died very young from cholera. His widow, Carolina, *née* Friedlaender, and her six children were supported by Jacob's brother, who had 10 children but was prosperous enough to take care of both families. Fritz Haber's great grandfather was Pincus Seelig Haber, a merchant in Silesia. Siegfried's brother Julius became baptized in order to become a high-ranking judge.

Fritz Haber's mother Paula was the youngest of the 10 children of his uncle Julius. The marriage with Siegfried was a happy one, but short-lived. She died in childbirth; the child, Fritz, survived. Fritz was nine years old when his father married Hedwig, *née* Hamburger. Siegfried and Hedwig had three daughters. Siegfried had a great love for his second wife and her three daughters. Toward Fritz, he was severe and sometimes even harsh. The relationship between Fritz and his stepmother and three half-sisters was excellent and very affectionate; it remained so throughout his life. Hedwig came from a thoroughly assimilated family. Although the family was not baptized, no Jewish customs or festivals were observed. Since Siegfried too had no relationships with the Jewish community, Fritz grew up without any knowledge of Jewish rites and traditions.

Fritz attended a humanistic gymnasium. In his school the emphasis was on Greek, Latin, literature, history, and philosophy. Little attention was paid to science. Fritz became an avid reader in his high-school years.

Goethe was his favorite poet, Kant his favorite philosopher. During his school years he was a member of an academic literary society, in which the young people devoted their time to reading literature, poetry, Greek classics, and the like. He liked to write verses, particularly hexameters. He kept this passion of expressing himself in verses, preferably in hexameters, throughout his life. Many of his postcards, even letters of congratulation on various occasions, were written in this form. He had also wanted to be an actor. During his vacation trips he became acquainted with many art treasures, particularly on his frequent visits to Italy.

After graduation he studied chemistry and physics for one semester at the University of Berlin, which had become a great center in both fields, as mentioned before. Helmholtz was a poor teacher, but A. W. von Hofmann was inspiring and Haber became interested in chemistry. In the following semester he went to Heidelberg, where Bunsen headed the chemistry department. Bunsen, with his tendency to emphasize physical methods in chemistry, strongly influenced Haber, whose interest turned to physics and mathematics. In 1889 Haber entered the *Technische Hochschule* in Berlin, at that time an outstanding center. Karl Liebermann (1852–1914), a member of a highly respected and influential Jewish family, was an outstanding organic chemist there, a pupil of Adolf von Baeyer, and it was under his guidance that Haber started with chemical research. It was essentially devoted to artificial dyestuffs, a branch of chemistry rapidly developing in Germany. He received his Ph.D. in 1891. At that time he became interested in physical chemistry. After some unsatisfactory periods in industrial laboratories, which he found not suitable for creative work, in 1892 he joined the *Eidgenössische Technische Hochschule* in Zurich. There he studied several subjects related to science, particularly to its many applications to technology. It was during this period that he became familiar with a wide range of chemical technology. This intimate and unusual knowledge became an important factor in his future career and is reflected in his great achievements in his most creative years.

Haber then joined his father's company, but the difference in their temperaments and viewpoints did not lead to a good relationship. After two frustrating years in his father's business, Haber spent a short time at the University of Jena and then joined Professor Bunte at the *Technische Hochschule* in Karlsruhe. This institute was destined to become the place of Haber's greatest scientific achievements in the following two decades. He became *Privatdozent* in 1896, *Extraordinarius* in 1898, and, at the age of 38, *Ordinarius* in 1906. In Karlsruhe he found his way to physical chemistry and there started his meteoric rise toward becoming one of the greatest leaders in the world in science and technology. Many of his fundamental contributions originated during this time. In addition, it was not only the most creative but also the happiest period of his life.

Scientific Achievements

Haber's first research at Karlsruhe dealt with the decomposition by heat of hydrocarbons, chemicals formed by carbon and hydrogen. No other two elements combine in so many different ways. Gasoline and fuel oil belong to this group, and coal is to a large extent formed by hydrocarbons. Haber established some basic principles of the processes of combustion of these compounds. He summarized his research work in a book entitled *Experimental Studies on the Decomposition and Combustion of Hydrocarbons*. It appeared in 1896, when he was 28 years old. His studies were a milestone in the investigations of a very important process and played an important role in the development of the oil industry. For this book he received the *venia legendi*.

Haber's long-standing interest in physical chemistry received a new and great stimulus by Hans Luggin. Luggin was a pupil of the Swedish scientist Svante Arrhenius, considered one of the founders of modern physical chemistry together with Jacobus Henricus van't Hoff (1852–1911), Wilhelm Ostwald (1853–1932), and Walther Nernst (1864–1941); all four were Nobel laureates (van't Hoff in 1901, Arrhenius in 1903, Oswald in 1909, and Nernst in 1920). Haber was fascinated by the new perspectives that this new and rapidly developing area offered. Under the influence and inspiration of Luggin, he soon became fully familiar with the field, and within a few years he was one of its most prominent leaders.

Haber's investigations at this time centered on reactions called *oxidation–reduction processes*. Such reactions betwen molecules involve transfer of electrons, mentioned previously. The net loss of an electron is oxidation; acquisition of an electron is reduction. Oxidation–reduction reactions play an important role in chemistry, particularly electrochemistry, and they are also crucial processes in living organisms. For instance, the ultimate source of energy in the human body is oxygen, which is activated in the cell, as we know today and as will be discussed later, through a whole chain of oxidation–reduction processes. Haber established the essential role of the force of electric currents at the junction, between the electrode and the surrounding solution, in this type of reaction. He developed a great variety of different types of electrodes. He proposed a mechanism that would permit scientists to calculate the force of oxidation–reduction reactions. These and many other observations were described in a book entitled *Outline of Technical Electrochemistry on a Theoretical Basis*, published in 1898. The title again indicates how Haber tried to combine basic and applied problems. Electrochemistry increasingly became his passion. Like many other scientists of that period, he was particularly interested in fuel cells, in which electric currents are produced by atmospheric oxygen. When energy is obtained from the burning of coal, less than 25% of the potential heat energy is used; the rest is lost. A much more efficient way of preserving energy is its direct conversion into electricity without the intermediate production of heat. This may be achieved by using certain types of fuel cells. The significance of such studies hardly needs to be emphasized in the present era, now that the importance of energy conservation has become so apparent. Haber's work in the field was greatly helped by the establishment of an institute of physical chemistry and electrochemistry in Karlsruhe. In 1905 he published a book entitled *Thermodynamics of Technical Gas Reactions*. This book, his third, was an immediate success. It was praised by the greatest authorities in the field all over the world as a milestone remarkable by its depth, competence, and vision. He was by this time established as one of the great figures in science and technology and in 1906 became, at the age of 38, *Ordinarius* and director of the new institute.

The Ammonia Synthesis. At the beginning of the twentieth century it became apparent that mankind was approaching starvation because of the growing scarcity of fertilizers containing nitrogen. As Liebig had recognized half a century earlier, this element is essential for proper plant growth but is usually not available in sufficient quantities in soil. Nitrogen-containing inorganic compounds called *nitrates*, available in large quantities primarily in Chile, were the only rich source of nitrogen then known. At the turn of the century Chile provided two-thirds of the nitrates used as fertilizers in the world. But in view of the continuously increasing rate of consumption, these natural nitrate deposits were becoming rapidly depleted. Finding a new source of nitrogen was vital for the survival of mankind.

Nitrogen is plentiful in the atmosphere; it constitutes about 80% of the air. Fixation from the air thus appeared to be the obvious solution. But the problem was to find a process by which this source could be effectively used for the production of large amounts of fertilizers. Many attempts were made by quite a few scientists in various countries. None of the procedures proved to be satisfactory.

About 1905 Haber began to work on transforming nitrogen into ammonia, which is a combination of nitrogen (symbol N) and hydrogen (symbol H). One molecule of nitrogen is formed by two nitrogen atoms (thus it is written as N_2), and one molecule of hydrogen is formed by two hydrogen atoms (thus, H_2). When one nitrogen molecule combines with three molecules of hydrogen ($3H_2$), two molecules of ammonia are formed: $2NH_3$. The formula of this process, which has become a famous symbol of Haber's work, is $N_2 + 3H_2 \rightleftharpoons 2NH_3$. The two arrows in two directions indicate that the process goes easily in two directions. Thus, the problem arose of how to achieve a condition favorable for one direction, the production of ammonia.

About the same time, Walther Nernst, who had moved from Göttingen to Berlin in 1906 to become professor of physicochemistry, also became interested in the problem of nitrogen fixation from the air. Nernst was at that time one of the most outstanding and celebrated physicists and physicochemists, and by his dynamic and brilliant personality played a great role in the scientific life of that era. He was liberal and free of any prejudice against Jews. In fact, many of his close associates and pupils were Jewish.

Nernst, like Haber and other scientists already mentioned, was one of the founders of modern physical chemistry. His theory of the galvanic cell has been mentioned before. This work revealed Nernst's extraordinarily deep insight and his superb command of thermodynamics. Thermodynamics increasingly became his chief interest and his name in history is associated with this field. His book, *Theoretische Chemie*, which first appeared in 1903, saw many editions. It was the standard book on thermodynamics for more than two decades, indispensable for all students in chemistry and physicochemistry. After Otto Meyerhof introduced thermodynamics in biochemistry by his epoch-making lecture on the energetics of living cells in 1913 (see IV, D), it also became essential for biochemists interested in physicochemistry and thermodynamics.

When Heinrich Rubens died, Nernst became, in 1924, his successor as *Ordinarius* of physics. In 1920, Nernst received the Nobel Prize. The Gibbs–Helmholtz equation had shown the difference between the free energy of a compound or a system and its total energy—the total energy is that contained in a compound or a system, and the free energy is that which can be readily used at a given temperature and under certain conditions. Nernst postulated that, on a thermodynamic basis, the difference between free and total energy must decrease with temperature, the two energies becoming equal at absolute zero, $-273°$ C. This idea became known first as *Nernst's theorem* and later as the third law of thermodynamics. Actually, as it turned out later, the physical quantity that changes at absolute zero is the entropy of the system—the quality introduced by Clausius in his formulation of the second law of thermodynamics.

The notion of entropy subsequently became greatly clarified and its significance well defined. During the last twenty years, its great importance in molecular biology, a field to be discussed later, was established. Biological systems are never in a static condition, but always in flux; they are in a continuously changing dynamic state on a molecular level, the πάντα ρεῖ (everything is in flux) of Heraclitus. Classical thermodynamics applies essentially to equilibrium states and is therefore only of limited use in studying living systems. A new type of thermodynamics, referred to as nonequilibrium thermodynamics, has emerged since the 1930s. One of the most brilliant and foremost pioneers of the application of nonequilibrium thermodynamics to biological systems was Aharon Katzir-Katchalsky. In the midst of his work on startling new ideas in this field he was killed in the massacre at Lod in 1972. Few events have so deeply shocked the scientific community all over the world as this murder of one of the most brilliant minds of a generation, a leader in the field of modern biology. His death at the height of his creative activity was an irreplaceable loss, not only for Israel but for mankind.

Nernst's dynamic and brilliant personality attracted many outstanding pupils not only in Germany, but from all over the world, and his institute became a famous scientific center. Among his many outstanding Jewish pupils were Franz Simon and Kurt Mendelssohn, great authorities in the field of low-temperature physics. Both are direct descendants of Moses Mendelssohn (personal communication of Sir Hans Krebs). Their work greatly contributed to the field of low-temperature physics, which first attracted their interest in connection with Nernst's concepts. This field later became greatly important in technology and industry. Other pupils were two brothers, Charles and Frederick Alexander Lindemann from England. F. A. Lindemann became Nernst's favorite pupil. He in turn was a great admirer of Nernst and they remained personal friends until Nernst's death. Lindemann became professor of physics of the Clarendon Laboratory in Oxford, and was a close friend of Winston Churchill during World War II; he later became Viscount Cherwell. When the Nazis came to power, Lindemann was only too happy to accept Franz Simon and Kurt Mendelssohn in his laboratory in Oxford. The Clarendon Laboratory became another great center of physics, whereas before it was in no way comparable to the Cavendish Laboratory in Cambridge, which through the activity of J. J. Thomson, Lord Rutherford, and others had become the outstanding laboratory in physics in England. When J. J. Thomson opened a meeting of physicists in Oxford in 1936, he opened it with the words, "Heil Hitler! Thanks to Hitler we have today with us Franz Simon, Kurt Mendelssohn, Rudolph Peierls, and many others."

Nernst had two sons and three daughters. The two sons were killed in World War I. Two of his daughters were married to Jews and forced to leave Germany under the Nazis. Nernst retired in 1933. Thanks to the generosity of Lindemann he was invited to Oxford in 1937, so that he could see his family. He died in 1941. As for countless other German scientists, the Nazi period was for him a personal disaster in many respects.

Let us now return to the production of ammonia from nitrogen and hydrogen. When the two elements are added to a container, virtually no ammonia is produced; ammonia decomposes as fast as it is formed. The equilibrium does not favor the production of ammonia, and the reaction proceeds to the left in the formula. Two factors may influence this equilibrium in favor of ammonia production. The first is the temperature: low temperature favors the production of ammonia. But there is a limit to how much the temperature can be lowered without undesirable side effects. The other factor is pressure. Since both elements are gases, pressure changes also affect the equilibrium. Increased pressure favors the formation of ammonia. Nernst tried high pressure, but the results were unsatisfactory. He informed Haber of his failure. Although Nernst was at that time the greatest authority in physicochemistry and thermodynamics, Haber did not accept Nernst's skepticism. Haber, with his associate Robert Le Rossignol, the son of an English physician, had obtained some positive results with extremely high pressures and low temperatures, but it was apparent that these two factors would not provide a satisfactory procedure for large-scale industrial production. Nernst, whose main interest was theoretical problems and whose main approach was based on thermodynamics, abandoned the problem. An additional factor may have been his dynamic personality. He had a tremendous drive to obtain results rapidly and had little patience if the experiments proved to be disappointing. Haber had a large amount of experience in chemistry and particularly in industrial chemical processes. Moreover, he was not the type to give up easily when a solution to a problem would have such far-reaching importance.

It was known that many chemical reactions could be influenced by catalysts. They accelerate the process without being affected themselves. In living cells virtually all chemical reactions require catalysts: the enzymes, which are proteins. Haber started to look for catalysts and found two that proved to be efficient: uranium and osmium. The results were impressive. Haber then turned to industry for help with large-scale production. Badische Anilin and Soda-Fabrik, after some hesitation, accepted his idea and Karl Bosch, a brilliant research engineer, was instrumental in developing the large-scale production of ammonia. The technical difficulties were tremendous and solving them required great ingenuity. The process is therefore frequently referred to as the Haber–Bosch process. After several years, the production of ammonia on a large scale became a reality. Today many hundreds of huge ammonia-producing plants are spread all over the world.

In 1931 Karl Bosch received the Nobel Prize, and in 1935 he succeeded Planck as president of the Kaiser Wilhelm Society. He was appalled by the destruction of German science by Hitler and was one of the first who had the courage to explain personally to Hitler the disastrous consequences of the loss of Jewish scientists. But like Max Planck, who tried the same thing a few months later, he only aroused Hitler's rage.

It may be mentioned that Nernst recognized that he abandoned the problem too early. It is perhaps only natural that the triumph of Haber, who proved that Nernst was wrong in a field in which he was the great authority, created some coolness in Nernst's feelings, but their public relations always remained correct and friendly. However, an intellectual giant like Nernst could not help admiring the brilliance of Haber's achievement. In a lecture in Munich, in 1913, on the importance of nitrogen for life he gave Haber full credit for his success, as one would expect from a person of the stature of Nernst. In the 1920s, he attended many of the Haber colloquia in Dahlem and became one of the board members.

Life in Karlsruhe

The years in Karlsruhe were the happiest in Haber's life. He had magnificently equipped laboratories in the new institute. His reputation and fame attracted many brilliant students not only from Germany, but also from the United States, Britain, France, Australia, Japan, and elsewhere. He had more than 40 research students; some of them were outstanding collaborators. To English and French students he spoke in their native language. He was for all of them the kind and fatherly friend and all felt a great affection for him. His attitude to students, which also prevailed later in the Kaiser Wilhelm Institutes, was extremely generous in every respect. He never objected to differences of opinion, his attitude was that young people did not always have to listen to older ones. The affection of his students was strikingly demonstrated in a two-day farewell party in Karlsruhe in 1912, after he had accepted the offer from the Kaiser Wilhelm Institutes in Berlin. Some 75 former pupils, some from distant lands, came especially to attend this celebration. The *Zeitschrift für Elektrochemie* dedicated an issue to Haber, to which former and present members of the institute contributed. The issue contained many cartoons, verses, and stories that gave a lively picture of his personality.

On the initiative and under the sponsorship of van't Hoff, Haber visited the United States in 1902 to obtain firsthand information about research and education at the universities and industries in the fields of chemistry and physicochemistry. He visited a great number of universities in various parts of the country and some representative industrial plants. He was greatly impressed by the vigorous, dynamic, and enterprising spirit, the courage and the pride of the people. He greatly admired New York and the beauty of California. He was extremely well received by his colleagues and many scientists invited him to their homes. He thus received a good insight into the lives and activities of his American colleagues. He also was a special guest of the American Electrochemical Society at a meeting at Niagara Falls.

On his return to Germany, at that time in the process of a tremendous industrial expansion, he was repeatedly invited to present his impressions of the United States, since it was known to have a dynamic and rapidly developing industry. The interest centered around the role of electrochemistry in technology, a field with which Haber was particularly familiar. He found the training of chemists in the United States practical, but superficial as to the theoretical basis. American educators and scientists fully agreed with his reports, which were considered extremely competent presentations; their critical but constructive nature was much appreciated.

In spite of his passion and devotion to his own research and the time and efforts spent for his students and collaborators, Haber maintained a genuine interest in other fields of science. He was particularly attracted to biology. In addition, he continued to spend much time on literature, art, and philosophy. His strong constitution and willpower permitted him to have long working days. He liked to have friends, and his genuine kindness, warmth, and magnanimity, his wit and humor, his cheerful and outgoing character attracted many people. One of his closest and earliest friends in Karlsruhe was August Marx, a teacher of Greek and Latin in a gymnasium in Karlsruhe. He was the first Jew to become a professor at that school. Together they went sightseeing and mountain climbing, and they frequently dined out. Other friends were a group of scientists, writers, artists, and educators. They liked to dine together and formed a group of friends who called themselves a *Tischgesellschaft*. Their meetings were devoted to storytelling, serious discussions, humor, Sunday excursions to the Black Forest, and so on. Haber was famous for his stories, and he loved to express himself in verses. Once he took part in a play as an actor. With his strong personality and wit, his *joie de vivre*, his buoyancy, his broad knowledge in science, art, and humanities he soon became the dominant figure in the group. Willstätter, a close friend and greater admirer of Haber, recalls one of Haber's stories in his book *Aus meinem Leben*: On a very hot summer day Haber was hiking in the mountains. He became very thirsty. Discovering a well, he immersed his head in the water. He did not notice the ox approaching the well from another side. When they withdrew their heads, they found them interchanged. That was in 1906, and Haber has just been offered the position of *Ordinarius*. Now he said he felt ready for this position.

In 1902 Haber married Dr. Clara Immerwahr. They had met in Breslau, when Haber was 18 and Clara 15 years old. They became very close friends and wanted to marry, but their parents objected. It was not the custom at the time to marry before the man could support himself and his wife, and the engagement was broken. They kept in touch, although with the passage of time their interests developed in many different directions. Clara studied chemistry; she was the first woman to receive a doctorate at the University of Freiburg. Early in 1901 they met at a scientific meeting. He was now 33, and she was 30; they married in the summer of 1901. When Haber left for the United States she could not go with him since she was pregnant. Their son, Hermann, was born in 1902. The marriage was not a happy one. Clara wanted to continue her career, which became almost impossible when the child was born. She disliked being a housewife. Haber was not a good husband. He was completely absorbed in his work and his many other interests; he was very absent minded and neglected her. The marriage slowly disintegrated.

The Period in the Kaiser Wilhelm Institute

Although Haber's institute in Dahlem did not open until 1912, he moved to Berlin at the end of 1911 and devoted much time to the planning and the layout of the building. His aim was to have eight independent divisions, one of them devoted to physiological chemistry, an indication of his passionate interest in the problems of the biological sciences. When he finally moved in, all the work and plans were soon disrupted; a year and a half later the war broke out.

World War I. Haber soon became involved in work for the war. He detested war, with its unavoidably high cost in human lives, and the destruction and tragedies inevitably associated with it. Up to the last minute he hoped the war would be avoided. The actual outbreak was a terrible shock for him. He only reluctantly signed the manifest in which many leading German artists and scientists protested against Germany's responsibility for the war. But his attitude was that in time of peace scientists should work for all of mankind, and in time of war they were obliged to put their energy and their knowledge at the service of their country. He was a great German patriot; his love of Germany was deep-rooted. He was convinced, as many Germans were, that Germany was the victim of deliberate aggression and had no other choice than to defend itself. Actually many scientists in other European countries on both sides involved in the war had a very similar attitude.

The German General Staff, like those of all the other countries involved in the war, had counted on a very short war. It soon became apparent that the war might last for years. Germany faced an almost impossible situation; ammunition, many essential raw materials, and food were in short supply. The country was almost completely cut off from outside supplies. Much to the annoyance of the military, only civilians were able to cope with the situation. Walther Rathenau was selected for the economic organization of the war effort, a job he performed with extraordinary vision and brilliance. Haber became the primary force in mobilizing the scientific and industrial resources of Germany. Nitrates, produced from ammonia, are essential not only for fertilizers but also for explosives. Through Haber, their production was accelerated. Without them, the German army would probably have been forced to surrender in 1915. It is impossible not to think in this connection of the similar role played by Chaim Weizmann in the mass production of acetone, essential for producing the kind of explosives required by the British army and navy. It is useless to speculate whether an early end to the war might have been a blessing for all of Europe, saving the millions of young people who were sacrificed on both sides.

Although Walther Nernst was also asked to help the war effort, he was devoted essentially to fundamental problems and did not contribute much to the war effort. He was moreover convinced early in the war that Germany had lost. Haber's resourcefulness and his knowledge of a wide range of industrial problems qualified him in a unique way to help his fatherland by developing substitutes for many essential materials that were in short supply.

One chapter of Haber's activities may be briefly mentioned. He directed efforts to develop chemicals that could be used for chemical warfare. Chemicals have been used in warfare since ancient times; they were used in the war between Athens and Sparta. They were used in the American Civil War. At a conference in The Hague, in 1899, many nations agreed not to use asphyxiating gases. Interestingly, the United States refused to sign the agreement. The Russians and the French had used chemical warfare, but without success. Haber believed that the use of gas might drive the enemy troops out of the trenches, permitting a mobile war that might be more quickly decided, thereby saving many lives. The story of Ypres is well known. The chlorine used led to 150,000 casualties in the British army; one-third of them died. Later, stronger gases— mustard gas and phosgene—were used by both sides. Actually, the casualties due to chemical warfare were an extremely small fraction of those due to conventional weapons. The idea of chemical warfare is repulsive, but so is the mass killing of men by any other means. Haber was despised and hated by many colleagues, particularly in France and England, for his role in this type of warfare. But the question may be raised of whether this weapon is more criminal than the wholesale bombing carried out in World War II by both sides (and in Vietnam or Cambodia in the 1960s) or the use of the atom bombs in Japan, killing 200,000 people and resulting in additional victims who suffered severely for a lifetime. Scientists have since time immemorial contributed to the development of new weapons (e.g., Archimedes, Leonardo da Vinci, and Galileo). The development of the atom bomb was supported by Einstein and Szilard in their famous letter to Roosevelt because they were afraid that Germany might develop it. Many outstanding scientists worked hard on the development of the atom bomb. They did not foresee its use for the massacre in Japan. But in any event, few will consider the scientists who worked on the production of the bomb as criminals.

In fact, it is pathetic that a man who hated war as intensely as Haber did should have aroused so much hostility because he promoted chemical warfare. For Haber all deaths were a cruelly felt experience, whether by illness, by conventional war weapons, or by chemical warfare. An explanation of this hostile reaction may be the complete ignorance of his real motivation—that is, his firm conviction that chemical weapons would lead to an end of trench warfare, permit a rapid decision by a mobile war, and save countless lives. It took several years for this dislike of Haber to die down. In fact, J. E. Coates (1939), an outstanding English chemist, rejected, in his Haber Memorial Lecture, the assertion that Haber's work was a violation of international law. A fairer and more realistic judgment was also expressed by many other experts, such as the Americans Major West and Benedict Crowell, the French chemist Turpin, and the English chemist Hartley. However, some people never forgave him. Rutherford, for example, refused to meet him at a party in 1933 when Haber was in Cambridge as a refugee, an old, sick, broken, and depressed man. One wonders how Rutherford would have reacted toward many of his closest colleagues and friends, such as Niels Bohr, who worked on the atom bomb. Some scientists even supported the use of the atom bomb against Japan, for instance James B. Conant (1893–1978), for a long time the president of Harvard University, as was recalled in his obituary in the *New York Times.*

Haber's work on chemical warfare led to a personal tragedy. His wife implored him not to participate in producing such a terrible weapon. They had several heated discussions. He refused to accept her arguments. The same day he left to go on with this work his wife committed suicide. When he learned the news Haber was deeply shaken. The tragedy left a deep scar; he never got over this frightful experience.

The defeat of Germany and the many humiliations the country suffered for years afterward were a terrible shock and a deep personal tragedy for Haber. He had devoted all his energy to the war effort, even at the expense of his health. He became ill with diabetes insipidus, a disturbance of the pituitary gland, and many thought he would not survive for long. The discrimination against German scientists in France, England, and even in the United States, the expulsion of illustrious scientists from many societies, and the odium attached to his name for his role in chemical warfare contributed to a deep depression. His name was on a list of several hundred war criminals and he had to escape to Switzerland. Haber had received the Nobel Prize for chemistry in 1918. When he and four other German Nobel laureates—Max von Laue, Max Planck, Richard Willstätter, and Johannes Stark—went to Stockholm in 1919, Frenchmen awarded the prize did not go there because they refused to be in the same company with Haber.

The Postwar Period. After a period of a year or so, Haber regained his energy and dynamism, although he never regained the buoyancy and happiness of his younger years. He knew that the strength of Germany had been built upon science and technology. It appeared to him vitally important for the future of Germany that the rebuilding of the universities and research institutes should have top priority. Because of Haber's achievements in peace and war he enjoyed a tremendous prestige with the government, in industry and agriculture, and among scientists at the universities. Thus he was in a good position to use his influence to achieve his aims. He emphasized that Germany was unable to supply food for more than half the population. The rest then depended of necessity on industry. For maintaining its superiority industry required intensive scientific research of high excellence and it was thus essential for the survival of the country to go on with the support of research. But the rebuilding of science was not an easy task in view of the political turmoil and the social, economic, and financial upheavals in the postwar years. He and Nernst, together with Schmidt-Ott, were instrumental in creating the *Notgemeinschaft der Deutschen Wissenschaft* ("Emergency Association of German Science"). Many special committees were formed to help research proceed with maximum efficiency. Contributions were obtained not only from the government and industry, but even from big industrial concerns in foreign countries, such as the General Electric Company in the United States. The *Notgemeinschaft* played a decisive role in the rebuilding of German science.

In 1919 Haber started with the reorganization of his institute in Dahlem. He was one of the pioneers to foster a multidisciplinary approach to research. The era in which a scientist performed his research work in his own field, alone or with the help of his students and assistants, was coming to an end. Haber recognized that the rapidly expanding knowledge in many fields, the development of highly refined techniques and methods, would increasingly require close cooperation between different disciplines —physics, chemistry, physicochemistry, and physiological chemistry. Such cooperation would increase the potential and the efficiency of research in many areas. He therefore envisaged eight separate departments in his institute with more or less completely independent heads. Physics was led by James Franck, colloid chemistry by Herbert Freundlich. Among the other departments planned were theoretical physics and pharmacology. For several months Albert Einstein spent a few hours every day in the institute acting as adviser in theoretical physics.

But Germany's increasing financial troubles, the inflation, and the reparations imposed made systematic planning and real development difficult, if not impossible. The Allies had imposed on the exhausted German economy $33 billion in reparations to be paid over a period of 120 years. Haber estimated this sum to be the equivalent of about 50,000 tons of gold. Svante Arrhenius, the famous Swedish scientist, had once spoken to Haber about the occurrence of gold in the oceans and had suggested that it might be possible to extract large amounts of gold from the sea. There were some reports about the content of gold in seawater varying in different parts of the globe. The amounts were small, but even if the estimates were too high, extracting considerable amounts of gold appeared to be possible. Many colleagues warned him not to get involved in this adventure, but they were unable to change his mind. Maybe he was influenced by his success in obtaining wealth from the air; with sufficient energy and ingenuity he hoped to extract wealth from the sea.

He started his search for gold with the help of a senior colleague, Johannes Jaenicke (1935). With the help of industry, laboratories were installed in several ships for tests to be performed in different parts of the ocean. But the project was a complete failure; the estimates of the gold content in seawater reported were orders of magnitude too high and Haber was forced to give up his plan. It was a great disappointment, but he believed that the work was at least a help to oceanography, which it was indeed. Moreover, quite a number of useful analytical techniques were developed; among them was a method by which infinitely small amounts of gold could be determined. Other useful techniques made it possible to extract bromine and magnesium from solution.

During the inflation, which ended with 1 billion marks being converted to 1 Rentenmark, all scientific activities were drastically reduced; it was extremely difficult to maintain even a limited staff. But after the stabilization of the currency, Germany's economy began slowly to recover, although a large fraction of the middle class was completely ruined. Germany was greatly helped by large loans from the United States government and by investments made by United States industry. Haber resumed his work on the plans of his institute. James Franck had left in 1921; he had accepted the chair of physics in Göttingen (see II, G). Rudolf Ladenburg became the head of the physics division; Michael Polanyi, of physical chemistry; and Herbert Freundlich, of colloid chemistry.

Scientific activity of high quality resumed with full intensity in the institute once the inflation ended. Haber's fame attracted many students and collaborators from all over the world. Since he had an appointment as professor at the university, he was entitled to train students for the Ph.D. His institute became completely international in character. About half of the scientific staff of about 50 to 60 people working in the institute came from abroad; many came from the United States, but there were also quite a few from England, France, Japan, and other countries. Before accepting students, Haber had long interviews with them to find out their intelligence, their devotion to science, their honesty. These personal impressions of the qualities of a candidate were more important to him than letters of recommendation or previous records. To the students whom he accepted he was a fatherly friend, as he had been in his younger years in Karlsruhe. He had long discussions with them that sometimes lasted until late at night, since he had no sense of time. The discussions were a great inspiration because of his many ideas and broad perspectives. The students were trained to respect facts and not to adhere to preconceived and dogmatic views. In the evaluation of their papers he was very generous; they did not always have to agree with his own views. In examinations, thinking ability and originality were more important than memory. He did not mind when a student replied that he did not know. But he was irritated and angry when a student tried to guess. Once he asked a Ph.D. candidate to describe a method for preparing iodine. After some hesitation the student said it was obtained from a tree. Haber, undisturbed, asked him to describe the tree. The student thought he had guessed correctly and described the height and the shape of the tree and the type of leaves it had. Asked where the tree grew, the student replied that it grew in India and Brazil. Finally, Haber asked when the iodine blossoms appeared, and the student answered, "In the fall." Haber put his arm around the student and told him, "Well, my friend, I will see you again in the fall when the iodine blossoms."

With his genuine kindness and understanding, Haber provided his students, not only with scientific guidance, but with help and advice in difficult situations and aided them financially on occasion. In some cases he paid doctor bills without letting them know who was responsible. He often lunched with them or even took them to his home. Sometimes when he had famous visitors for a seminar and invited them later to his home, he also invited a few young students to join them. One of these visitors was Niels Bohr. Haber knew how much it would mean to the students to meet this illustrious man. It shows his warmheartedness and his personal concern for his students. He greatly admired Bohr for the application of the quantum theory to the hydrogen atom and compared him to the great prophets of the Bible.

Haber was always anxious to assist his students in finding jobs and he continued his support to further their careers, whenever there was a possibility. His love for students and assistants was deeply appreciated. They called him affectionately "der alte Fritz" after Frederick the Great.

Among Haber' senior associates, in addition to those already mentioned, were Karl-Friedrich Bonhoeffer, who had received his Ph.D. with Nernst, and the two brothers A. and L. Farkas. In fact, in his discussions with Weizmann in 1933 (see below) about his move to Palestine, Haber considered Ladislas Farkas as a most suitable candidate to be his associate in building up his laboratory there. During the Hitler period the latter spent some time in Cambridge and then joined the Hebrew University in Jerusalem. He made brilliant contributions to the development of Israel until his tragic death in an airplane crash. Among those who visited Haber for shorter periods were J. von Neumann, E. Wigner, L. Szilard, and Henry Eyring. The personal relationships between Haber and his senior associates—Polanyi, Ladenburg, Freundlich, and Bonhoeffer—were not only cordial but very close; they were for him real friends.

Many factors contributed to making Haber's institute a superb place for training young scientists: the many scientists there of great stature, many highly qualified collaborators, frequent visitors, excellent facilities, and the friendly and relaxed atmosphere important for devotion to research work. His experience, his insight, and his knowledge in many fields, and above all his dynamic, forceful, and lively personality created an extraordinary atmosphere. He was the most dominant figure of the whole complex of the Dahlem institutes when, in the second part of the 1920s, they were at the acme of their glory. Many people in Berlin identified the Kaiser Wilhelm Institutes in Dahlem with Haber.

One of the most characteristic and central features of the scientific and intellectual life in Dahlem became the famous Haber colloquia, which took place every second Monday afternoon. These colloquia had started in 1919. Colloquia were an old tradition. Magnus had introduced them in the physics department of the university and there this tradition was continued. But most colloquia consisted in a presentation of work performed in the institute, followed by a discussion in which the whole staff participated. The Haber colloquia had an entirely different and extraordinary character. There were many famous guest speakers in different fields and many of the department heads or senior scientists of different institutes in Dahlem presented papers. Everything "from the helium atom to the flea" could be the topic. The colloquia provided an ideal opportunity for Haber to realize his aim of breaking down the barriers between physics, chemistry, physical chemistry, and the biological sciences. The presentation of a paper was followed by vigorous discussion, deliberately stimulated by Haber, in which scientists of different disciplines participated. Thus the discussions automatically stimulated interests beyond one's own field and frequently suggested ways for trying a new approach. Young people were encouraged to participate. Controversies were considered an essential element for clarifying a subject. Haber greatly disliked dogmatic views. The young people taking part in these colloquia became immunized against authoritative and dogmatic thinking. Science in all fields was in a rapid expansion. Nobody could claim to have the right answers. For the author the colloquia were one of the most striking and valuable experiences in his scientific formation. It was an illuminating experience to listen to the opposing views of so many illustrious people, a fascinating demonstration of the limitation of our knowledge, a warning to be flexible and to avoid rigidity of views. Moreover, none of these great people hesitated to admit errors. The author vividly remembers how Polanyi once wrote a long mathematical formula on the blackboard. A young postdoctoral fellow got up and objected to one of the figures. Haber took Polanyi's side and a vigorous discussion followed between him and the young fellow. Suddenly, Haber said: "Donnerwetter, Sie haben Recht." This spirit of searching for the truth, not worrying about reputation or prestige when one was wrong, was one of the most extraordinary features of the colloquia. There were many illustrious visitors who presented papers, such as Niels Bohr, Richard Willstätter, Peter Debye, E. Newton Harvey, and Selig Hecht. Many other papers were given by the leading scientists of the institutes in different fields, among them Hermann Mark, Otto Meyerhof, Otto Warburg, and Richard Goldschmidt. Warburg was perhaps the only exception to the prevailing attitude. He disliked any kind of criticism of his work, as will be discussed later (IV, C). His relationship with Haber was rather cool. Their characters were too different.

Haber would usually sit in the front row and he frequently started the discussion. In these discussions Haber's extraordinary qualities became manifest. Frequently he would get up, excusing himself and saying that he was getting old and might not have followed the presentation correctly. For his own benefit he would summarize briefly the meaning of the lecture to be sure that he had understood it. This he did in a few minutes with amazing clarity, bringing out the essential points and their importance. These comments were invaluable. Sometimes he could be very critical, but sometimes also very generous in praise, especially of young people. Haber's comments revealed his capacity of grasping the essence of the subject presented, even if it was removed from his own field, his ability to perceive the fundamental problems beyond the details, and to outline perspectives with remarkable insight. Polanyi once presented in a paper some notions that were critically received by Einstein and Haber. These objections led him to rethink the whole problem and had a great influence on reshaping his ideas.

The colloquia, by the wide scope of their topics and the exciting discussions, contributed to create a unique atmosphere and were an unforgettable experience. They became an intellectual center in the lives of scientists of various fields, outlooks, and interests for exchanging views and learning about developments in other fields. Among them were Einstein, von Laue, Hahn, and other leaders of the different institutes. Many outstanding members of the University of Berlin were frequent visitors, among them Nernst, Leonor Michaelis, and Peter Rona. But the fame of the colloquia attracted scientists from other universities and even from abroad. People traveled to Berlin just to attend a Haber colloquium. This was at a time when travel was not as easy as it is today.

It may be mentioned that W. Nernst's department of the University of Berlin also had colloquia in which an exciting atmosphere prevailed and in which vigorous discussions and controversies took place. Sitting in the front row were Max Planck, Albert Einstein, Max von Laue, Edwin Schrödinger, and occasionally visitors such as Peter Debye. Students sat in the back and for them it must have been an experience similar to that felt at the Haber colloquia. However, these colloquia were strictly limited to physics. In most seminars at universities one had no opportunity to hear opposing views. It was usually inadmissible to oppose the view of the *Geheimrat*. But even today, in countries with much more liberal attitudes, such controversies as were the rule of the Haber colloquia are a rare exception. In Anglo-Saxon countries, controversies seem to be considered "unpleasant"; A. V. Hill (1965), one of the leading scientific figures of this century in England, actually very openminded and far from being dogmatic, wrote in his memoirs, *Trails and Trials in Physiology*, that "controversies are distasteful." The Haber colloquia were a particularly great stimulus for the remarkable development of biological sciences in the Kaiser Wilhelm Institutes. Physics, chemistry, and physical chemistry are the foundation on which modern biological sciences are based; the living cell obeys the same laws. The revolutionary developments in these fields were brought to the attention of biologists, thus providing them with new insight and later with important tools for the analysis of the processes taking place in a living cell. The many unique features of the colloquia were an important factor in the progress achieved there in the 1920s. A new era was beginning in biochemistry, the field that is crucial to all others in biology, whether genetics, enzyme chemistry, immunology, cell growth, or cell differentiation. Haber's contributions to these colloquia can hardly be overestimated. He must be credited with having a direct and essential part in stimulating new ideas and developments in biology. For the author, who attended these colloquia quite frequently, many of Haber's comments, especially those referring to biological sciences, were most inspiring and he feels that they had a great influence on his scientific thinking and his research. In 1963, at the opening of the Ullman Building of Life Sciences in the Weizmann Institute in Rehovot—a project to which the Adenauer government had generously contributed considerable funds—it was decided to put up a commemorative plaque recalling Adenauer's help and giving the names of six German-Jewish scientists who had inspired the development of biological sciences, but had left Germany and had died in exile, victims of Nazi persecution. The author was asked to propose the names. He suggested Fritz Haber, Richard Willstätter, Otto Meyerhof, Carl Neuberg, Max Bergmann, and Rudolf Schoenheimer.

In the mid 1920s the Kaiser Wilhelm Institutes became one of the world's greatest scientific centers, particularly in chemistry, physical chemistry, and the biological sciences. In physics there were centers of great attraction, especially in atomic physics at the University of Berlin, with Planck, Schrödinger, von Laue, and Nernst; at the University of Göttingen, with Franck and Born and their intimate collaboration with the famous mathematicians there; at the University of Hamburg, with Otto Stern; at the University of Munich, with Sommerfeld; and at the University of Leipzig, with Werner Heisenberg. In spite of the lost war Germany had become one of the leading countries in these fields.

Haber's great scientific fame, his brilliant achievements in theoretical and applied science, his services to the country in peace and war, his impressive personality and the admiration of his colleagues, all these qualities contributed to making him one of the most powerful and influential men in postwar Germany. Government and industry, universities and research institutes frequently turned to him for advice.

Internationally, too, Haber became one of the most celebrated scientists. On a trip around the world with his second wife (see below) he participated, in 1924, at the one-hundredth anniversary celebration of the Franklin Institute in Philadelphia as one of six Nobel laureates. He gave a lecture entitled "Technical Results of the Theoretical Developments of

Chemistry," in which he explained his ideas about the relationship between pure and applied science. He praised the people of the United States for their creative genius, their effectiveness, their industriousness and ingenuity. He stressed the great potential of science for the future benefit of mankind.

He and his wife then crossed the Pacific Ocean and visited Japan. Haber had several students in the country and was the guest of many distinguished and important people. Everywhere he was received with exceptional honors. Some of the leading industrialists were anxious to get his advice on problems of developing industry. He found the laboratories well equipped, but was not impressed by the factories. He greatly admired the paintings and the sculptures. He urged closer relationships between Germany and Japan and suggested the establishment of joint institutes in Tokyo and Berlin for the exchange of scientific knowledge. A German–Japanese institute was indeed opened in Berlin in 1926 and in Tokyo in 1927.

In 1928 he attended the fiftieth anniversary celebrations of the Spanish physical and chemical societies in Madrid. He was on the board of directors of the International Union of Chemists from 1929 until 1933. In 1931 he became vice-president of the union and was elected to become president in 1934. He became an honorary member of many societies and prestigious academies in the United States, England, France, and elsewhere. In 1930 the *New York Times* called him "the clue to Germany's survival."

Haber was a fascinating speaker and a master at portraying famous scientists. An especially superb performance was his lecture on Justus von Liebig in 1928 on the occasion of the Liebig–Wöhler celebrations. Haber greatly admired Liebig. The two men had many similar qualities and attitudes. Both were outstanding in the promotion of theoretical and applied science. Liebig had always had a great interest in agriculture and had been the first to emphasize the importance of fertilizers. He was, like Haber, a brilliant speaker, an impressive personality, always full of ideas, endowed with a rich imagination, and very internationally minded. He was concerned with inspiring a young generation of scientists and used to say that young people are not receptacles to be filled, but fires to be kindled; that was exactly the same attitude that Haber had. But above all Haber praised Liebig's idealism and the greatness of his soul. "No one deserves to be counted among immortals—no matter how great his power and inspired his mind—if he lacks greatness of soul." These words on the importance of magnanimity as a criterion of the greatness of a scientist deeply impressed Willstätter (1949) as a characteristic of Haber himself.

In France, Haber once gave a lecture in French about the great French chemist Louis Berthollet (1748–1822). His presentation was received by the audience with great enthusiasm. In view of his special ability to give a lively picture of a personality and his achievements, he was in great demand and frequently invited to give lectures at celebrations of famous people.

On Haber's sixtieth birthday in 1928, the scientific community expressed its admiration for him. Many articles in numerous scientific journals were devoted to the celebration of this event. *Naturwissenschaften* (1928) published a number of commemorative articles by Richard Willstätter, James Franck, George de Hevesy, Otto Stern, Herbert Freundlich, and others. Significantly, *Biochemische Zeitschrift* dedicated a whole volume as *Festschrift für Fritz Haber*. His colleagues at the institute planted an oak tree to commemorate his birthday. A circular stone with the inscription, "In honor of Fritz Haber," was placed around the tree. At the ceremony many representatives of government officials, universities, and scientists were present. Haber celebrated his birthday by making a trip to Egypt and going on a boat up the Nile River. He was accompanied by several of his relatives and friends. Among them was his second wife, although at that time they were already divorced, his son Hermann and his daughter-in-law Margarete, *née* Stern, the daughter of the physician Dr. Rudolf Stern (see below), and Haber's oldest sister Else Freyhan, with whom he had always the closest relationship. Among the friends was Richard Willstätter.

In the two following years Haber was still active and together with Karl-Friedrich Bonhoeffer made outstanding contributions to the mechanism of autoxidative reactions. They succeeded in demonstrating by the use of spectrophotometry a number of very quickly disappearing intermediary radicals leading eventually to the end product. These processes, first studied in gas reactions, were later extended by Haber and James Franck to autoxidation in solution. These observations led to the work on biological oxidation that Haber continued in the last months of his life with his associate Joseph Weiss in Cambridge.

However, in the last few years his health began to deteriorate. In 1927 he had the first attack of angina pectoris. After that he always carried a bottle of nitroglycerine with him. He started to suffer from insomnia. In view of his drive and willpower it was not easy for him to slow down with his activities, but he was just unable to keep up with the huge amount of obligations and work to which he had been accustomed. When the Haber colloquia were moved to a new club and guest house, the Harnack House, he stopped attending them regularly. After 1929 he came quite infrequently and was not as active as he had been. The colloquia thereby lost much of their splendor; people sometimes telephoned before coming to inquire whether Haber would attend. But the Wall Street crash at the end of 1929 and the start of the economic depression greatly disturbed him and he began to worry about Germany's future. Haber was one of the few scientists who was interested in politics. He was a Democrat and was frequently consulted by members of the Democratic party, even by members of other center parties. He was friendly with Friedrich Ebert, the first president of the Weimar Republic, although he was critical about many aspects of the republic, especially about the lack of a strong and competent leadership. He realized the necessity of many social adjustments in the new era. Now the economic deterioration and the rapid rise of the Nazi movement in 1930–1931 deeply depressed him and probably affected his health adversely. Once, when he talked to some high government officials in 1931, he was appalled about their ignorance of many facts and their lack of insight into the dangerous situation.

Family Ties and Social Life

Two and a half years after the tragic end of his marriage to Clara Immerwahr, Haber married again. He met at the Deutsche Club a young, witty, alert, and intelligent lady, Charlotte Nathan, in 1917. She was less than half his age. The marriage went well for a couple of years, but then began to disintegrate. His wife belonged to another generation. She had an entirely different style of life and different interests. She loved parties, dancing, night clubs, and cabarets, activities utterly disliked by Haber. The marriage ended in divorce in 1927. They remained, however, friends. But although Haber was a poor husband and had no happy married life, he was actually a family man who had a great affection and close ties with the members of his family, his stepmother and his three stepsisters, especially the oldest, Else, who was married to a physician, Dr. Freyhan. Haber was extremely helpful financially and in other ways, to many other relatives as well, whenever there was an occasion. He was very fond of the family of his son Hermann and his wife and loved to play with his two granddaughters. Although he had lost his fortune (as most members of the German middle class did) by the inflation, he never had real financial worries. He inherited considerable fortunes soon after the inflation. He still received royalties. He had his big house on the Faraday street that belonged to the Kaiser Wilhelm Society and was rentfree. He committed serious mistakes in investing his money and suffered great losses, but he was always able to follow his genuine generosity and to help his family and his students.

Haber was also a warm and faithful friend; the close relations with his senior colleagues have been mentioned. In view of his many interests in a variety of fields he had many other friends. One of his closest friends was Richard Willstätter, whom he had first met in 1911 in Zurich. But when both moved to Dahlem and lived in two neighboring houses on the Faradayweg, a very intimate personal relationship developed between them (see also IV, B). Usually they went for a walk in the evening before retiring. But even after Willstätter moved to Munich, the close friendship remained. They spent many summer vacations together. In addition to the trip to the Nile mentioned before, they traveled to the Engadine, to Rome, to the Canary Islands and Madeira, to BadGastein, and other places, or spent their vacations on an estate in the country. Willstätter was a great admirer of Haber. On several occasions Willstätter presented superb and lively descriptions of Haber's dynamic and fascinating personality and the greatness of his scientific achievements, the wide range of his interests, his wit and humor, his theatrical sense for acting and showmanship, for expressing himself in verse (Willstätter, 1928, 1949). There are also very amusing descriptions of the summer trips. Among his close friends was his personal physician, Dr. Rudolf Stern and his wife Katherine, the daughter of a professor of pathology in Breslau. Stern had met Haber in Pontresina, in 1913, just after he had passed his *Abiturientenexamen*. Haber was surrounded by eminent professors of science, medicine, and law. Stern was amazed by the wide scope of different fields in which Haber appeared to be competent. Stern met Haber again in 1921, when Haber was in a deep depression in a sanitarium. By that time he was already a physician and he tried to be as helpful as possible. Haber encouraged him to spend some time in scientific research; this would be most valuable for his medical future. He arranged for Stern to spend two years in his institute in Dahlem with Herbert Freundlich. From that time on an increasingly intimate friendship developed, and Stern, who greatly admired Haber, made every effort to keep an eye on Haber's health. They frequently spent vacations together and went on several trips; these are described in an article in a Year Book of the Leo Baeck Institute (Stern, 1963), in which Stern gives a vivid picture of many facets of Haber's personality, his character, his warmth, and his charm. Some revealing details are given about his political views, his remarkable intelligence and flexibility in discussing the political and economic problems facing Germany.

Haber liked to entertain in his home, not only colleagues and friends, but also writers, artists, diplomats, and others. Although he was not a great music lover, he was a good friend of Siegfried Ochs, the founder and conductor of the Berlin Philharmonic Orchestra. He was a most charming host, a famous storyteller and a witty and charming raconteur, describing his many adventures frequently extended by his imagination and fantasy, not taking too seriously the distinction between *Dichtung und Wahrheit*, as was easily realized.

Jewish Attitude

Haber had been brought up in a family without any connection or ties to Judaism. His Jewish descent was thus completely meaningless to him. His father Siegfried had received his education in a gymnasium. The name Siegfried was frequently chosen by assimilated Jews who wanted to identify their children with Germany and German history. Haber's stepmother too came from a completely assimilated home; she had many Jewish and Christian friends. The attitude was in line with the tendency of a large sector of German Jewry to become completely assimilated, especially in the second part of the nineteenth century. Thus it is not surprising that Haber had no ties to his Jewish heritage. Like many educated people, he knew the Bible well, but he did not associate himself with the people whose ancestors had written it thousands of years before.

In his early twenties he became baptized as a Protestant. He was not motivated by the hope of facilitating his career. He was influenced by Mommsen's writings and attacks against Treitschke's anti-Semitism. Mommsen accepted that those Jews who were Jews out of conviction should remain so. But he pleaded with the Jews who were Jews just by convention, without any real ties to the Jewish community, to consider the danger of a dualistic life. Since Christianity links the different civilizations and societies of Europe together, the best solution seemed to him a mass conversion of Jews for secular reasons. It was not out of any anti-Semitic feelings that he gave this advice, but out of concern that the Jews avoid tragedy. These views of Mommsen have been well described by Hans Liebeschütz (1962) in his article, "Treitschke and Mommsen on Jewry and Judaism."

It was under the influence of Mommsen's writings that Haber decided to accept baptism. He took this step at a period when he did not think of an academic career. It was for him just a formality without any significance. His rapid climb was not based on this step. It appears unlikely that it facilitated his career in a period of rapid expansion of German science during which the enlightened leaders looked for people of high quality and talents. The story of the Kaiser Wilhelm Society mentioned before documents this situation. It is interesting, however, that his baptism became for him a problematic act only in 1933. Apparently, according to Stern (1963), he was more tormented by the idea of having taken this step than by the loss of property and position. It seems that he associated this action in retrospect with a lack of dignity.

It is noteworthy that Haber's closest friends were nevertheless mostly Jews: August Marx in Karlsruhe, his associates in Dahlem (Franck, Ladenburg, Polanyi, Freundlich). An interesting story shows how he instinctively still identified himself with Jews. During his visit to the United States in 1902 a Jewish friend took him to Atlantic City. In the hotel he noticed a sign: "No Jews." Haber was shocked; it was for him a new and unexpected experience. He felt that they could not stay there. But his friend assured him it was not meant for their type of Jews; they could ignore the sign.

Even after the war Haber ignored the rising wave of anti-Semitism. Like most Jews, he considered these forces as a temporary phenomenon, bound to disappear with the improvement of the economic and political situation and the continuously increasing assimilation of the Jews. Were

Jews not in prominent positions in many fields in greater numbers than ever before, in far higher proportion than their percentage of the population? He believed that the Zionist movement had no legitimate function in Germany. It might provide a solution for Jews in some eastern states, but not in such a highly developed country as Germany. There it might have only adverse effects. He disapproved of Einstein's trip with Chaim Weizmann to the United States in 1921 in support of the Zionist movement. An interesting story is reported by Stern (1963). After a discussion with a young Zionist about the many problems facing Palestine, in 1921, Haber said, "If there are enough Jewish people with his enthusiasm, Palestine will have a great future."

But the rise of the violent and increasingly terrorist anti-Semitism and the victories of the Nazis in the late 1920s and early 1930s had a strong effect on Haber and gradually changed his mind. In contrast to many Jews he began to realize that these strong irrational forces were a reality and could not be ignored or fought by reason. It should perhaps be recalled that the realization and the analysis of the power of emotions and irrational forces have greatly developed in this century, starting perhaps with Freud's emphasis on the unconscious. We are much more aware that even in advanced democracies irrational elements and emotional factors may play a decisive role and readily prevail over reason, no matter how much in the end the whole population suffers. The Nazis shrewdly played the bitterness and hatred felt because of Germany's humiliation and the disastrous conditions created by the Versailles treaty against latent anti-Semitism. They associated the Jews with the defeat in the war and the weak resistance of the "Jewish dominated" Weimar Republic to the harsh conditions imposed. They thereby succeeded in inflaming large sections of the population to the extreme form of anti-Semitism.

Haber saw the signs on the wall more clearly and earlier than most of his scientific colleagues and friends. He identified himself increasingly with the Jews and his whole attitude toward the Jewish problem began to change. In 1921 he had opposed Einstein's trip with Weizmann to the United States. When Weizmann visited Berlin in 1932, Haber invited him for lunch and showed him his institute. Weizmann, who had not forgotten Haber's attitude in 1921, was surprised to find Haber now greatly interested in his work in Palestine and frequently discussing scientific problems that might be of interest in the development of the country.

Disaster struck soon. Only a few weeks after Hitler became chancellor, the civil service laws appeared, driving most of the illustrious Jewish scientists out of Germany, as described in Chapter II. In April Polanyi and Freundlich voluntarily submitted their resignations although they could have kept their positions. Haber tried to restore the situation, but without success. According to the civil service laws, those Jews who were active in World War I could keep their positions. Haber was therefore entitled to continue as director. He indignantly refused and submitted his letter of resignation in April 1933, although he was willing to continue his duties for a few months in order to arrange an orderly transition to a new directorship. In his letter he stressed that although the laws entitled him to stay in his office in spite of being a descendant of Jewish parents and grandparents, he could not accept that the selection of collaborators should be based on racial origin and not on scientific and personal qualities, the traditional principle he had followed all his life. He had served his country for his whole life, but his pride did not permit him to accept the new attitude. He left Germany in the early summer.

He arranged a meeting with Chaim Weizmann in London in the summer of 1933 through his son Hermann, who was employed in an enterprise of Joseph Blumenfeld, an industrial chemist in Paris and a brother-in-law of Weizmann. It is hard not to see the historic perspective, the dramatic symbolism of this encounter of two giants of the Jewish people. Haber had been almost an uncrowned king in German science by his glory, his great achievements, his power and influence. Weizmann was in the eyes of mankind the incarnation of the Jewish dreams and aspirations for a Jewish homeland, what is today Israel. Both were representatives of two opposing views. Weizmann, the Russian Jew, now a British subject, was devoting his life to save Jewry of central and eastern Europe from the disaster he saw coming, although he could not know how great the disaster would be. Haber's life was devoted to the service of his country. Assimilation was for him a natural and unavoidable process. But he now fully recognized that Weizmann had been right in recognizing the reality and he had been wrong, and he offered his services to Weizmann, who was already contemplating a chemical research institute in Rehovot. He hoped to be able to offer opportunities to at least some of the German refugee scientists. He was deeply shocked to see Haber as a refugee in exile, almost penniless. Sir William J. Pope, a former associate during the Karlsruhe days, had offered him space in his laboratory in Cambridge. But that was not really a solution. It was decided that Haber would take some rest and that they would meet again later in the summer in Switzerland.

Weizmann spent the summer in Zermatt. He refused to attend the Zionist Congress in Prague in view of political disagreements. After a vacation in Switzerland and on his return from Spain, where he had given a lecture in Santander, Haber visited Weizmann against the advice of his physicians. In view of his serious heart condition, Zermatt (about 5000 feet high) was at a dangerous altitude. Haber implored Weizmann to go to Prague. He reminded him of his own contributions to Germany, of his powerful position there; now at the end of his life he was bankrupt and in exile. Weizmann's work would stand as a shining monument in the long history of the Jewish people. Even if he would have disappointments in Prague, he should not abandon his efforts. Weizmann was appalled and moved by this encounter. He most generously offered Haber a laboratory and able assistants in Rehovot to work in peace and dignity.

It would be a return home. Haber accepted the offer with enthusiasm. When Weizmann inquired about the personal accommodations he would need, Haber asked for two furnished rooms, one for his sister Else Freyhan, and one for himself.

In view of his poor health he asked for a delay of a few months to rest in a sanitarium. On his return trip he suffered a heart attack in Brigue, but recovered after a few days. He had originally planned to return to Germany to arrange his emigration in an orderly fashion. On the advice of his friends he abandoned the trip, primarily because his health was so poor that he could not expose himself to the agony that such a trip would hold. He went to Cambridge for a short time to complete some research. He started his trip to Israel early in 1934. He wanted to spend a couple of weeks in the Tessin to rest. When he arrived in Basel at the end of January 1934, he died there a few days later. His friend Willstätter came to the memorial service and gave a stirring farewell speech.

An interesting collection of letters from Haber to Weizmann during that period is reproduced in the Year Book following the article by Stern (1963). In one of his letters Haber mentioned a discussion he had with his friends Einstein and James Franck. He wrote that Franck was a great admirer of Weizmann. Should Franck's son-in-law go to Istanbul, he would visit him there and might go to Jerusalem from there for a visit.

It may be mentioned that Max Planck, on his return from the south, was shocked by what had happened. He turned first to Bernhard Rust, the Minister of arts and sciences, who told him that he was through with the Jew Haber ("Der Jude Haber ist für mich erledigt"). Planck's interview with Hitler has been described. In the spring of 1936 Planck had the courage to praise Haber publicly for his great contributions to Germany in peace and in war.

In the eulogies after Haber's death, particularly in the Anglo-Saxon countries, in spite of former feelings of hostility, Haber's personality and achievements were highly praised. His tragic end in exile was considered a disgrace and noted with general disapproval. The Kaiser Wilhelm Society, as well as other scientific societies, decided to commemorate the first anniversary of his death on January 29, 1935. In spite of Rust's vigorous protests and warnings, more than 500 scientists, students, industrialists, and educators gathered in Dahlem to pay tribute to Haber. Max Planck opened the ceremony by recalling Haber's great contributions. One of several other speakers on the program was Karl-Friedrich Bonhoeffer. The Nazis prevented him from attending the meeting and his speech was read by Otto Hahn. A service was held in the Daniel Sieff Institute in Rehovot, to which Haber had given his library. Weizmann and two former students of Haber, Ernst David Bergmann and L. Farkas, spoke. Weismann kept two photographs on the desk of his office in the institute: one was of Haber and the other was of Willstätter.

It was not until 1952, several years after the end of the Nazi regime, on the occasion of the fortieth anniversary of the foundation of the Kaiser Wilhelm Society, by then the Max Planck Society, that a memorial was erected, inscribed with the words Max von Laue had written for *Naturwissenschaften* in 1934:

> Thermistocles
> entered history not as the exile at the court of the Persian King, but as the victor of Salamis.
> Haber
> will live in history as the brilliant discoverer of that method of combining nitrogen with hydrogen which is the basis of the technical fixation of atmospheric nitrogen, as the man who created, as was stated on his receipt of the Nobel Prize, "an exceedingly important means of advancing agriculture and the welfare of mankind," who obtained bread from the air and won a triumph in the service of his country and mankind.

B. RICHARD WILLSTÄTTER (1872–1942)

Richard Willstätter was one of the giants in organic chemistry who, at the turn of the century, had laid the foundation for the spectacular developments of biochemistry in this century. He belongs to the extraordinary and ingenious group of Adolf von Baeyer's pupils, who paved the way for the new era in biological sciences by providing the knowledge of the structure of the most essential components of animal and plant cells, a prerequisite for the analysis and understanding of chemical cell reactions.

One of his most brilliant contributions was his work on the structure of chlorophyll; it is one of the classic works in organic chemistry. His many achievements in solving this problem were summarized in collaboration with his associate Arthur Stoll in *Investigations on Chlorophyll* (Willstätter and Stoll, 1913). These contributions are a milestone in the exploration of photosynthesis. For this work and for his investigations on the structure of the pigments of flowers and fruits, the anthocyanins, he received the Nobel Prize in 1915. His lifework was of great importance not only for biochemistry, but also for medicine and technology.

However, as he writes in his autobiography (Willstätter, 1949), two factors dominated his life: his chemical observations and the anti-Semitic movements, turning later to a *Sturmflut* ("storm flood"). His deep and genuine attachment to Germany, combined with a great awareness of his Jewishness, led to deep tragedy and dilemma, perhaps more stirring than in many other cases. Because of the combination of the two features, being a brilliant scientist and caught by anti-Semitism in a most dramatic, pathetic and, in some respects, unique way, this chapter on Willstätter forms a particularly interesting part of this book.

Ancestry

It is characteristic of the strength of his feelings of being a Jew that he starts his autobiography with the statement, "My ancestors were Jews." The family had lived since 1720 in Karlsruhe and later in some villages in the vicinity. The great-grandfather of his grandfather was Rabbi Ephraim, who immigrated from Willstätt, an old fortress in Alsace, and accepted the name Willstätter. At that time only nine Jewish families lived in Karlsruhe. Two years later they received privileged positions. Willstätter adds the interesting remark that his ancestors may just as well have come to Germany with the Roman legionnaires at the time of the Roman emperors, thereby apparently implying that he was just as legitimately a German as anyone. Among his ancestors were rabbis, teachers, and one physician. Efforts for material gains were rare in the family; modesty and altruism were marked.

The son of Ephraim, Rabbi Elias Willstätter, died in Karlsruhe in 1795. He left five sons and five daughters. His eldest son Ephraim (1761–1817), Richard's great-grandfather, wrote some notes about his oldest son, presenting a lively picture of the family and the prevailing atmosphere. His father never talked about his family; memories seemed to him of no importance. Ephraim was a teacher for 15 years in Frankfort on the Main, but had to return to Karlsruhe to take up business in order to take care of his large family. Jews at that time had to combine learning and teaching with business activity to earn a living, since academic careers were virtually closed to them.

Richard's grandfather was a child from the second marriage of Ephraim, with the daughter of Rabbi Juda Oppenheimer. From this second marriage there were many children; eight of them survived the grandfather. He devoted all his efforts and money to the education of his children. Three of Ephraim's sons were scientifically trained. At that period there was a liberal tendency in Baden, and the so-called *Judenedict*, in 1809, recognized the Jewish confession as "constitutionally tolerated." But it took another 53 years for a "law about equal civil rights of Israelites" to be pronounced.

Richard's grandfather Meier, called Maximilian, was born in 1810 in Karlsruhe. He attended the gymnasium and received many distinctions there. He studied medicine in Heidelberg and, after another year of training in Munich, practiced in Karlruhe. He married Johanna Paula from Darmstadt, the daughter of manufacturer Michael Kaula and his wife Esther. Her parents belonged to a family well known in Jewish circles in southern Germany. The grandfather was modest, always ready to help, a devoted physican who never took vacations or relaxed. He did not write bills and life was not easy for the family, especially when, in 1847, a serious economic crisis occurred. Because of his strenuous life he aged prematurely and both grandparents died before Richard's birth. The three brothers of the grandfather were decent and dignified people, although typical representatives of the lower middle class. The youngest brother, Benjamin, became grandrabbi (the chief rabbi) in Baden. He had a very progressive and liberal attitude, and was a brilliant lecturer.

Richard's father Max (1840–1912) was the oldest child of five. After an apprenticeship in the textile business in Karlsruhe he spent some years in England, but had to return to Karlsruhe to take over the family business. In 1870 he married Sophie, daughter of Moritz Ullmann, a wealthy textile businessman in Fürth. She was nine years younger than he. The Ullmanns' ancestors had been businessmen in Augsburg for centuries, although they were not permitted to live in the city. The records go back to Simeon Günzburg ben Ullmo, one of the richest Jews of his time, who died in Pfersee on January 9, 1585. Richard's great-grandfather Beer (called Bernhard) Ullmann (1751–1837), whose house still existed in Pfersee at the time the autobiography was written, had studied at the University of Prague and was a respected banker. To attend to his business, which was more modest than that of his ancestors, he still had to go to Augsburg. Since his first marriage remained without children, his wife persuaded him to remarry and helped him in many ways. In the second marriage he had six children. They all lived very long and had big families. Richard knew three of the sons quite well. One of them was his grandfather Moritz Ullmann (1814–1890). He was trained as a banker, but founded a big textile business in Fürth (Bavaria). He married Doris Kleefled, the daughter of a prominent businessman whose family also could be traced back to the seventeenth century. Richard's mother Sophie was the oldest daughter, born in 1849. She had six brothers and two sisters. She was outstanding among all the other children by her intelligence, her broad education, her warmheartedness, and her ambitions.

Although Willstätter devoted no less than 14 printed pages of his autobiography to the history of his ancestors, these few excerpts may suffice to illustrate that he was descended from old and prominent Jewish families, among them many rabbis, scholars, and respected and successful businessmen. His extensive report of the long line of ancestors who lived for centuries in Germany and played a productive role in the country in many ways may have been unconsciously motivated by his desire to emphasize his legitimate right to be considered a German citizen, as much as anybody else, a right denied by the Nazis. The fact that his ancestors were Jews, which he does not want to ignore, is considered simply a matter of history, an aspect that has relatively little importance. The problem will be discussed in the context of his Jewish attitude and his reactions to anti-Semitism throughout his life.

Willstätter spent the first 10 years of his life in Karlsruhe and it was a happy time for him. The city was beautiful, with many parks, open-air spaces and playgrounds; he frequently walked with his mother in the lovely park of the castle. When he started to go to school, at the age of six, he had his first encounter with deep-seated anti-Semitic feelings, and Karlsruhe was a relatively liberal place. When he walked alone in the streets, other young boys threw stones after him and shouted insulting remarks about the *Judenjunge* ("Jewish boy," in a derogatory sense).

When he was 10 years old his father decided to go to the United States for a few years in order to accumulate enough money to ensure a comfortable life for his family. The business in Karlsruhe did not go well. Moreover, he was actually bored by the way the business was run. Several relatives had gone to the United States and had been highly successful in their businesses. Instead of 2 to 3 years, as was originally planned, he was gone for 17 years. When he left, in 1883, his wife moved the family to Nuremberg, where her two married sisters lived. For Richard this move was like an expulsion from paradise. He disliked the narrow streets; there were no trees, no parks, no playgrounds, no opportunity and time to be in the fresh air. The living conditions were much less comfortable. He was much too young to appreciate the beauty of the city's monuments and churches, as he realized only much later. However, the worst was the much stronger anti-Semitism that prevailed in the population, particularly in his school. Most of the pupils there came from the lower middle class, usually much more prone to anti-Semitism than the sons of public officials and army officers who were his schoolmates in Karlsruhe. He suffered considerably from this anti-Semitic atmosphere and it appears likely that some of his later reactions were based on these unhappy experiences of his youth. At home the family had a liberal attitude toward religion. But to break the ties to Judaism was inconceivable. The attachment to Judaism, to the ancestors, was strong; one remained a Jew as a matter of course; baptism was unthinkable.

Willstätter's mother, an extremely strong person with great willpower, had a great influence on Richard. She seems to have been overprotective and worried that he would not take his work seriously enough. She watched all his activities carefully, not only during his school years but in his student years, in fact up to his marriage, and even up to the time he received the Nobel Prize. An illustration of her influence is his report that, although an avid reader of literature, he did not read the work of certain authors, such as Zola, Tolstoi and Dostoevski, when he was a student in Munich because his mother objected!

In school he did not do too well in Latin, whereas he was excellent in mathematics and geology. But in view of his relative weakness in Latin, the family decided that he should attend a *Realgymnasium*. Early in his school years he became attracted to science. He does not remember what influenced his inclinations, since science was never discussed in the family. He did exceedingly well in school and got highest marks. His schoolteachers were outstanding and inspiring, many of them real scholars. Some of them later turned to academic careers. However, the atmosphere was somber and serious; cheerfulness, joy, careless hours of pleasure or relaxation did not exist. Everything was concentrated on work and studies. Willstätter was liked by his teachers and his performance became increasingly excellent. This superiority created jealousy, unfriendliness, and even hatred among his schoolmates. The reaction to superiority in sports, he comments, is everywhere admiration and acclaim; but intellectual achievements, in science or humanities, are usually disliked, although they contribute so much more to the well-being of the society than does physical performance.

In the middle of his school years he contracted diphtheria and was so severely ill that for some time there seemed no hope for his survival. There were many complications, including effects on the central nervous system. It took more than half a year for him to recover. Soon afterward he caught scarlet fever, complicated by a kidney infection; again he took several months to recuperate. But he fully recovered from both diseases, and they had no influence on his progress in school.

He disliked social life and drinking. For a short while he participated in such activities under the influence of some relatives, but was so antagonized and bored that he tried to avoid them as much as possible. He felt most happy when he was alone. Shortly before leaving school, he was permitted to go alone on a trip. This was for him a most happy and wonderful experience. He particularly enjoyed the beauty of the Bavarian mountains.

Of interest in his description of the *Abiturientenexamen* is the topic he was given for discussion in the *Religionsunterricht* (religious instruction); Jewish pupils had a separate education in the field of religion. The question was, "What does Judaism teach about the immortality of the soul?" He wrote in his paper that since the Jewish religion is not a faith, but a law, one can say very little that is definitive about this topic. His general performance in this examination was so outstanding that the rector of the gymnasium proposed him for the Royal Maximilianeum, an institute in Munich for particularly gifted students. He was turned down because he was a Jew.

Academic Career

After the *Abiturientenexamen* the question arose as to his future. He had an excellent offer from a cousin to enter a prosperous textile business. He flatly rejected the offer. He wanted to study physiology and medicine. However, because this was a profession that would take many years of studies and training, the family concluded that the expenses involved would be too high for the modest means available. He therefore decided to study chemistry. But he always maintained a great interest in biology, and many of the problems on which he worked were motivated by the desire to contribute to biology and medicine by exploring the chemical structure of substances important for these fields.

He started at the *Technische Hochschule* in Munich, but was dissatisfied with the standards and pressures, which did not permit him to acquaint himself with the field on a level acceptable to him. He therefore transferred to the University of Munich, where Adolf von Baeyer held the chair of chemistry. After having passed the preliminary examinations, he wanted very much to work on his Ph.D. under the direct guidance of Baeyer. But this was hopeless, since Baeyer had at that time too many students. So, under Alfred Einhorn, he started his thesis on the structure of cocaine and related compounds, a problem on which he continued to work after he received his Ph.D. in 1894. However, during a short presentation on some aspects of his work, prior to obtaining his degree, Baeyer was present and was greatly impressed by the work as well as by its presentation. Soon a personal relationship developed between the famous teacher, a world-wide recognized and celebrated authority, and the young and inexperienced student. This relationship became increasingly close and intimate and permitted free and uninhibited discussion. Finally, a warm friendship developed that lasted through life, although there were occasional strains.

After the promotion Baeyer accepted him as a private assistant, thus opening to Willstätter the possibility of an academic career. He was even permitted, an exceptional favor, to continue his own work on cocaine, since Baeyer realized the great importance of the work and Willstätter's outstanding abilities, originality, and critical judgment. To the surprise of Willstätter, Baeyer proposed him for promotion to *Privatdozent* in 1896, when the position became available because of some changes in the staff. This was a great distinction, since he was only 24 years old, and a decisive step for his academic career. Willstätter was jubilant. He went in high spirits on a vacation trip and subsequently made his first trip to the United States to visit his relatives. He had an opportunity to see New York, Baltimore, Philadelphia, and Washington. On his return he started immediately with great eagerness, energy, and enthusiasm to continue his work, which meant everything to him and to which he devoted all his time and efforts.

Unfortunately, the facilities at his disposal were not improved by his promotion. He had to carry out his investigations at the same bench he had had before in the student laboratory, a cluttered, noisy, crowded room that lacked many adequate installations. On the other hand, he now had, in addition to his own research, a rather heavy teaching load. He took teaching most seriously and considered it a great and important duty of every scientist. Transmission of knowledge was for him just as important as the contribution of new facts and insight. Thus his workload was greaty increased and he had less leisure time for his research problems without the compensation of better research conditions. On the other hand, he had a much greater opportunity to meet colleagues and interesting scholars from all faculties, since every *Privatdozent* automatically became a member of the association of *Privatdozenten*. Whereas before he was completely isolated, his membership in this association offered him the chance to meet colleagues in a great variety of fields outside chemistry, such as physiology, zoology, geology, mathematics, and history. This was a great stimulus for his general development.

Only three years later he seems to have been seriously considered for promotion to *Extraordinarius*, since Baeyer was increasingly impressed by the brilliance and rapid progress of his research. Instead, apparently at the last moment and quite unexpectedly, Baeyer's son-in-law, Piloty, a pupil of Emil Fischer, was offered the position. One day Baeyer told Willstätter that he should get baptized. Willstätter flatly and abruptly rejected this suggestion, perhaps in a slightly impolite manner. Baptism to facilitate his career—that is, out of opportunism—was for him inconceivable. This attitude will be discussed later. When he was offered a highly paid position as a director in an industrial enterprise, he turned the offer down and showed a copy of the letter to Baeyer, who was a little unhappy about this rejection. He was worried that Willstätter as a non-baptized Jew would have great trouble with an academic career. Willstätter assured him that he was not at all interested in an academic career; all he wanted was to be able to continue his research.

Nevertheless, Johannes Thiele, who had been *Extraordinarius* since 1893 and was a forceful and enthusiastic teacher who had greatly improved the teaching in organic chemistry, was offered a position as *Ordinarius* of the chemistry department at the University of Strasbourg. He accepted the offer and in 1902 Willstätter, then only 30 years old, became his successor and was appointed *Extraordinarius*. His situation and his possibilities changed quite considerably, far beyond his expectation. He had a private laboratory with adequate space, facilities, and technical help for his own research. He was able to work at any time, in the evening hours, on Saturdays and Sundays, completely undisturbed. This was for him invaluable and he was very happy about this great improvement. He was able to accept more students for work on their doctoral theses and also postdoctoral and visiting fellows who wished to do research and to collaborate with him. Soon he had quite a number of excellent co-workers.

One of the foreign scientists who came to work with Willstätter in Munich was Ernest Fourneau, later one of the leading chemists in France and for many decades director at the Pasteur Institute. French chemistry was at that period no longer at the high level at which it had been for much of the nineteenth century. Fourneau had already worked in three German laboratories, one of them that of Emil Fischer, before he joined Willstätter. The collaboraton went very well and soon a friendship developed between them. They took many walks and excursions together, since there was little time in the laboratory for personal discussions and meetings. Fourneau learned much from Willstätter about chemistry, and in return he taught the latter a great deal about France and particularly Paris, about scientists such as Pasteur and Berthelot, about French culture, literature, and art. He even tried to teach Willstätter French, but had to give up, since Willstätter had too much difficulty learning languages. Willstätter found the strong nationalistic feelings of his French friend surprising. Fourneau greatly suffered because France had fallen behind Germany in chemistry. This kind of thinking never occurred to Willstätter. When he gave a lecture in Paris shortly before the outbreak of World War I, he greatly enjoyed Fourneau's hospitality. He had great difficulty understanding the many hostile anti-German remarks Fourneau uttered in the postwar period.

At the turn of the century Baeyer's institute was a great center, famous all over the world. There were many distinguished visitors. Because of his great fame Baeyer was usually treated not as a colleague but as a prince. Chemists from many countries came to work there. Among the scientists who worked there in Willstätter's time were Vladimir Ipatieff and Moses Gomberg, both from Russia; they later moved to the United States and became famous and successful there. Willstätter met both of them again in the fall of 1933 when he was a guest of the American Chemical Society. The opportunity to meet many interesting people proved quite valuable to Willstätter, sometimes more in a cultural respect than in chemistry. After his promotion to *Extraordinarius*, he began to attend scientific meetings and congresses more frequently and he participated in them quite actively. His research advanced extremely well and his reputation grew rapidly; Baeyer, on several occasions, emphasized the brilliance and importance of Willstätter's achievements.

Only three years after his promotion to *Extraordinarius*, Willstätter became, in 1905, *Ordinarius* of chemistry at the *Eidgenössische Technische Hochschule* in Zurich, one of the great scientific centers in Europe with a glorious past, particularly in chemistry, but also in other fields, and still today a leading institute. Two of the former heads of chemistry were Nobel laureates. The institute is supported by the federal government, in contrast to Swiss universities, and thus always has had the necessary means to provide for the large expenses research requires. The differences between working conditions in Zurich and in Munich were striking. The chemical institute was relatively modern; it had large laboratories with good lights, big halls, and modern facilities. It was one of the first institutes in which the temperature was regulated. The private laboratory was very large, with space for three to four private assistants. It was not a difficult decision for Willstätter. Here he would have the opportunity to proceed freely with his research and to build up a large and excellent school.

At the farewell party Baeyer gave a warm and moving speech and the students and collaborators expressed their feelings toward the beloved and admired teacher in many ways. A telegram arrived announcing that Switzerland had renamed the *Vierwaldstättersee* the *Vierwillstättersee*. Several of his best collaborators went with him to Zurich. It was there that his creative work reached its zenith. Although he had developed the main concept of the chlorophyll structure before leaving Munich, it was in Zurich that the experimental work was performed, as well as the work on some flower pigments to be discussed later. Among the large group of collaborators who were trained there he mentions with great pride Professor Michael Heidelberger, the founder of modern immunochemistry. Heidelberger spent decades at Columbia University until his retirement in 1957, although he is still fully active, at the age of 90, at New York University. In 1977 he and the first student to get his doctoral degree with Heidelberger, Professor Elvin A. Kabat of Columbia University, received the highest award of the Columbia University Medical Center, the Louisa Gross Horwitz Prize, a great personal satisfaction for the author, their colleague at Columbia University since 1942 and a great admirer of their achievements.

In the summer of 1911, in the midst of his very productive activities in Zurich, Willstätter received an offer to join the Kaiser Wilhelm Institutes in Berlin–Dahlem as head of the organic chemistry division in the chemistry building. This building and that of physical chemistry, of which Fritz Haber would be the director, were the first two to be ready and opened in the fall of 1912. Willstätter turned the offer down. He was happy where he was and did not want to move. Emil Fischer came to Zurich to persuade him to reconsider. Willstätter was a great admirer of Fischer; he considered Fischer the most brilliant and most outstanding organic chemist of the whole era. Fischer's vision of the plans to build one of the greatest scientific centers of the world in Dahlem, which he described in glowing terms, made a deep impression on Willstätter. Fischer succeeded in convincing him to change his mind. He gave Willstätter all kinds of assurances concerning his possibilities there, of his complete freedom and independence. A decisive factor in finally accepting the offer was his desire to return to his beloved Germany, to his *Heimat* (native country). He moved to Berlin–Dahlem in 1912.

On a short visit to Berlin he was informed at the ministry that in addition to his position in Dahlem he would become the successor of van't Hoff with the rank and title of *Ordinarius* at the University of Berlin.

The inauguration of the two institutes in the fall of 1912, the magnificent ceremonies in the presence of the kaiser, many ministers, leading scientists and industrialists, have been described earlier. Willstätter was particularly happy that President von Harnack stressed in his speech the paramount importance of adequate space, excellent facilities, and the best possible equipment for the future of chemistry, which now played a great role in the life of the society. He also greatly enjoyed Paul Ehrlich's lecture, given in the presence of the kaiser, on chemotherapy, since he had personal and scientific contacts with Ehrlich. Willstätter even managed to have the kaiser look at his beautiful crystals of chlorophyll, of which he was extremely proud. His laboratories were indeed superb in every respect. There was much space, room for 12 assistants, a coldroom at −10°C, a darkroom, a special room for preparations, and an excellent library. No less than 10 collaborators came with him from Zurich to Dahlem. These and others who joined him there came from many different countries and contributed to the creation of the international atmosphere that was typical of the Kaiser Wilhelm Institutes, except for the war years. He was happy to have Otto Hahn and Lise Meitner in the same building.

The greatest experience of his time in Dahlem became the close friendship with Haber, which lasted for life. They both lived on the Faradayweg; their two houses were next to each other, separated only by Hittorf Street. As he writes in his autobiography, the two men who had the greatest impact on his life were Baeyer as his teacher and Haber as his friend. He met Haber too late in his life for his scientific work to be influenced, however.

Willstätter had a most happy, productive, and stimulating time in Dahlem. The "Casino" in Haber's institute, where one could eat, offered an opportunity to meet many colleagues. Restaurants did not exist in Dahlem at that time. A clubhouse, called Harnack House, in honor of the President, was built by the Kaiser Wilhelm Society much later and opened in 1929. It was an elegant and superbly decorated building with restaurants, a magnificent auditorium, all kinds of facilities, even for chamber music, and a few guest rooms for distinguished visitors. In Berlin Willstätter met many famous and interesting people whom he had not known before, and saw again old friends and colleagues. One of the first to invite him for dinner was Nernst, who had been very active in the planning of the center. He kept Willstätter after the other guests had left and pointed out that they all had high hopes that he, Willstätter, would soon greatly contribute to the fame and prominence of the new center, although it could not be of such great importance as the heat theorem (the third law of thermodynamics of Nernst).

Soon, in 1913, a third institute was opened, with August von Wassermann, a pupil of Ehrlich, as the director and Carl Neuberg as head of the biochemistry division. The Kaiser Wilhelm Institute of Biology was opened in the spring of 1915, delayed by the war, with Karl Correns and Hans Speemann, and later Richard Goldschmidt, as directors. Otto Warburg and Otto Meyerhof had their laboratories in this institute at a later time. Willstätter noticed that a strikingly large fraction of the leading scientists were Jews. He recalls the story that Emil Fischer, when asked why he was so completely free of anti-Semitic feelings, replied, "We who come from the Rhineland are not so stupid that we have to be afraid of the Jews." Emil Fischer invited Willstätter to dinner, but like the latter, he had no inclination and time for social life. He still worked at his bench, had a tremendous teaching load, many collaborators, and administrative duties. Because of his eminence as one of the world's leading scientists and his forceful personality, his broad perspectives, he had many other obligations and was frequently consulted by government and industry. It was an unbelievable tragedy that he ended his life at the age of 67, in 1919, by taking cyanide because his physician diagnosed cancer, erroneously, as the autopsy showed.

Although Willstätter tried to avoid social life, his friend Haber drew him into it, at least to a limited extent. They visited the banker Koppel who had given the money for Haber's institute and whose house was a real palace. Koppel picked up Willstätter for an occasional walk in the Grunewald. He visited Franz Oppenheim, director general of the aniline manufacturing company, several times. Oppenheim lived in Wannsee, a charming suburb located on the lake. He was very fond of Oppenheim's wife, who was a great art enthusiast. They both frequently visited him later in Munich.

The happy time in Dahlem came to a sudden end on August 1, 1914, with the outbreak of the war, which hit Willstätter and many of his colleagues totally unexpectedly, like a catastrophe of nature. They thought the war threats would disappear as they had before. Most of Willstätter's colleagues were politically disinterested and uninformed. They did not seriously consider the problem of political responsibility. This question came up much later. Willstätter's ideas will be discussed subsequently. In the beginning, most thought that Germany was the victim of an attack and that the war was fought to defend the country. The war's effects on scientific work were disastrous. Many collaborators left. Many precious and irreplaceable preparations were abandoned and perished. Many laboratories were empty, or later used for war work. Research was not completely abandoned, but drastically reduced. Willstätter and Haber were elected to membership in the Prussian Academy of Sciences. In view of his reduced scientific activities Willstätter attended the sessions quite frequently. They gave him the opportunity to meet many distinguished members of the University of Berlin and of other research institutes. His seat was next to that of Albert Einstein. Max Planck and Emil Fischer were quite active in the academy. He met many colleagues from the humanities as well. After the sessions many members walked to a nearby cafe, *Unter den Linden*, in which a special room was reserved for the members of the academy. Thus the meetings offered some compensation for the paucity of scientific activities. In the beginning most people were optimistic and hoped for a short war, but the mood changed gradually and gave way to considerable skepticism and pessimism. Haber did not participate, since he was completely absorbed in his activities for the war.

In the year and a half he spent in Dahlem before the war Willstätter, who had a much easier life than before since he was no longer teaching and burdened with administrative tasks, more frequently accepted invitations to lecture, in Germany as well as abroad, and attended meetings. He was particularly pleased to attend the celebrations in honor of Paul Ehrlich's sixtieth birthday and thus pay tribute to his genius. He got offers from several universities and received, in 1915, the Nobel Prize, although the award took place only after the war.

In the academy he learned that he had been proposed, *primo et unico loco*, as the successor of his teacher Adolf von Baeyer, then 80 years old. When the Prussian ministry learned of it, it tried to prevent Willstätter from going to Munich by offering him Wallach's chair in Göttingen. Actually he would have preferred for many reasons to go to Göttingen, but a telegram from Baeyer influenced his decision to accept the chair in Munich. President von Harnack was anxious to keep Willstätter in Dahlem. He wrote him a letter and paid him a personal visit in the hope of persuading him to stay on. During this visit he noticed a great and beautiful portrait of Justus von Liebig in Willstätter's office. He asked him, "Do you know that I am a grandson of Liebig?" Willstätter replied that he knew it and that he wanted to become another grandson of him.* Harnack then realized that there was no hope of keeping him. Willstätter was appointed in September 1915, but he moved to Munich in 1916. His only condition was a new building, since the old one was completely outdated. When King Ludwig III signed the appointment, he stated that this was the last time he would sign the appointment of a Jew. It was the last time, since the Wittelsbach monarchy, which had lasted seven centuries, was toppled two years later. But the anti-Semitic feelings under which he had suffered so much in his youth, forgotten in Zurich and Dahlem, soon reappeared in a particularly strong and violent form after the defeat of Germany in the war.

One of the main contributions to the war effort, carried out on the urgent pleading of Haber, was the development of a mask for protection against poison gases. This mask proved to be extremely successful and saved many lives, as was stated in a letter from the secretary of war in 1917. He never mentioned this letter, but in view of the events after 1933 he published it in full in his autobiography. He received as recognition of his great contribtuion the Iron Cross in May 1917.

A quarter of a century after he had entered the University of Munich as a student, he returned as *Ordinarius* of the famous chair in chemistry

---

* In Germany one used such a term in reference to the succession of chairs. Liebig was the predecessor of Baeyer.

held by his admired and beloved teacher Baeyer, who died in 1916 at the age of 81. But the conditions under which he took over the chairmanship were frightful and depressing. The institute was completely obsolete and in a state of dilapidation; so were his living quarters (*Dienstwohnung*). All had to be renovated and rebuilt, especially the auditorium. It took until 1920 for the auditorium to be completed. But the lost war had other devastating effects, owing to the political, economic, and social turmoil. Shootings, street fights with machine guns, assassinations of leading personalities, student revolts by extremists, the overthrow of governments by force, made life miserable and dangerous. It was not until 1921 that Willstätter was able to start teaching seriously. In view of the great number of students returning from the army, the teaching load was extremely heavy. Moreover, from 1921 up to the end of 1923, all research activities were severely hampered and restricted, although still possible to a limited extent, by the runaway inflation in which prices went skyrocketing every day, until the mark was finally stabilized.

There was, however, some compensation. In the faculty of the university were many distinguished people; some of them were about 80 years old, since there was no forced retirement age. Willstätter soon developed warm and cordial relationships with many prominent colleagues. Among them were the zoologist Richard von Hertwig, the botanist Karl von Goebel, the mineralogist Paul von Groth, and the physicist Arnold Sommerfeld. He was particularly pleased by the friendly and warm personal relations he had with Wilhelm Röntgen, whose revolutionary discovery of X-rays had always fascinated him and whom he had always considered as a forerunner of a new era. He highly valued all these friendships, which greatly increased his perspectives and enlightened him in many new areas. The mathematician Alfred Pringsheim, who came originally from Berlin, the father-in-law of Thomas Mann, was his neighbor. A warm friendship soon developed with him and his wife. Willstätter was greatly pained when Pringsheim, a Jew, had to emigrate at the age of 90.

In his institute Heinrich Wieland was *Extraordinarius* and the head of the organic chemistry division. He soon became *Ordinarius* at the *Technische Hochschule* in Munich, and Kurt H. Meyer became his successor. Meyer, a half-Jew, was the son of Hans Horst Meyer, the distinguished pharmacologist of the University of Vienna. He, Hermann Staudinger (1881–1965), Nobel laureate in 1953, and Hermann F. Mark (1895–   ) were the founders of polymer chemistry, which plays such an important role in industry and in modern life, in addition to its theoretical importance. Meyer left the institute when he was offered a great industrial position in Ludwigshafen, but later accepted, before the Nazi regime, the chair of the Department of Chemistry at the University of Geneva.

Willstätter was fully aware of the great importance of physical chemistry and the chemistry of radioactive compounds for the future of chemistry. His friendship with Haber and his close contacts with Otto Hahn and Lise Meitner had certainly greatly strengthened this conviction. He tried to get support for a special new building for physical chemistry, but this was impossible in view of the war condition. The only possibility was the establishment of a special laboratory in his own institute, a step for which he needed no authorization. He invited Kasimir Fajans to join him in Munich. Fajans, a pupil of Georg Bredig in Heidelberg and Lord Rutherford in Manchester, was *Privatdozent* at the *Technische Hochschule* in Karlsruhe, but had difficulties, since he was a Russian Jew, not yet a German citizen. He came to Munich as *Privatdozent*, but Willstätter soon promoted him to *Extraordinarius* and provided him with a special division when the new building was finished after the war. It was years after the resignation of Willstätter that Fajans became *Ordinarius* in an excellent new building established with funds from the Rockefeller Foundation. Just when the new building was finished, the Nazi regime came to power and forced him to leave Germany. He was invited to join the University of Ann Arbor in Michigan.

As to Willstätter's co-workers, several of the Swiss associates who went with him to Dahlem followed him to Munich. He was particularly happy that Arthur Stoll stayed with him, so he was able to continue the work on photosynthesis. It was in the context of this work that he had come across the action of some enzymes. In the 1920s the enzyme work moved increasingly into the center of his research. This aspect will be discussed later.

In 1919, after the tragic death of Emil Fischer, the University of Berlin tried hard to get Willstätter back to Berlin. But in view of the many warm and friendly relationships he had established in Munich, the urgent pleadings of his friends, of the faculty, the administration and the ministry officials, he decided to remain in Munich.

Shortly after the end of the inflation, and just when everything began to look more promising and cheerful, came the dramatic turning point that put a sudden end to his meteoric academic career.

Appointments to professorship of three Jews named Goldschmidt did not come through for anti-Semitic reasons. The first one was that of Stefan Goldschmidt, *Extraordinarius* at the *Technische Hochschule* in Karlsruhe, whom Willstätter wanted as head of the organic chemistry division in his institute. The second was that of Richard Goldschmidt, second director of the Kaiser Wilhelm Institute of Biology in Dahlem as successor of R. von Hertwig. Goldschmidt had been Hertwig's most outstanding pupil. Hertwig objected on the basis that Goldschmidt was planning an extended research expedition. Later Hertwig admitted privately to Willstätter that he would have preferred Goldschmidt, that the reason given was a pretext, and that the real reason for his objection was the desirability of taking into account the anti-Semitic feelings of the population and the students. The third one was Victor H. Goldschmidt in Oslo, who had been proposed as successor of Paul von Groth, primarily because he was the most brilliant and outstanding candidate in the field. Instead, the faculty voted for a rather unknown and mediocre scientist, allegedly because Goldschmidt was a foreigner. But the real reason was again anti-Semitism. The dean at that time, who was actively preventing the appointment, was Wilhelm Wien, the famous physicist. Willstätter had been instrumental in Wien's appointment to the University of Munich.

The last case was too much for Willstätter to take. Immediately and rapidly, the same day, without much consideraton, he decided to resign. This was in June 1924. He wrote a letter of resignation pointing out that this decision was final and that nothing would change his mind; he explained his motives. This action had the effect of a bombshell. Friends, colleagues, the whole faculty pleaded urgently and fervently with him to reconsider. Friedrich von Müller, the outstanding professor of internal medicine, and the famous surgeon Sauerbruch visited him and took him under some pretext to his auditorium. There was a huge crowd of students. He got a most moving ovation. They expressed in strong terms their admiration and affection for their inspiring teacher; they implored him not to abandon them in such a serious situation, when the university needed the forceful leadership that only he could offer. A letter was read at this meeting and presented to him, signed by 337 students. He was for them not only a great and ingenius scientist, but the devoted and dedicated teacher, unmatched in his ability to raise their enthusiasm for science. They paid tribute to the greatness and magnanimity of his personality. The country, in this bitter and depressing condition, had few leaders of his stature. They were deeply stirred and upset by the possibility of losing his leadership.

Willstätter was greatly moved and excited. In his answer he explained to the students what motivated his decision and he pointed out the principles that should prevail in the interest of a truly qualified leadership and of the standard of German universities. He received a letter from the rector who, in his name and on behalf of the senate and the faculty, expressed sadness about his intention to resign from the famous chair, the most brilliant successor of his great teacher, unanimously elected by the whole faculty. His worldwide fame and his admirable leadership of one of the most important institutes was an invaluable asset for the university. They expressed their admiration for his genius, his willpower, his success in helping the country during the war and the postwar years, the deep affection of his colleagues and students, his integrity. In the name of everybody they appealed to him to reconsider his decision.

Willstätter's decision was final and irreversible. When the news of his resignation became widely known, he got a dozen offers from all over the world, among them three American universities, one of them the University of Chicago. Another offer for a high position came from

Madrid. He did not consider any offer, since he did not want this event to be viewed as a matter connected with career or position. He also turned down a new offer from the Kaiser Wilhelm Society in Dahlem, proposed in 1926, to build a new institute for him designed completely according to his wishes. Several German universities tried to attract him as *Ordinarius*. Heinrich Wieland, who became, on Willstätter's proposal, his successor, pleaded forcefully with Willstätter in a letter to think of his responsibilities and of what Baeyer and Emil Fischer would have felt about his intention.

In his letter imploring Willstätter to reconsider his resignation, Wieland wrote that there was no anti-Semitism in Germany. This statement is almost pathetic. It shows how many German scientists, and educated people in general, were unable to recognize the powerful forces, no matter how irrational and emotional, that were behind the anti-Semitic feelings. But they were a reality that could not be ignored. To many Jews and non-Jews it was simply inconceivable that the barbaric forces of an age long past should take hold again in a highly civilized and advanced country such as Germany; the attitude is similar to that already discussed in the chapter on atomic physics.

Heinrich Otto Wieland (1877–1957). A brief description of Willstätter's successor is indicated. Born in Pforzheim, he had studied chemistry first in Berlin and then with Baeyer in Munich. In his research he was always passionately interested in the structure of biologically active compounds. His achievements and his personality have been superbly summarized by one of his brilliant pupils, B. Witkop (1977), in an article in *Angewandte Chemie*, on the occasion of Wieland's hundredth birthday anniversary. Witkop is a half-Jew and came to the United States after the war; at the National Institutes of Health he continues with outstanding success the glorious tradition of Wieland in the field of animal toxins, compounds that have become an increasingly important tool in biological research. Only a few of the most important contributions of Wieland will be briefly mentioned.

One of Wieland's major interests was the mechanism of oxidation. He recognized very early that hydrogen should be just as activated as oxygen by a reducing enzyme, a dehydrogenase. He was bitterly criticized by Otto Warburg, but as is fully discussed in the section on Warburg, Wieland was right. Wieland made a whole series of outstanding contributions to our knowledge of steroids. He became interested in bile salts, which are derivatives of cholesterol, an important compound on which he did much pioneer work. The final knowledge of the sequence of the intermediary steps in the formation of cholesterol was obtained only after World War II, when isotopes as markers became readily available. This ingenious work was performed, independently and simultaneously, by Feodor Lynen (1911–   ) and Konrad Bloch (1912–   ); the two shared the Nobel Prize in 1964 for this work. Lynen was a pupil of Wieland's and his son-in-law, having married Wieland's daughter Eva. Lynen was also instrumental in elucidating the manifold roles of "activated" acetic acid, which plays a central role in many important metabolic pathways (Hartmann, 1976). The studies were initiated in Wieland's laboratory. Bloch, born in Neisse, Germany, began his studies at the *Technische Hochschule* in Munich in 1930. As a Jew, he had to leave Germany, and he received his Ph.D. at Columbia University. He worked there with Rudolf Schoenheimer (1898–1941), one of the great pioneers in the studies on cholesterol. Today he is professor of chemistry at Harvard University.

A research area greatly stimulated by Wieland and his school and continued by his pupils is the structure of alkaloids, such as strychnine, lobeline, calabash–curare, and several others. Amanitine, the toxin of one of Europe's most dangerous mushrooms, was crystallized in the early 1940s. The structures of several other toxins, such as bufotoxin, a venom with a steroid skeleton produced by the toad, were successfully explored. The area of animal toxins is a wide and promising one; the number of toxins being discovered increases continually.

Some of Wieland's basic research led to important developments of great practical value. One example is the use of steroid hormones as anti-fertility agents, which may have far-reaching implications in an age threatened by a dangerous population explosion, even if at present it may not yet provide a satisfactory answer. Wieland received the Nobel Prize in chemistry in 1927.

Looking back on Wieland's work one cannot help admiring the wide range of his outstanding accomplishments. They reveal an extraordinarily great imagination and intuition. His personality attracted many highly talented pupils and he built up a great school. He was an artistic type and, like many of the scientists mentioned before, a passionate piano player. Some features of his personality must be mentioned in the context of this book. He was extremely liberal. His liberalism went so far that his children were not baptized and did not attend religious instruction, because the church was for him too reactionary and he wanted to bring his children up in a truly liberal spirit, free of prejudice. He was deeply shocked and appalled by the Nazi regime, but just as helpless as the atomic physicists described before, and unable to act politically for the same reasons. When he was asked to remove the bust of Willstätter that was in his institute, he simultaneously removed the busts of Baeyer and Liebig. In 1938 the author met him at a banquet of the Chemical Society in Paris. After the dinner the members dispersed in small groups to different rooms. Then Wieland's hatred of the Nazis exploded and he vigorously expressed his contempt for the "Nazi criminals who are ruining Germany." Probably only the author realized that Wieland risked his life; if there had been a Nazi in the group, he could have denounced him. In

1943 one of Wieland's students, Hans Carl Leipelt, was arrested by the Gestapo and hanged. At the trial Wieland, present as a witness, had the incredible courage to defy the Nazi in charge of the trial by walking to Leipelt and shaking his hand to express his sympathy and approval.

Willstätter left the institute in September 1925 and never entered it again. He moved to a house on the outskirts of the city, located a considerable distance from the institute. He accepted the plea of the minister of arts and sciences, who asked him as a personal favor to remain on the faculty and to take part in some activities. He used the time he had for writing up and summarizing his research results of the last years, particularly on enzymes, in which many distinguished scientists had participated. He gave quite a few lectures at universities, meetings of the Chemical Society, and elsewhere, in Germany and abroad. In 1928, Dr. Margarete Rohdewald, a former student of his who went with Richard Kuhn to Zurich to obtain her doctorate there, returned to Munich and started to work with Willstätter as his private assistant. Wieland gave her space in his institute. She did the experimental work on enzymes, but discussed her results on the phone with Willstätter every evening. This unusual collaboration went on for a period of 10 years, until 1938. Willstätter realized how unsatisfactory such an arrangement was. A chemist working only on his desk was, in his own words, like a fish on a dry beach. But he was just unable to detach himself completely from the research problems that preoccupied his thoughts. Despite the difficulties, it gave him some satisfaction since he had the feeling that the work made some valuable contributions.

When Hitler came to power Willstätter stayed on in Germany. He made several trips abroad, received great honors, was offered several positions. In Germany he suffered many humiliations and hardships. He witnessed the terror, the voluntary or forced emigration of Jewish scientists, many of them his friends. He was extremely active in helping friends, relatives, and colleagues who had difficulties in settling down after their emigration. But he himself stayed on; nothing was able to move him. In September 1938 his collaborator Stoll invited him to Basel. It was meant to be a no-return trip. He did return. After the pogroms of November 10, when many Jewish shops and synagogues were destroyed, it became apparent that the lives of all Jews were in danger and their situation untenable. He finally decided to accept Stoll's offer to come to Switzerland. He became desperate after the treatment he received and the difficulties raised by the Gestapo (*Geheime Staatspolizei*, the "secret state police"). In a moment of despair he tried to escape across the Bodensee, but was caught by the Gestapo. Through the efforts of the German embassy in Berne, he finally crossed the border early in March 1939. Arthur Stoll gave him a lovely villa, the Ermitage, in Locarno–Muralto where he could display the few works of art he was able to take out of Germany. Stoll took care of his beloved teacher in a most moving way. The last year of his life he suffered greatly from a heart condition and passed away on August 3, 1942, shortly before his seventieth birthday. It was in Locarno that he wrote his autobiography, a fascinating document not only of Willstätter's work and personality, but of the whole era of his lifetime.

Marriage and Family Life

During a vacation in Wiesbaden, in the spring of 1903, Willstätter met Sophie Leser, her father, and her two brothers. The father, Emanuel Leser, was originally a historian; he was very young when he became professor of *Nationalökonomie* at the University of Heidelberg. He was a greatly respected scholar; his main field of interest was the history of England's economy. His colleagues and friends had great affection for him. Sophie was a lady of the world, of great beauty and intelligence, very elegant, lively and warmhearted. She was highly educated and had studied for several years, especially literature, the history of art, and social sciences; she was an ardent reader. Her two brothers were law students. The family loved social life; their house in Heidelberg was, because of their warm hospitality, a great meeting place of interesting people. At their parties there were always animated discussions about literature, politics, and law; they were always gay and lively. Leser loved stories and anecdotes.

Leser was born in Mainz; his wife Ida, *née* Rohr, came from Silesia, from a village not far from the Polish border. After World War I the village became Polish. She was, when Willstätter met Sophie, already in poor health. She was a deeply devoted mother. Leser loved to travel and Sophie had frequently traveled with him to Italy, England, and Holland. He died at the age of 65, shortly before the outbreak of World War I, deeply mourned by the university and by his many friends.

Willstätter rapidly fell in love with Sophie in Wiesbaden. A few weeks later he visited her in Heidelberg. They soon became engaged and married in August 1903. He tried to introduce her to the problems on which he worked, and he succeeded in arousing her interest. She made great efforts to understand the problems and frequently attended his main lectures. It became a perfectly happy marriage; Sophie was an ideal and true partner and comrade of his life. Their son Ludwig was born in Munich in October 1904. He was a happy, lively, and gifted child. After two years in Munich the couple moved to Zurich, where they had a beautifully located house with a lovely view on the lake. At the door was written the verse of Horace: *Ille terrarium mihi praeter omnes angulus ridet.* Close by were forests where they took walks. In Zurich their second child, the daughter Margarete, was born. Sophie was a devoted mother and spent most of her time with her children.

Social life was quite limited. They met only with a few families. Among them was an outstanding colleague of Willstätter, Alfred Werner, from Alsace, who was professor of inorganic chemistry at the university. His main contributions were theories of the configuration of inorganic compounds and complexes. His laboratories were completely inadequate, but a new institute was built for him, and at a dinner during the inauguration ceremonies, in 1909, Willstätter was sitting next to Haber. It was the first time that he met Haber, who made an unforgettable impression on him.

Willstätter made a few trips with Sophie, one of them to Paris, where they had a wonderful time. That was shortly before the disaster struck. In May 1908 Sophie fell ill. It took the physician a day until he diagnosed appendicitis. The following day she was taken to the hospital, but the surgeon postponed the operation until the next morning. By then it was too late. The operation was performed 36 hours after the perforation. She had a high temperature for weeks. Everyone tried to hide it from her, but one day a substitute physician revealed the secret and she knew she was going to die. In the last weeks of her life she expressed her deep feelings and her happiness about their marriage. After seven weeks she became unconscious and died in the night. It was an indescribable disaster for Willstätter. He realized what the greatness of a single life may be. He never married again. He immediately tried to do everything in his power for the children and took care of them as much as possible in order to replace their devoted mother.

When they moved to Berlin–Dahlem, in 1912, the children grew up under favorable conditions. The suburb was full of parks and gardens, and the air was clean; the climate of Berlin is exceptionally good. When Ludwig was 10½ years old, in 1915, Willstätter noticed one day that the boy looked tired and that he was thirsty. The physician was called, but came only on the following day. He found the boy to be in excellent condition. But Willstätter remained worried and the next day he tested the urine; he found a large amount of sugar. He urgently called the physician, who promised to come the next day with prescriptions. When the physician arrived, the child was already in a coma; he died that night. Willstätter had lost his son, whose charming, cheerful nature and great talents had raised in him great hopes. Willstätter wrote in his autobiography that there are wounds that cannot heal and that he did not want to heal. These two tragedies were certainly among the wounds he had in his mind.

His daughter Margarete married a young chemist, Dr. Ernst Bruch, who later became a physician. When the Nazis came to power, they emigrated to the United States and settled in Winnepago, Illinois. It was on the occasion of the birth of his grandson, Ludwig W. Bruch, that Willstätter decided to write his autobiography.

Scientific Achievements

Willstätter made a vast number of important scientific contributions that made him one of the outstanding leaders of his generation. They are summarized in his autobiography and fully described in his books and his many publications. Here only a few outstanding achievements will be mentioned to illustrate the nature of his research and the direction of his interests.

His first great achievement was the elucidation of the structure of cocaine. This alkaloid, which occurs in plants, was the first local anesthetic, discovered in 1860. Its pharmacological actions were described by B. von Anrep (1879), who suggested that it might be clinically useful as a local anesthetic. But nobody paid attention to this observation. Sigmund Freud independently discovered, in the 1880s, the anesthetic action of cocaine and suggested to a friend, an ophthalmologist, that he test this action. Another ophthalmologist, Carl Koeller, took up the idea and made the decisive experiments on the eyes of animals; he demonstrated the anesthetic effects at a congress in Heidelberg.

Willstätter started to work on the structure of cocaine as the topic of his doctoral thesis and made valuable contributions. But when he finished his thesis, the problem of the structure was far from being solved. He became so involved in this question that he decided to continue his work on the structure of cocaine and many related compounds. One of the split products of cocaine and the closely related atropine is tropine. The structure of tropine had been proposed by A. Ladenburg, a distinguished chemist and professor of chemistry at the University of Breslau, to be a six-membered ring, formed by five carbon atoms and one nitrogen atom (a pyridine ring) with a side chain. Professor G. Merling later proposed a modified structure, which came closer to the real structure: a pyridine with a bridge formed by a two-carbon compound between two carbon atoms of the pyridine ring. But it was Willstätter's work that elucidated the real structure of cocaine and of atropine. They are six-membered carbon rings in which two of the carbon atoms are connected by a bridge containing a nitrogen group (NH_3).* The difference between cocaine and atropine is due to different side chains.

Willstätter and his collaborators succeeded in the synthesis of tropine by using a whole series of reactions, starting with a seven-membered

---

* The skeleton structure of the two compounds, leaving out the hydrogens and the side chains, is as follows:

```
C — C — C
|   |   |
    N
    |
C — C — C
```

carbon ring (suberone). He used many novel and ingenious methods in this work. It was a true masterpiece of the classical method of approach. It earned him the admiration of Baeyer and he became his preferred pupil and the pride of the institute. The imaginative methods applied to the synthesis of tropine opened new ways that were very useful for the chemical and pharmaceutical industry; they were particularly valuable for the synthesis of new anesthetics. The stimulus produced by an achievement is the best indicator for its significance. In 1917 Willstätter succeeded in the synthesis of cocaine.

Cocaine blocks nerve conduction, but has many undesirable toxic side effects; it produces euphoria. Willstätter's teacher, A. Einhorn, who originally suggested the work on the structure of cocaine for the doctoral thesis, became interested in the development of local anesthetics by modifications of the structure. He developed, in 1905, procaine (or novocaine), for a long time one of the most widely used local anesthetics. Today a large number of local anesthetics are available, such as butacaine, tetracaine, and dibucaine. Many of them are widely used in medicine and dentistry as local anesthetics. The molecular mechanism, by which nerve conduction is blocked, was elucidated only in the 1960s by the author and his associates, when information about the chemical and molecular basis of nerve impulse conduction began to accumulate (see for summaries, e.g., Nachmansohn, 1969, 1971).*

A second contribution, which was achieved in Zurich, resulted from the degradation of an alkaloid that is found in the pomegranate tree and is similar in structure to tropinone. A series of hydrocarbon compounds were obtained. Willstätter tried, with his associate Ernst Waser, joined later by Michael Heidelberger, to elucidate the structure of the hydrocarbon C_8H_8, seemingly similar to the benzene (C_6H_6), but completely different in behavior. It turned out to be cyclooctatetraene. This work

$$
\begin{array}{c}
\text{CH}=\text{CH} \\
\text{HC}\diagup\qquad\diagdown\text{CH} \\
\parallel\qquad\qquad\parallel \\
\text{HC}\qquad\qquad\text{CH} \\
\diagdown\text{CH}=\text{CH}\diagup
\end{array}
$$

---

* The structure of local anesthetics is analogous to that of acetylcholine, a compound that has a control function in nerve conduction. This function is associated with four proteins, which form the "acetylcholine cycle." By its action on one of the four proteins, the receptor protein, acetylcholine initiates the reactions that lead to increased ion permeability across the excitable membranes that surround nerve and muscle fibers. The resulting rapid ion fluxes are the carriers of bioelectric currents that propagate impulses along nerve and muscle fibers. Whereas acetylcholine activates the receptor protein, its analogue, the local anesthetic, also react with the receptor protein, but prevent its activation; they act as inhibitors or "antimetabolites," thereby blocking conduction.

was also a major accomplishment. It required great resourcefulness and the application of a great variety of newly developed catalytic methods; with the previously existing methods the elucidation of the structure would have been impossible. In particular, it was extremely difficult to find out the number of saturated and unsaturated carbon–carbon bonds, since the compound is rather unstable.

The significance of this work became apparent when it was used as the starting material for the synthesis of a variety of other compounds and the methods applied became valuable tools. The difference of behavior between the six- and eight-membered hydrocarbon compounds found an explanation only much later in terms of resonance theory (see II, F).*

But Willstätter's most outstanding achievement, which established his great fame and with which his name will remain associated in the history of chemistry, is the elucidation of the structure of the green coloring pigment in plants (chlorophyll). All free energy consumed by biological systems arises from solar energy. The latter is trapped in plants by the process of photosynthesis, in which the carbon dioxide of the air is transformed by a series of reactions that lead to the release of oxygen necessary for respiration, that is, for animal life. Chlorophyll plays a key role in this process, since the first step in photosynthesis is the absorption of light by a chlorophyll molecule. The production of oxygen in photosynthesis was discovered, in 1780, by Joseph Priestley (1733–1804). Antoine Lavoisier (1743–1794) started to work on the basic process of photosynthesis, but almost two centuries passed before the mechanism was elucidated (see IV, C).

For the exploration of the mechanism of photosynthesis it was essential to know the structure of chlorophyll. Thus its elucidation by Willstätter represents a landmark; it opened the way for the investigations on the mechanism of one of nature's most important processes. Although Willstätter had some notions of how to proceed before he left Munich, the main work was carried out with a great number of collaborators from 1905 to 1911, and still continued, when he moved in 1912 to Berlin. Many of these collaborators went with him from Munich to Zurich, where they were soon joined by others. However, his most outstanding associate, who joined him in Zurich, went with him to Berlin, and stayed with him when Willstätter moved back to Munich in the difficult war years, was Arthur Stoll. The elucidation of the structure required a huge amount of imaginative work. Chlorophyll in the form of crystals was obtained in 1907. These crystals had been described by the Russian botanist J. Borodin in

---

* In the eight-membered carbon compound, the carbon atoms are not arranged in the same plane, as is the case in benzene. The transformation of one form into another would require a shift in the position of carbon atoms. Since resonance occurs only by electron shifts, not by movement of atomic centers, the cyclooctatetraene is incapable of achieving stabilization through resonance.

1882, who found them in microscopic studies. But it was many years before the structure was finally established.

Willstätter's work has shown that chlorophyll consists of two chemically closely related forms: chlorophyll a and chlorophyll b:

The R group in chlorophyll a is a methyl (CH_3); in b, a formyl (CHO). (The carbon, hydrogen atoms, and the side chains are not shown in the structure.) The pigments are bound to proteins. The chlorophyll is a substituted tetrapyrrole. A pyrrole is a five-membered heterocyclic* ring with one nitrogen and four carbon atoms. The four nitrogen atoms of the pyrroles are coordinated to a magnesium (Mg) atom. Thus chlorophyll is a magnesium porphyrin, whereas the heme in hemoglobin is an iron porphyrin. The latter compound will be discussed in the following section on Warburg (see IV, C). Chlorophylls are very effective photoreceptors because they contain a network of alternating single and double bonds. The two kind of chlorophylls, a and b, complement each other in absorbing the sunlight since their maximum absorptions are at different wavelengths.

The vast amount of work necessary to unravel the structure of chlorophyll was accomplished within a period of nearly a decade. The various advances were reported in Liebig's *Annalen der Chemie*, and summarized in a book by Willstätter and Stoll on the structure of chlorophyll (Willstätter and Stoll, 1913).

Although many structural problems remained unsolved, Willstätter became interested in experiments on intact leaves, hoping to find out something about the process of photosynthesis and the absorption and transformation of carbon dioxide in the plant. After the outbreak of the war he and Stoll, who had a broad scientific outlook and was much more familiar with biological problems than Willstätter, tried a number of studies. In view of his old love for biology, Willstätter, although he realized his lack of competence, greatly enjoyed this excursion into a field that fascinated him. Although it was soon apparent that chlorophyll out-

---

* Heterocyclic rings are five- or six-membered ring systems that contain one or more atoms other than a carbon as ring members—either nitrogen, oxygen, or sulfur atoms.

side the plant is unable to assimilate carbon dioxide, a number of interesting observations were made and he and Stoll summarized them in a second book entitled, *Investigations on the Assimilation of Carbon Dioxide* (Willstätter and Stoll, 1918).

In the last years in Zurich Willstätter had also become interested in the pigments of flowers and fruits. These pigments are called *anthocyanins* (from Greek *antho*, "flower," and *kyanos*, "blue"). They are glycosides, compounds formed by a combination of various sugars with nonsugar components, called *aglycones*. The aglycones of anthocyanins are known as *anthocyanidins*. A large number of anthocyanidins were isolated and their structure elucidated. Most of them are formed by a two-heterocyclic-ring structure, to which a third ring is attached in a side chain.* Willstätter made the striking observation that the same pigment can give rise to different colors. For instance, cyanidine chloride is obtained by acid hydrolysis of the pigment of either the blue cornflower or of the red rose. Willstätter and his associates also initiated the synthesis of these compounds and the first anthocyanidin synthesis was achieved by his associate Dr. Laszlo Zechmeister on July 30, 1914, just before the outbreak of the war. Being Hungarian, Zechmeister had to return to Hungary immediately for military service. Although some work in this field was continued in Berlin–Dahlem, the conditions soon became prohibitive for research. Nevertheless, the structures of about 20 different pigments were elucidated and quite a few of them were synthesized. After the war the work on flower and fruit pigments was taken up by the brilliant chemist Sir Robert Robinson, Nobel laureate in 1947, and his wife G. M. Robinson. They determined the structures of a large number of additional pigments, using greatly improved new methods. Their research included, moreover, the analysis of the more complex compounds of natural flower colors. Their superb knowledge of reaction mechanisms permitted the synthesis of many glycosides. Their work represents an elegant completion of Willstätter's work. It is typical of Willstätter's character that he not only followed these accomplishments with great admiration, but also strongly expressed his genuine satisfaction and happiness at seeing his work continued and greatly improved and extended. Robinson had almost joined Willstätter in Zurich. Their collaboration did not materialize, but the two men later became good friends.

Before the war Willstätter had started work on a pigment of the heme series, etioporphyrin. This work was stopped by the outbreak of the war and was later taken up by Hans Fischer (1881–1945), his colleague at the *Technische Hochschule* in Munich. Fischer had become involved in work on chlorophyll and made many outstanding contributions long after Willstätter abandoned this line of research. Fischer later turned to investi-

---

* Among the many structures elucidated, some of which were synthesized, were those of the cyanins, flavones, and pelargonidin.

gations on the red blood pigment. His work on the structure and synthesis of iron porphyrins represents the most glorious and admirable achievements in that field. In 1930 he received the Nobel Prize.

After the war Willstätter felt that it would be preferable, instead of returning to old and abandoned fields, no matter how promising they were, to enlarge the scope and the methods of organic chemistry and to try to penetrate unexplored areas of neighboring disciplines. He became excited by the idea of trying out completely new pathways and possibly bringing some light into a hitherto dark field. These were the motives that drove him to plunge into enzyme research, in particular into enzyme purification, problems completely new to him and from which classical organic chemists had stayed away. His love of biology and medicine, going back to his early student years, the hope to make valuable contributions to these disciplines, may have been, unconsciously, a driving force.

Willstätter had first come in contact with some enzymes in 1910, when he worked with Stoll on chlorophyll. Moreover, his discussions with Haber had stimulated his interest in certain catalytic actions. But it was only after the war, in the early 1920s, that he started systematic investigations on the isolation of enzymes, their purification by adsorption and elution, their specificity, and so on. A vast number of various kinds of enzymes involved in the degradation of carbohydrates, proteins, lipids, and others were studied. A considerable degree of purification of a great number of enzymes was achieved and valuable information about enzymes accumulated. About 50 collaborators participated in this research during those years for shorter or longer periods. Just a few of the more well known may be mentioned: Richard Kuhn, Ernst Waldschmidt-Leitz, Kaj U. Linderstrøm-Lang, Arthur Stoll, and Wolfgang Grassman. But there were quite a few other highly qualified investigators. The results of this work were summarized in a number of publications, lectures, and books (see, e.g., Willstätter 1927, 1928).

A large part of the publications is devoted to the procedures used for purification, the degree of purification obtained, the specificity of the preparations, and so on. However, of special interest to biochemists and in particular to enzyme chemists are Willstätter's views and interpretations of the nature of enzymes. He realized that the enzyme preparations he had purified were not yet homogeneous substances but aggregates in which the enzyme and several other accompanying substances of various kinds were united. He also recognized that it was difficult to remove the protein from enzymes, but he did not consider this difficulty as conclusive evidence for the protein nature of enzymes. Being an organic chemist it seemed to him more appealing to attribute the catalytic function to a small active group so tightly bound to a protein that it could not readily be dissociated. He stubbornly adhered to this belief up to the end of his life—although at that time the erroneous character of this idea was generally recognized—and he considered his enzyme studies as one of his most important contributions.

Much has been written about Willstätter's enzyme work and his views in this last major field of his endeavors. A number of biochemists and historians have taken a sharply critical view of his denial of the protein nature of enzymes.

To evaluate Willstätter's enzyme work fairly, one must consider several pertinent factors. This may clarify why one of the most celebrated chemists of that period went astray in his interpretation. When Willstätter started the enzyme work in his laboratory, it was performed under miserable and almost impossible conditions. The runaway inflation in the years 1922–1923 did not permit systematic work; essential supplies and equipment were unavailable; lack of funds prevented required repairs, and so on. Just when normal conditions started to prevail, in 1924, Willstätter resigned from his chair, as discussed earlier. Thus he removed himself from the laboratory, an indispensable element for the work of a chemist.

There are other serious factors to be considered. Protein and macromolecular chemistry in general, including enzyme chemistry, began its rapid development in the 1920s, just about the time Willstätter left the laboratory for the rest of his life. If he had accepted the offer of a great Kaiser Wilhelm Institute in Berlin–Dahlem, in 1926, built according to his wishes, his work may have taken a different turn. There Warburg, Meyerhof, and Neuberg carried out their brilliant enzyme studies (see the following sections). In 1925, Meyerhof isolated and subsequently purified the glycolytic enzymes of muscle; in the 1930s Warburg crystallized nine of the most important enzymes in respiration and glycolysis. Willstätter would have been exposed to continuous discussions with these colleagues and with his friend Haber, familiar with these advances. It seems hard to believe that the exciting and inspiring atmosphere in Dahlem would not have had a deep impact on such a brilliant mind as that of Willstätter. The "heroic" era of enzyme chemistry, to use the term applied to the great era in quantum mechanics, started at about the time Willstätter had totally abandoned work in the laboratory and lived in a villa on the outskirts of Munich. The telephone discussions with Dr. Rohdewald, who from 1928 on performed experiments in Wieland's laboratory, cannot possibly be considered a surrogate for laboratory work in the best possible conditions, surrounded by competent collaborators and outstanding colleagues in the field.

In 1926 J. B. Sumner at Cornell University in Ithaca crystallized an enzyme, urease, for the first time, and from 1930 on John Northrop and Moses Kunitz (Northrop, 1939) obtained several enzymes in crystal form. It is sometimes claimed that Sumner's Nobel Prize was delayed for several years because his finding was not accepted by the great authority of Willstätter. This seems an open question. But the statement by Leicester

(1974), that Willstätter's rejection of the protein nature of enzymes delayed the understanding of enzyme chemistry for several years, appears hardly justified. Enzyme research went on at full speed, and if there were factors preventing a still faster growth rate, it was not the question of its nature—that is, whether it was a protein or not—but the slow realization of many biologists of the vital role of enzymes in cellular functions and of the insight that enzymes might provide into the mechanisms of living cells. These developments gained momentum only in the 1940s, and the schools of Warburg, Meyerhof, and Neuberg played a crucial role, as will be discussed later. Another great center of enzyme chemistry developed, especially in the 1930s, in Cambridge, England, under the leadership of Sir Frederick Gowland Hopkins.

Willstätter's work nevertheless stimulated the interest in enzyme research and greatly contributed in a variety of ways to its progress. Many of these achievements were outstanding. Perhaps they were not commensurate with his great genius, but this was, certainly to a great part, the result of the tragic conflict in his life. He could not find an answer to the dilemma facing him: his deep love of Germany and his determination to maintain his dignity as a Jew. When the Nazis rose to power, many illustrious contemporary scientists faced a similar dilemma: their deep-seated love for Germany versus their hatred of the Nazis, their horror of Nazi anti-Semitic terror. But they reacted, sometimes under much more trying circumstances, in a way that permitted the continuation of their creative work. Willstätter's extremist reaction amounted to the self-destruction of a great genius.

Personality

One of the most striking features of Willstätter's personality was the rare nobility, the high ethical and moral values and principles, that dominated his aims, his actions, and his feelings. He was in this respect the true scion of a family in which these principles were a deeply rooted tradition for generations. Spiritual, intellectual, and ethical values were more important than the accumulation of money and wealth. The altruism and idealism of several ancestors illustrate the family's values.

His noble character is demonstrated by the driving force in the two main activities to which his life was dedicated: research and teaching. His devotion to his work was deep and genuine. It provided him with great personal happiness to have achieved something important or outstanding. But it is typical of his attitude that career and recognition were not primary and compelling elements. In fact, he cared very little about career, wanting only adequate conditions for his work. What gave him the greatest inner satisfaction was the feeling of having made a contribution of eternal value. Truly significant or outstanding findings become the property of mankind, even though the name of the scientist is usually soon forgotten. The great artists who were the anonymous architects of the magnificent Gothic cathedrals must have had the same attitude. Science is in its nature progressive and corrective, as has been frequently stressed by scientists and historians of science. There are no final answers; as Willstätter said, good contributions end with a question mark and not with an exclamation mark. Some of these views, their philosophical basis, were discussed earlier. It was not the glory that counted for Willstätter, but the personal happiness of having been so fortunate as to be able to increase human knowledge. Real progress usually requires the work of generations. The elucidation of such a simple structure as urea resulted from the work of three generations: Liebig, Baeyer, and Emil Fischer. What a difference is seen when one compares Willstätter's attitude with that of scientists for whom career, honors, and material success are primary aims, sometimes more important than the happiness of having made scientific achievements. As to theories, Willstätter expressed his view in his lecture in Chicago in 1933:

> It is not important for the scientist whether his own theory proves the right one in the end. . . . It is not inglorious at all to have erred in theories and hypotheses. Our hypotheses are intended for the present rather than for the future. They are indispensable for us in the explanation of the secured facts, to enliven them and to mobilize them, and above all, to blaze a trail into unknown regions towards new discoveries.

Willstätter fully endorsed, as most scientists do, J. J. Thompson's view that a theory is a tool and not a creed.

According to Willstätter's well-known and frequently quoted statement, scientists need five G's for achievement: *Geschick, Geduld, Geld, Glück,* and *Gehirn* (skill, patience, money, luck, and brains). By "brains" he meant more than just intelligence and knowledge; he fully realized the importance of intuition and imagination for great discoveries. But the statement omits two important additional qualities required, which he had: enthusiasm and idealistic motivation. In his modesty, he explained that he needed all his time to concentrate on his work, since he was not of the high caliber of other scientists, to whom achievement comes easily. Actually it is common knowledge that all creative work, be it in art or in science, requires a great amount of hard work, determined efforts, concentration, inner drive, and devotion. The great artists and scientists were usually hard workers who excelled not only by quality, but also by quantity. Willstätter graduated in 1894 and resigned in 1924; in addition, during the last decade, from the outbreak of the war in 1914 until his resignation, research was greatly impeded and restricted; it was sometimes almost impossible because of the many factors described. One must admire not only his brilliance, but the vast number of his achievements in a relatively short period.

The man who had the greatest influence on Wittstätter's scientific thinking was his beloved and admired teacher, and later close friend, Baeyer. As he pointed out in his autobiography, it is not enough to attend lectures and read books; the personal effect of the scientist on his students and associates is much more important than the lectures. What really counts is to watch the teacher in his experiments at his bench or to be with him at his desk when he evaluates the results obtained, to follow his thinking and reasoning, his imagination and the ability to integrate his own experience, that of many contemporaries, and that of the previous generation. These are factors that are decisive for the young pupil. The observation of the personality of the teacher, with modesty and respect, is much more helpful than reading literature.

This attitude and the relationship to Baeyer were important factors in his teaching. The moral responsibility of teaching students, to build up a new generation that would be equipped to advance the field and enlarge our knowledge, was for him one of the noblest and most important aims of a scientist. His enthusiasm for teaching was combined with a great ability to express himself clearly and easily. He was a superb lecturer and he devoted much time to supervising and helping students and associates. But he also discussed with them the experimental data; when a paper was written, they had to be at his desk and participate in evaluating the results and in formulating the presentation. It is thus not surprising that he became a forceful leader, admired and loved by his students. The truly remarkable ovation given by his students when his resignation became known and the expression of their feelings in the letter presented are superb testimony to his extraordinary qualities as a teacher. He created a large school, and his pupils played a great role in many countries all over the world. This explains the warmth of the ovations he received wherever he went and was certainly a factor, in addition to his scientific achievements, in his receiving a remarkable number of honors from all over the world; in addition to his Nobel Prize he was awarded some of the most coveted medals, became a member of about 20 academies, and received many honorary doctorate degrees.

His integrity shows up also in his principles in publications. He would never sign a paper as coauthor unless he had really contributed to the work. Moreover, everybody who had worked on a problem had to be coauthor on the publication. He was extremely careful to give proper credit to preceding work. He detested people who omitted to mention pertinent contributions of others. He wrote, in his autobiography, "Miserable creatures [*armselige Geschöpfe*] are the scientists who are afraid to decrease their own glory when they do justice to their predecessors." The author was once informed that in the Talmud there is a statement that no theft is worse and more despicable than that of intellectual properties and ideas. For Willstätter ethical behavior, generosity, decency—that is, character and personality—are equally important attributes in the evaluation of a scientist as his contributions. He fully endorsed the statement of his friend Haber: "Nobody deserves a place among the immortals, no matter how great his strength and how illuminated his mind, if he lacks greatness of soul."

While Baeyer had the greatest impact on Willstätter as a teacher, Haber had it as a friend. When they lived next to each other in Berlin–Dahlem, they became close friends. Haber soon suggested that he and Willstätter call each other *Du*, an expression of a close personal relationship, instead of the formal *Sie*. This would facilitate the frank expression of differences of opinion. Haber loved vigorous and frank discussion, as discussed in the preceding section. He greatly admired Willstätter's competence and brilliance in chemistry. For Willstätter, who was almost completely absorbed by his work and was a rather quiet and even withdrawn person, the dynamism, the vitality, the ebullience and joie de vivre of Haber, his broad perspectives, his wit and humor, his wide range of interests in addition to his great achievements that had revolutionized science, industry, and agriculture, were a fascinating attraction. Moreover, both had the same high ethical standards and shared compassion for the young, always trying to encourage and to help them. They both despised lack of integrity, decency, and magnanimity; bad character was for both repulsive.

When the happy days in Dahlem ended and Willstätter moved, during the war, to Munich, the close friendship was maintained and the two friends spent many vacations together and took wonderful trips, as mentioned in the preceding section. They are vividly and with humor described in Willstätter's autobiography. His superb description of Haber's great scientific achievements in the *Festschrift* published by *Naturwissenschaften* on the occasion of Haber's sixty-fifth birthday has been mentioned. The expression of his admiration for Haber's many great human qualities throws as much light on his own personality as on that of his friend. The similarity of their attitudes was probably the basis of the affinity that cemented the friendship of these two giants.

When Haber died in Basel, in 1934, Willstätter was asked to speak at the memorial service. In the stirring and deeply moving tribute to his friend he stressed again the exceptional human qualities, the greatness and warmth of his personality, his rare nobility. He was deeply pained not only by the death of his friend, but by the fact that this great servant of his country, whose unique genius had contributed in a decisive way to Germany's strength and power before, during, and after World War I, who had promoted the well-being of mankind, should have died in exile. According to Plutarch, Aristides' answer to the question of what had hurt him most in his exile was this: "That my fatherland by my expulsion disgraced itself." This obviously expresses Willstätter's own feelings toward Haber's exile, to the ruthless, barbaric, and brutal forces ruling Germany.

Like most scientists in Germany, Willstätter lived, until the outbreak
of World War I, in an ivory tower. He was completely detached from
politics, believing that science by its achievements would steadily contribute to the progress and the improvement of society. The war came as
a total surprise. Because of the war, the disastrous defeat, and the postwar
conditions, Willstätter and scientists in general began to realize their political responsibilities as citizens. In discussing the question of guilt for the
war, he puts the chief blame on the citizen. When Bismarck established
his autocratic regime in the *Kaiserreich*, with a Parliament with little
influence and power, he did not realize the dramatic effects of science
and technology on society. These changes necessitated the active cooperation, advice, and responsibility of a great number of experts, leading
citizens in many fields. Bismarck also did not foresee the danger that a
successor might inherit the power, but might not have the great statesmanship that he had. In any event a modern society requires the participation
and responsibility of its citizens. German citizens had failed to fight for
a government capable of coping with the drastically changed conditions
of modern life. Thus the citizens were to blame in the first place for the
disaster. Willstätter was throughout his life an ardent monarchist, but a
monarchy as the symbol and dignified presentation of its country, with the
power exercised by Parliament. In view of his love of England, its constitutional monarchy was apparently his preferred form of government.

As a student Willstätter was a great fan of opera; he was enthusiastic
about Wagner. He loved the theater. Later his artistic interests turned
increasingly to painting and sculpture; he was particularly fond of French
impressionists. In his home in Munich he had an exquisite collection of
paintings, sculptures, and pieces of furniture. When he left Munich, in
1939, he had to leave behind most of his property. Only a few pieces
was he permitted to take to his villa in Locarno.

Looking at his colored portrait in his autobiography, one is struck by
the deep sadness revealed by his eyes.

Jewish Attitude

Of all the Jewish scientists mentioned in this book, Willstätter presents
the most complex, most puzzling, and probably most tragic picture in his
attitude toward his Jewishness. Although a really competent analysis is
extremely difficult, and the author feels unable to provide a satisfactory
answer, some comments appear appropriate in view of Willstätter's scientific greatness and the context of this book.

As Willstätter said, the two dominant factors in his life were his work
and anti-Semitism. The first question that arises is that of his own attitude
toward his Jewish heritage, his feelings about being a Jew. Judaism was
for him purely history, allegiance to his ancestors. It was not the exercise

of directions and rules prescribed thousands of years ago, not the Jewish laws, that is, not conservative Judaism, not to speak of faith. Adherence to old religious customs that, by the cultural changes that had taken place over milleniums, had lost much of their meaning, appeared to him as rigidity and weakness, dangerous for the survival of Judaism in highly civilized countries. His religious faith was the same as that expressed by Einstein; he believed in the God of Spinoza, manifesting itself in the harmony and beauty of nature, not in a personal God who cares for the fate of individuals. Divine is the unexplored miracle of creation. He greatly admired the ethical values in the New Testament, which he considered the greatest advancement of humanity. Unfortunately these values were frequently ignored by the institutions of the church, particularly when one recalls the history of the cruelty, the persecutions, the torturing and killing in the name of Christianity. He was horrified and deeply shaken and depressed by the outrageous anti-Semitic statements of the Nazis and their ruthless and brutal persecution of Jews. In 1934 a book was planned with a collection of articles about Jews in German culture from the period of the Enlightenment until Hitler. Willstätter was asked to write a preface to the book. In it he gave a few illustrations of the great achievements of Jews, their outstanding contributions to Germany. Among the examples selected were those great scientists with whom he had had personal relationships, such as Baeyer, Ehrlich, and Haber. The book was never published.

In view of his emphasis on his family background, of his awareness of Jewish history, it is surprising that he did not realize their effects on his personality. Here one is obviously struck by his failure to understand the implications of these factors. How could he not see and deny that his illustrious ancestry, traced back for centuries paternally and maternally, must have had a strong impact on his personality, on his genetic endowment? Even if the spectacular achievements of the molecular basis of genetic factors took place after his death, their role in the formation of personalities, of ethnic groups, of nations was well known and accepted long before. Moreover, how could he fail to realize the role of family environment, of education and upbringing, the effect of the high ethical and moral values, a characteristic feature of Jewish heritage and a long-standing tradition in his family, on his own personality and attitude? These are questions; as to the answers one can only speculate.

In spite of the strong anti-Semitism to which he was exposed from the age of six, he went on his way with iron willpower and determination. Although a career appeared to him irrelevant as he stressed in his writing and by his actions, his success was absolutely spectacular: At the age of 43 he was unanimously elected by the University of Munich to the most famous and coveted chair of chemistry in Germany, held before him by two of the greatest German chemists of the nineteenth century, Liebig and Baeyer. He reached this pinnacle without making any efforts or any concessions incompatible with his high ethical and moral standards. His deep and genuine love of Germany, shared by the overwhelming majority of German Jews, was after all also based on the impact of German civilization, of its art and science, on his personality. Almost all of the scientists mentioned elsewhere in this book also were attached to Germany. The formation of their personalities was influenced by Kant and Goethe, Mozart and Beethoven and so on. This, and the European civilization, influenced their thinking, their feelings, emotions, their pleasures of life. Even if we are unable to explain the precise connection between the factors mentioned and scientific thinking and work, just as we are unable to trace the ability and traits of intuition or imagination, we cannot deny or ignore these factors. The influence of German, and European, civilization was so strong compared to that of Jewish tradition and history, that most scientists, even those proud and conscious of their Jewish heritage, were deeply attached to Germany.

Willstätter's flat rejection of Baeyer's suggestion of baptism was mentioned before. Baeyer belonged to a group of a few distinguished Jewish families that were totally assimilated in the nineteenth century. Baptism and mixed marriages were common and not even considered serious problems. Heinrich Heine's dilemma, his cynicism, his unhappiness about the price of the "ticket of admission" to European civilization was an exception, not a widely shared feeling in this group. For Willstätter baptism, especially for opportunistic reasons, was inconceivable. It meant denying the long and proud history of his family, a dissociation from his ancestors.

His sudden resignation, in 1924, in view of the opportunistic view of some colleagues, that was based more on fear of public opinion and student unrest than on really strong anti-Semitic feelings, came as a total surprise to all. People were stunned and perplexed. All the great efforts to change his mind failed, as described before. When Chaim Weizmann learned the news he felt that this reaction might be an indication of Willstätter's attachment to Judaism. As an organic chemist, Weizmann was a great admirer of Willstätter. He believed that interesting Willstätter in the Zionist movement would be a great asset, especially at a time when very few prominent Jews were interested in Zionism. Many such Jews were even hostile to Zionism; this movement was for most of them perhaps an answer to the problems of persecuted Jews in Eastern Europe, but not in the Western democracies. Weizmann asked Kurt Blumenfeld to visit Willstätter in Munich. Blumenfeld had won over Albert Einstein to active support of the Zionist movement; he was known for his remarkable ability to attract the interest of intellectuals, artists, scholars, and eminent and prosperous businessmen. Blumenfeld visited Willstätter. On his return he told the author: "I can give you in writing, this man will never understand the Jewish problem." This diagnosis has been borne out by the subsequent events.

Willstätter's sudden resignation may have been motivated by a long accumulation of repressed experiences and emotions going back to his childhood. They may have led to a sudden irresistible explosion, his refusal to reconsider his action despite the extraordinary ovations of the students, and the urgent appeals of all his friends and colleagues. But it is even more difficult to understand several of his later reactions. What was the basis of his refusal, two years later, to join the Kaiser Wilhelm Institutes in Dahlem, when he was offered a special institute, to be built entirely according to his wishes? In Dahlem, as he knew, there were a great number of distinguished Jewish scientists. Haber was a close friend. He knew Einstein, Lise Meitner, and several other Jewish scientists. The biochemists Warburg, Meyerhof, and Neuberg (see the following sections) were, in 1926, working in the field of enzymes, which was then his primary interest. In Dahlem there was certainly no anti-Semitic atmosphere. Moreover, in Berlin many Jews were prominent in art and literature, in the theaters, opera, and music; many Jewish artists were adored and acclaimed by the public at that very time. Professor Schlenck, the *Ordinarius* of chemistry at the University of Berlin, the successor of Emil Fischer, had offered him his own chair and was willing to resign, quite a remarkable gesture. Since Willstätter was attached and devoted to Germany, here was a great opportunity to use his creative genius under the most favorable and pleasant conditions and thereby contribute to the future and the well-being of his country. Instead he chose isolation, which meant for an active scientist, at the age of 52, self-destruction. The strange arrangement of participating in scientific work by telephone with Dr. Rohdewald was clearly an indication of how much he must have suffered from being unable to continue the research that meant everything to him. It reveals his frustration, his deep urge to find some way to be active. Actually, it was more an illusion, which may have given him some satisfaction and the feeling of accomplishment, than a reality. Certainly it was far from the superb opportunity he would have had as the director of an institute in Dahlem, surrounded by many excellent coworkers, and many illustrious colleagues and friends. In such an atmosphere he would have been able to make great contributions and his ideas and his life would have received a great stimulus.

However, the most puzzling and perplexing aspect is offered by his attitude and behavior during the Nazi period. His greatly admired friend Haber, a baptized Jew, had voluntarily resigned a few weeks after Hitler took over; he had left Germany and had offered his help to Weizmann. Many brilliant Jewish scientists, Nobel laureates, and many famous scholars and artists resigned their positions and left voluntarily or were soon forced to resign and to emigrate. It is true that in the beginning many people, Jews and non-Jews, liberals, leaders in their fields, did not realize the true nature of the Nazi regime. They were convinced that it would be a short episode, and that the regime would be replaced in a few months by another type of government. The meaning of the absolute dictatorship established by Hitler early in April 1933 was not realized in Germany and still less, for several years, in France, Britain, and other countries. The Nazis were able to crush every opposition by terror, persecution, assassinations, and concentration camps. Many of the victims were liberals, scions of prominent old German families, some belonging to the old nobility. Scientists, many of whom had lived in an ivory tower in the era before Hitler, became aware of the frightful reality and the criminal nature of the regime only when it was too late to act. The courageous attitude of Planck, von Laue, and Heisenberg has been mentioned before. Even non-Jewish scientists left Germany, some immediately, as Schrödinger did, others later, as Max Delbrück, Hans Gaffron, and many others did. The factors that led to this situation are discussed in another chapter.

In 1933 Willstätter visited England and the United States. He was received enthusiastically everywhere and honored in many ways. He was offered several excellent positions. He turned them down and stayed on in Germany, even when the situation became threatening and humiliating for the Jews. It seems difficult to understand the psychology of a man who had so vigorously and forcefully reacted to the veiled anti-Semitism he had sensed in the faculty in 1924. In 1934 he was invited by Weizmann to be chairman at the opening ceremonies of the Daniel Sieff Institute in Rehovot, Israel. The institute was the first building of the huge complex completed after World War II, the Weizmann Institute of Sciences. In his autobiography of 400 pages, only two paragraphs are devoted to this trip to Palestine, now Israel. The ceremonies in the Sieff Institute, built by the "meritorious Chaim Weizmann," are referred to very briefly. Otherwise he speaks about the beautiful view and the magnificent scenery seen from Mount Scopus, and he makes a few comments on the chemicals of the Dead Sea. No feelings are expressed about the great achievements of the Jews in transforming a desert into a green garden country, no emotions about the miracle of Jews returning after 2000 years in exile to their homeland and building it up with incredible creativeness, devotion and enthusiasm. A man conscious of his illustrious Jewish ancestry, who had suffered all his life from anti-Semitism, who witnessed the horrible persecutions of the Jews prevailing in Germany, quite obviously remained quite untouched by the achievements he saw, achievements that moved and deeply impressed even non-Jewish visitors. When Weizmann asked him, "Professor Willstätter, why do you not leave Germany?" he replied that "one does not leave his mother even when she behaves badly."

As mentioned before, he continued to stay in Germany until the end of 1938 and decided to leave only when the conditions became unbearable and his life was threatened. When he crossed the border, in March 1939, he wrote that others were jubilant when they crossed the border and escaped the Nazis; he felt like crying about leaving his beloved Germany.

It is fortunate that thanks to the efforts, the devotion, and the generosity of Arthur Stoll, Willstätter was able to spend the last few years of his life in a beautiful place, in most pleasant and comfortable conditions, in the solitude he wanted. During the last year he suffered from poor health, but received the best possible care. Thus the end of this scientific giant, of this great personality, whose life was exceptionally full of triumphs and tragedies—the latter to quite an extent self-inflicted—came peacefully.

Before turning to biochemisty, it may be useful to remind the reader that this book is not intended to be a survey of German-Jewish pioneers in science, not even in the three topics selected. In view of the agreed-upon space limitation just a few examples have been chosen to illustrate the wide-ranging importance of the contributions of some of those pioneers, the spiritual and cultural atmosphere, and the intensity of the collaboration between Jewish and non-Jewish scientists. Therefore it was necessary to omit many outstanding German-Jewish chemists and physicochemists, among them even two Nobel laureates. One is Otto Wallach (1847–1931), whose contributions to the chemistry of terpenes and volatile oils greatly stimulated the industry of scents and perfumes. He received the Nobel Prize in 1910. The other is Gerhard Herzberg, born in 1904 in Hamburg. He started his research in Göttingen in 1928, shortly after quantum mechanics had been developed. He applied the new knowledge of electronic structure to the analysis of diatomic molecules, and even to some polyatomic molecules, a field in which he soon became recognized as *the* outstanding pioneer. Because of events during the Hitler and postwar period he became involved in astronomy. His brilliant analyses of many spectra of the outer planets brought him great fame, and many scientists consider him to be an astrophysicist. Although the roots of his scientific formation were laid in Germany, his most important contributions were accomplished after he left Germany, especially in Canada. He received the Nobel Prize in 1971.

However, one illustrious chemist and physicochemist, Hermann F. Mark, must be mentioned for a number of reasons. He was born in Vienna in 1895 and studied chemistry there, receiving his doctoral degree in 1921. As a graduate student he synthesized a number of compounds belonging to an important group of compounds, organic free radicals, which remain uncombined even in the solid crystallized state. Although certain features of his work could not be explained at that time, the observations became important for the theory of resonance in organic molecules later established by Linus Pauling. This theory now plays an important role in the understanding of important biochemical reactions. In 1922 he moved to the Kaiser Wilhelm Institute for Fiber Research in Berlin–Dahlem, where he remained until 1927. His X-ray studies there led to the first complete elucidation of several relatively complicated substances, such as hexamethylenetetramine and urea. The aim of these studies was to establish the values of interatomic distances and valence angles in organic molecules. His X-ray work greatly contributed to the physics of X-rays in terms of clarifying the wave–particle duality problems.

In 1925 Mark started his work on polymers, a field in which he became one of the world's outstanding leaders. He, Hermann Staudinger (1881–1965), and Kurt H. Meyer (1883–1952), are considered cofounders of this field, the great importance of which is known to every layman. Mark began his work with studies of cellulose, rubber, and starch and elucidated the molecular structure of these important high polymers within a few years. They consist of long chains in which the free—or almost free—rotation about the single bonds of the backbone chain leads to the tendency to assume random conformations. This explains many properties of high polymers. Mark then developed, on the basis of kinetic studies, a general theory of the polymerization processes. His work on the relationship between structure and properties of macromolecules opened the way for the designing of new polymers with properties to suit specific requirements. Thus he not only laid the groundwork for the basic understanding of polymers, but in addition used his competence and ingenuity to apply these principles to industrial problems, and so became one of the great promoters of the gigantic growth of the polymer industry.

In 1927 Mark left Berlin and joined the I. G. Farben in Ludwigshafen, where he continued his research. During those years he also lectured at the *Technische Hochschule* in Karlsruhe. After 10 years in Germany, he became, in 1932, professor of physical chemistry at the University of Vienna. Since he was a half-Jew (his mother was Jewish), he was forced to leave Vienna after the *Anschluss*. He joined the Polytechnic Institute of Brooklyn in 1940 and became the director of its newly founded Polymer Research Institute. Because of his scientific stature, his inspiring and dynamic leadership, the institute soon became a most active scientific center and attracted a great number of brilliant chemists, physicochemists, and biochemists, since the polymer studies were extended to include biopolymers (proteins, nucleic acids, starch, and the like). The biopolymers, because of their central role in cellular function, had moved into the center of most essential biological problems. Among Mark's many brilliant associates were the two brothers Aharon and Ephraim Katzir-Katchalsky(i), two of the most eminent leaders of the Weizmann Institute. Mark also frequently arranged symposia covering a wide range of topics and involving outstanding speakers, which contributed to the fame and the attraction of the institute for New York's scientific community. Two of these meetings are of interest in the frame of this book. One, in honor of Otto Meyerhof, featured three Meyerhof pupils (Severo Ochoa, Fritz Lipmann, and the author) as the main speakers. The other one was a meeting in memory of Aharon Katzir-Katchalsky, and included a number of brilliant speakers; the meeting was attended by scientists not only from the New York area, but from many distinguished institutes of the East Coast.

Mark is one of the many outstanding examples of illustrious scientific leaders who became victims of the Nazi regime, but turned out to be a great asset for the countries that accepted them, as well as for the whole of mankind.

Two of Mark's activities are of particular interest for this book. By the end of World War II several rich Jews, under the shock of the Holocaust, pledged to provide funds for a great research center in Rehovot for the benefit of Israel and of humanity (see V, E). In view of his great scientific stature, his superb judgment, his ability and experience as an organizer, Mark became Weizmann's chief and most influential scientific adviser in the planning and the establishment of the Rehovot Institute, which opened in 1949. He played an important role on the board of governors in the development of the center in the following two decades. Another great service to the world's scientific community was his untiring effort to restore the disrupted collaboration between German and Jewish and later Israeli scientists. He had the courage to invite, very early, German scientists to his institute in New York. He fully accepted the duty and responsibility of scientists to fight in the forefront to create bridges between nations. Scientific collaboration, especially between Germans and Jews, may sooner or later help the coming generation to overcome the nightmare of a horrible past and lead to constructive steps in political, economic, and other fields of mutual interest.

C. OTTO HEINRICH WARBURG (1883-1970)

The next two scientists to be described, Otto Heinrich Warburg and Otto Meyerhof, must be considered as two of the chief architects of modern biochemistry and biology. Their work and their ideas decisively influenced the development of these fields in the first half of this century. The schools they created provided some of the most prominent leaders of this century.

Warburg's fame in history is inseparably associated with the greatest breakthrough in our understanding of cell respiration—the mechanism of oxidation—since Lavoisier demonstrated that life requires oxygen. This process is the ultimate source of energy for cell functions. Warburg is the undisputed pioneer in the field of enzymes, which catalyze virtually all chemical reactions taking place in living cells. He introduced several new and highly refined methods which dominated the field for several decades, for the analysis of enzymic reactions. These methods were replaced or supplemented only since the 1950s by a great variety of still more refined methods and sophisticated instruments provided by the rapid advances of physics. They were not yet available at the earlier period.

A superb biographical article has been written by Sir Hans Krebs (1972). His analyses of Warburg's scientific contributions and complex personality are of unsurpassed clarity, presenting in a most elegant way both Warburg's strengths and his weaknesses. The author knew Otto Warburg for more than four decades, since 1926 when the author worked in the same institute with Meyerhof. He visited Warburg and had many discussions with him during his frequent trips to Berlin in the postwar period. He last saw Warburg in 1969. He fully agrees with Krebs' masterful and admirable evaluation. Krebs' article was written for biochemists and biologists. This section is essentially based on Krebs' article, but it is greatly condensed and kept on a level that may be accessible to a wider range of readers, those who lack the technical knowledge to understand Krebs' article but are interested in understanding Warburg's contribution within the specific context of this book.

Ancestry and Career

Otto Warburg was born in 1883 in Freiburg in Breisgau. He was the son of Emil Warburg (1846–1931), one of the leading physicists of his era and at that time professor of physics at the university. They belonged to the famous Warburg family, which included many distinguished businessmen, bankers, scholars, and philanthropists. Otto was the tenth direct descendant of Simon Jacob Warburg. Simon Jacob lived in the sixteenth century, in the town of Warburg in Westphalia near Cassel, and was apparently a banker to the ruler of Hesse-Cassel. When he was pressed to be baptized, he refused and moved to Altona near Hamburg, accepting the surname Warburg. The rulers of this area of Germany were quite exceptional in their liberal attitude in matters of creed, trade, and commerce. The banking house of M. M. Warburg was established in 1798. Among the many distinguished members of the family may be mentioned the art historian Aby Warburg (1866–1929), who assembled a unique library. It was fortunately possible to transfer the library to London after the advent of Hitler. Another distinguished member is Sir Siegmund Warburg in London, a warm friend of Israel, who has been extremely helpful to the country and to its various institutions in many respects. He is a member of the board of governors of the Weizmann Institute. Another family member, also named Otto Warburg (1859–1938), was a botanist and a well-known Zionist leader who joined the Hebrew University in Jerusalem. Finally may be mentioned the publisher Frederic Warburg, of Secker and Warburg, the firm that now publishes the Year Books of the Leo Baeck Institute.

Emil Warburg was baptized. He was married to Elisabeth, *née*

Gaertner, a member of an old German family that included several people who distinguished themselves as administrators, lawyers, and soldiers. A brother was a general killed in World War I. She was apparently a lady of great vitality and wit. Otto was the only son, but he had three sisters.

In 1895 the family moved to Berlin, when Emil Warburg was appointed *Ordinarius* of physics at the University of Berlin, the chair of Helmholtz and thus one of the most distinguished posts for a physicist in Germany. Otto received his education in a humanistic gymnasium. His home in Berlin was a stimulating cultural center, frequented by many of the most outstanding scientists and artists. His father was an accomplished piano player and frequently played chamber music with Albert Einstein. Among the illustrious scientists whom Otto met in his father's house were Planck, Emil Fischer, van't Hoff and Theodor W. Engelmann (1843–1909), a physiologist from whom he first learned about photosynthesis, one of the topics to which he later devoted many efforts in his research.

After the *Abiturientenexamen* he studied chemistry at the University of Freiburg, and then moved to Berlin. He received his doctoral degree under Emil Fischer in 1906. For his thesis he prepared optically active peptides* and studied their hydrolysis by enzymes. Otto had a great admiration for Emil Fischer, which he maintained throughout his life. A statue of Emil Fischer was in front of his institute in Berlin–Dahlem. Fischer spent most of his time at his bench, despite many other obligations. His dedication to experimental work, his high standard of reliability, and his personal integrity set for Warburg a model after which he patterned his own life.

Warburg was very ambitious. When he was a student he set a high aim for himself: to find a cure for cancer. After he received his doctorate in chemistry, he decided to study medicine and went to Heidelberg. There he started to do experimental work in the laboratory of the Department of Internal Medicine. This was a great center of medical research, since the director was Ludolf Krehl, a pioneer in stressing the relevance of basic science to clinical medicine. Warburg obtained his M.D. degree in 1911.

---

* Proteins are formed by amino acids; these can be linked together to form peptides, which may consist of two, three, or several amino acids. If light passes through a solution containing "asymmetric" substances, the light may be turned to the left or to the right. Asymmetric substances have the same composition, but some atoms are differently arranged. Compounds capable of rotating the light are optically active. Amino acids, peptides, and sugars are some of many substances that may be optically active.

Scientific Contributions

Oxidation. Warburg started his research work in 1908 with studies on respiration occurring during growth. He measured the oxygen consumption in growing sea urchin eggs after fertilization and discovered that the oxygen uptake increased up to sixfold. These observations already revealed his great skill in selecting the right kind of material and perfecting experimental techniques. The selection of the material frequently decides success or failure. In sea urchin eggs the mass of active matter is large in relation to the egg yolk mass, which is used for nutrition; moreover, the development of the fertilized egg is very rapid. He assumed correctly that chemical work is done when living matter increases with growth and that therefore the rate of energy supply—and that means oxidation—must rise rapidly. In addition, he improved the technique used to measure oxidation. Thus by selecting the right material he succeeded where previous observations had failed or had given inconclusive results.

Subsequently Warburg turned to the analysis of the mechanism of oxidation. He chose to investigate the nature of the respiratory enzyme (*Atmungsferment*), which enabled molecular oxygen to oxidize substances in neutral solution and at biological temperatures. Under the same conditions, without the enzyme, the substances were completely stable. Such an analysis was difficult with the methods available at that time, since Warburg found that the enzyme was attached to some structure, which he later referred to as "grana." The mechanism of enzymes in solution is more accessible to analysis than enzymes attached to a structure. The enzymes of fermentation had been obtained by Buchner in 1897, in solution. More than three decades later, the "grana" was found to be subcellular organelles called *mitochondria*. With electron microscopy, in combination with biochemical studies, it was found that oxidation reactions take place in mitochondrial membranes. In fact, a whole series of steps is involved and is referred to as electron transfer or the oxidation chain.

But despite the difficulties involved in studying a structure-bound-enzyme reaction, Warburg was not willing to give up his efforts. He found that a number of substances inhibit oxidations according to their surface activity. These substances are chemically indifferent and on their removal oxidation returns. Thus their action could be attributed to their adsorption to the surface preventing access of substrates to the enzyme. A different kind of mechanism was clearly involved in the action of cyanide, a powerful inhibitor of biological oxidation. Since cyanide readily reacts with heavy metals, Warburg concluded that a heavy metal must be involved in the catalytic process. He suspected it might be iron and found indeed that iron was present in biological material. He demonstrated, by "model" experiments, that iron salts accelerate the oxidation of several substances, supporting his hypothesis that the heavy metal involved was an iron compound.

It was at this stage that the period in Heidelberg came to an end. He was appointed "member" (head of an independent department) of the Kaiser Wilhelm Society, in 1913, at the age of 30; the appointment was to take effect on April 1, 1914. One of the main promoters of this appointment was Emil Fischer. In his six years in Heidelberg Warburg had an amazing record of achievement, as indicated by about 30 publications, despite the work for his M.D. degree. Among his associates of that period were Otto Meyerhof (see the following section) and Julian Huxley.

Interruption by the War. Warburg moved to Berlin before his institute in Berlin–Dahlem was ready and started to work in Nernst's laboratory. A few weeks after the outbreak of World War I he joined a cavalry regiment and served in the front lines. He rose to the rank of lieutenant, was wounded in action, and was decorated with the Iron Cross First Class. In March 1918 he received an interesting and moving letter from Albert Einstein imploring him to give up active service, which was endangering his life. Einstein indicated that Warburg was considered one of the few scientists of great promise in physiology, a field still poorly developed and mediocre compared to the great advances in other sciences. He suggested that Warburg seriously consider whether his present duty could not be equally well served by other people. Warburg accepted the advice of his father's friend and returned to Berlin to take up his work in the Kaiser Wilhelm Institute of Biology. The hopelessness of the war was by then obvious and the endless frustrations may have helped to make him receptive to Einstein's appeal, which may have saved his life. In an interview in 1966 he stressed that during his military service, the only interruption of his scientific life, he had learned many realities of life that would have escaped him in the laboratory.

After the war Warburg had his own department and he took up several lines of research. But let me first return to his efforts to elucidate the nature of the catalytic process by which oxygen is activated. His evidence that iron is involved in the process has been mentioned. In 1926 he added new evidence for this assumption. Since some iron-containing biological compounds, such as hemoglobin, readily react with carbon monoxide, he tested the effect of carbon monoxide on the respiration of yeast cells. He discovered that this substance indeed inhibits respiration. Just at this time A. V. Hill visited his friend Meyerhof and also paid a visit to Warburg. When he heard about the experiments he drew Warburg's attention to the light sensitivity of the carbon monoxide–hemoglobin association discovered by Haldane and Smith in 1896. Warburg immediately tested whether the carbon monoxide compound of the respiratory catalyst was also light sensitive and found that the inhibition of respiration was greatly decreased when the yeast suspension was illuminated. Warburg recognized that this effect opened the possibility of obtaining the spectrum* of the component in the catalyst that reacted with carbon monoxide. He illuminated the yeast suspension with monochromatic light (light formed by one wavelength) of known intensities and quantitatively measured the effect of light on the inhibition of respiration by carbon monoxide. From these observations he obtained the absorption spectrum of the component of the catalyst that competes for oxygen and carbon monoxide. The principle involved is simple. Since only light absorbed by the catalyst can be effective in removing the inhibition, the absorption spectrum of the catalyst is related to the action of light. But to devise and to carry out the experiments and to develop the mathematical analysis of the measurements required great ingenuity and exceptional experimental and theoretical skill. Warburg had to find sources of monochromatic light of sufficient intensity; he needed precise methods for the gas exchanges and light intensities; and finally he had to elaborate the theory for the quantitative interpretation of the data obtained. The spectrum obtained agreed quantitatively with those of iron porphyrins† such as are found in hemoglobin. This was a monumental discovery. Warburg presented the data in a lecture, in 1928, with extraordinary lucidity and clarity. The excitement was great. The author, who attended the lecture, and many people in the audience had the feeling that the results inaugurated a new era in enzyme chemistry. It was the first time that the active group of an enzyme action had been identified. Enzymes are very large protein molecules. That the action could be associated with a specific and relatively small molecular group promised to open the way for a real understanding of the mechanism of enzyme action. In fact, even today active groups of enzymes are a central field of enzyme chemistry. Actually, since the early 1960s, it received a tremendous stimulus when the tridimensional structure of crystallized enzyme was explored by X-ray analysis and the precise events taking place between substrate and the active group in enzymes could be established. Warburg's achievement was thus a milestone in this field, widely referred to as molecular biology. He received the Nobel Prize for this work in 1931.

There immediately were many critical objections, some justified, some unjustified. This is a common experience with all great discoveries in science, be it physics, chemistry, biology, or any other branch of science,

---

* When light waves pass through a prism they form bands of light on a screen, which are called a spectrum. With the help of spectrophotometers one can obtain spectra of compounds, molecules, and atoms that show characteristic bands.

† Porphyrins are complex organic structures formed essentially by 4 five-membered ring structures in which each of the rings is formed by four carbons and one nitrogen. The four nitrogens are all bound to a single iron. Hemoglobin, for example, is a protein, globin, that is linked to a porphyrin called heme.

since not a single achievement, however brilliant, gives all the answers and leaves no problems unsolved. Science by nature is progressive and accumulative. Half a century afterward, many of the greatest discoveries may look naive. Still they had a decisive influence on further progress. One serious problem for Warburg was that hemoglobin and cytochromes also contain iron porphyrins. Cytochromes are a group of electron-transporting proteins that contain an iron porphyrin as active group. The eminent biochemist David Keilin in Cambridge had discovered, in 1925, that cytochromes play a central role in respiration; at that time he had found that three different cytochromes, referred to as cytochromes a, b, and c, readily undergo oxidation and reduction (electron release or up-take) in living cells. The spectrum found by Warburg agreed quantitatively with the iron porphyrins of hemoglobin and the cytochromes, but it was not identical. There exist quite a few types of porphyrins that have the same fundamental structure but differ as to the side chains attached to the five-membered-ring structures. This discrepancy originated a prolonged controversy between Warburg and Keilin, which cannot be discussed here. The contradictions were resolved in 1939, when Keilin discovered that cytochrome a consisted of (at least) two components, one of which he named a_3 or cytochrome oxidase. This enzyme combines with carbon monoxide and cyanide and has the same spectrum as Warburg's respiratory enzyme (*Atmungsferment*). Today, more than four decades later, we know that there are five different cytochromes that play an essential role in the electron transfer system. The respiratory chain is an assembly of many compounds that form an integral part of the inner mitochondrial membrane. Mitochondria have two membranes, an outer membrane, in which the citric acid cycle takes place (see V, A), and an inner one.

Dehydrogenases. The energy used by the human organism is derived from complex molecules that are metabolized in the cells, such as carbohydrates, fats, and proteins. The breakdown products are then oxidized. But oxygen does not react directly with these breakdown products: They transfer electrons to special electron carriers. When these carriers accept electrons, they are reduced. The reduced forms of these carriers then transfer their electrons to oxygen. For all these reactions enzymes are necessary. The transfer of electrons to molecular oxygen is the last step in a long chain. At the beginning there must be a transfer of hydrogen molecules from the breakdown products to the carrier, a process in which electrons are added to the carrier. The enzymes performing this action are called *dehydrogenases.* The oxidation and reduction processes are inseparably linked to effect the oxidation of the metabolic end products. The action of the respiratory enzyme (cytochrome oxidase) alone thus cannot explain the whole oxidation process.

Dehydrogenases—hydrogen-activating enzymes—were long known to occur in biological material from the work of Torsten L. Thunberg, Heinrich O. Wieland, and others, beginning in the first decade of this century. They found that biological materials contain enzymes that catalyze the reduction of certain synthetic dyes (e.g., methylene blue) by a variety of compounds known to occur in the cell as metabolic products. The common feature of these substances is their ability to donate hydrogen atoms; that is, their ability to undergo dehydrogenation. In the 1920s there were many discussions about whether the catalytic activation of oxygen, as studied by Warburg, or the catalytic activation of the hydrogen atoms of the substrates, studied by Thunberg and Wieland, was the essential feature of biological oxidation. Warburg rejected the concept of the activation of hydrogen because it was based on the use of nonbiological compounds (dyes). At that time he did not realize that even if the reactants used were not biological, the enzymes catalyzing the reactions and the substrates providing the hydrogen were cellular components. But during a visit to Johns Hopkins Medical School in Baltimore, in 1929, he was greatly impressed by a demonstration of E. S. G. Barron with red blood cells and became convinced that the dye promoted clearly enzymic reactions. After his return to Berlin he investigated the chemical action of this dye (methylene blue) and found that oxygen was taken up after cytolysis of the red blood cell (i.e., in solution). Since he obtained the process in solution, it became possible to investigate the problem by standard methods of biochemistry. This was the beginning of a decade of intensive studies of the mechanism of dehydrogenases. During that period he made a number of ingenious and brilliant contributions that profoundly affected and greatly stimulated the whole field of enzyme chemistry. When Warburg dialyzed* the solution obtained from red blood cells containing the dehydrogenase, his results indicated that the hydrogen-transferring mechanism involved two components: the protein and a coenzyme. Many, although not all, enzymes require in addition to the protein a small molecule, a coenzyme. Some vitamins have been found to be coenzymes. Heating denatures the proteins, whereas coenzymes are heat stable. Warburg discovered that yeast extracts showed much greater activity under the same conditions than the solution obtained from red blood cells. His subsequent work was therefore carried out with yeast cells. The protein component was found to consist of at least two fractions. One of these was yellow and was referred to as "yellow enzyme."

---

* Dialysis is a procedure in which the enzyme-containing material is put in a tube of cellophane or other semiporous material; the tube is kept for some time in a solution containing only water and some salts for keeping the solution neutral (i.e., a buffer solution). Small molecules, such as coenzymes, penetrate through the pores of the tube to the outside, whereas macromolecules—the enzyme proteins—remain inside. Thus coenzymes and enzymes may be separated. The enzymes are inactive, but become active again on addition of the coenzyme.

The second component was colorless. When the yellow enzyme was denatured, by adding methanol, the colored component went into solution. This compound was obtained in crystal form by Warburg. It was found to be a luminoflavin.*

As was shown by Hugo Theorell, working at that time, in 1934, as a guest in Warburg's laboratory, the yellow enzyme consists of a specific protein (enzyme) and a phosphorylated riboflavin, in which a phosphorylated sugar (a pentose) is attached to the nitrogen of the middle ring of the three-ring structure. This compound is the active group, the coenzyme, which undergoes hydrogenation or dehydrogenation (i.e., accepts or gives up hydrogen atoms). Soon a variety of yellow enzymes, now called *flavoproteins*, were discovered. In some cases the active group is the same as in the first enzyme discovered—it is a riboflavin phosphate —but in the majority of cases it also contains a compound called *adenosine*, which is a nucleotide.† This compound was also discovered by Warburg. He called this type of coenzyme *flavin adenine dinucleotide*. Whereas the riboflavins accept hydrogen atoms directly from many organic compounds occurring in the cell, this is not the case with the dehydrogenase of red blood cells that Warburg had used originally. He immediately started to investigate the structure of this coenzyme. Using classical organic chemical methods of isolation, Warburg obtained the pure coenzyme of this dehydrogenase. The new coenzyme contained phosphate, pentose, and adenine, but in addition a compound not known before to occur in coenzymes: nicotinamide.‡ This was of the greatest importance

---

* A luminoflavin is a compound formed by three fused six-membered rings; one ring has six carbon atoms, and the two adjacent rings have four carbon atoms and two nitrogens. This ring system is called *isoalloxazine*. The rings may have side chains and the nomenclature is correspondingly modified.

† Nucleic acids have today become widely known, since ribonucleic acid, the famous RNA, and deoxyribonucleic acid, DNA, are two compounds that are essential in genetics. Adenine, one of the major components of nucleic acids, is formed by a six-membered ring, called *pyrimidine*, which consists of four carbon and two nitrogen atoms, fused to a five-membered ring called *imidazole*, which has three carbons and two nitrogen atoms. These two fused rings are called *purines*. They form the basis of many compounds, such as the stimulants in coffee and tea: caffeine and theophylline. Adenine is a purine to which an amino group (NH_2) is attached to a carbon of the ring. When adenine is linked to a sugar (a five-membered-ring pentose), it is called *adenosine*. Adenosine linked to three phosphate groups, adenosine triphosphate, usually abbreviated ATP, is the most common source of energy for metabolic reactions in living cells (see IV, D).

‡ Nicotinamide is a derivative of nicotinic acid, which is formed by a six-membered pyridine ring (five carbons and one nitrogen); to one carbon is attached a carboxyl group (COOH). In nicotinamide there is an amide group ($CONH_2$) instead of a carboxyl group. Nicotine is only remotely related to nicotinamide; it has the pyridine ring, but instead of the amide group a five-membered ring group is linked to the pyridine. In the coenzyme adenosine and pyridine are linked to each other by two phosphate groups, which are attached to the nitrogen of the pyridine and one

because eventually nicotinamide was found to be the catalytically active group that transfers hydrogen by undergoing hydrogenation and dehydrogenation. Subsequently, the coenzyme was found to be present in numerous other oxidation–reduction systems, especially those catalytically active on enzymes of fermentation. Today over 150 different dehydrogenases, acting specifically on a variety of substrates, are known in which the nicotinamide-containing coenzyme plays a key role as a hydrogen carrier.

In the study of nicotinamide dinucleotides and the enzyme reactions involved it was of paramount significance that the coenzyme showed, when reduced, an absorption band of 340 nanometers (i.e., one-billionth of a meter) in the spectrum. This provided an extremely elegant method of determining very minute quantities of the coenzyme with great precision and rapidity by using spectrophotometry. It permitted the determination of enzyme activities and those of metabolites involving the coenzyme in a way far superior to the tests available at that time. Warburg thus introduced into enzyme chemistry a method invaluable to biochemists even today. The method also gained practical importance in clinical biochemistry.

The existence of a coenzyme has been known since the pioneer work of Arthur Harden in 1906 on fermentation, but its function remained obscure. The elucidation of the structures of the two coenzymes associated with oxidation–reduction processes in cells as special carriers of electrons was a monumental achievement. It provided an essential key to the analysis of many important enzyme mechanisms. Nicotinamide dinucleotide is a major electron acceptor in the oxidation of many metabolic products. In the oxidation of a substrate a carbon of the nicotinamide ring accepts a hydrogen ion and two electrons. The hydrogen molecule is formed by two hydrogen atoms, each made up of a positively charged nucleus and a negatively charged electron. When a hydrogen atom loses its electron it becomes a positively charged ion (proton). When the nicotinamide ring accepts a hydrogen ion, the second ion goes into the solution. The other major electron carriers are the flavins. They also accept two electrons, but they accept in addition the two hydrogens lost by the substrate; both hydrogen atoms are fixed on the nitrogen of two different rings of the three-ring structure of isoalloxazine.

Crystallization of Enzymes of Fermentation. During the 1930s, thanks to the work of Harden, Neuberg, Meyerhof, Embden, the Coris, Parnas,

---

of the nitrogens of imidazole. Some coenzymes have a third phosphate group attached to the second pentose. Warburg called them *pyridine dinucleotides* and distinguished between diphospho- and triphosphopyridine nucleotides (DPN and TPN). This nomenclature was used for 30 years, but then the International Nomenclature Committee replaced it by the terms *nicotinamide adenine dinucleotide* and *nicotinamide adenine dinucleotide phosphate* (NAD and NADP).

Dorothy Needham, and Lohmann, the enzymes of the intermediary stages of lactic and alcoholic fermentations had been identified and characterized, and a number of them had been partially purified. None of them had been obtained in pure or crystallized form. The ultimate analysis of the mechanism of an enzyme depends on its availability in crystallized form and, as we know today, on the knowledge of its tridimensional structure, made possible with the use of X-ray analysis. Warburg and his associates approached enzyme purification with the new rapid and efficient procedures and succeeded in crystallizing, within a few years, nine enzymes active in fermentation.* Prior to the development of the new methods, assays of enzyme activity, essential in purification, were very laborious. The new optical tests could be carried out in minutes and required very small quantities. Another step, which could be tested by spectrophotometry and was useful in many experiments, was Warburg's discovery that the absorption band of enol in phosphopyruvate was 240 nanometers.

In the course of his work on enzyme purification, Warburg discovered a new key intermediate of glycolysis (1,3-diphosphoglycerate) that has escaped Meyerhof (see the following section). This discovery became possible only after the purification of the enzyme glyceraldehyde phosphate dehydrogenase. The purification of the enzymes led to the explanation of the observation of Lohmann and Meyerhof, that glycolysis and fermentation are inhibited by fluoride by specific block of the enzyme enolase. With the pure enzyme Warburg showed that the inhibitory substance is a magnesium fluorophosphate, which combines with the enzyme.

In the eighteenth century Antoine Lavoisier initiated what is widely referred to as the chemical revolution; he deeply influenced the thinking of scientists on respiration and fermentation throughout the nineteenth century. The achievements of Warburg (and those of Meyerhof described in the following section) during the first part of this century may be considered as the greatest breakthrough in these two fields since Lavoisier. In addition, they laid the groundwork for some of the most spectacular advances of biochemistry in the period following World War II. These two men and their brilliant pupils, several of them Nobel laureates, who became great leaders in the last decades, are striking examples of the impact of German-Jewish scientific collaboration on the twentieth century in the biological sciences.

Photosynthesis. All the energy of biological systems is derived from solar energy trapped by green plants in a process called *photosynthesis*. The energy of sunlight catalyzes reactions by which water (H_2O) and carbon dioxide (CO_2) form carbohydrate (sugar) and oxygen; on this

---

* According to the now accepted nomenclature, they were lactic dehydrogenase, enolase, aldolase, glyceraldehyde phosphate dehydrogenase, phosphoglycerate kinase, alcohol dehydrogenase, pyruvate kinase, α-glycerophosphate dehydrogenase, and triose phosphate isomerase.

oxygen, given off by plants, all animal life depends. Hence the paramount importance of the process and its interest to scientists. Joseph Priestley, the discoverer of oxygen, had already observed that an animal kept in a closed container with a green plant exposed to sunlight may survive because the air is improved. He and Lavoisier and many investigators after them worked on the problem of photosynthesis, but biochemistry was not yet sufficiently developed to provide an explanation of the mechanism.

Warburg became interested in the problem very early. Einstein had formulated, in 1912, the law of photochemical equivalence, according to which one quantum of light is involved per photolyte (i.e., the substance becoming activated after the absorption of light). Soon afterward Emil Warburg, Otto's father, provided experimental evidence for the law postulated by Einstein in experiments with inorganic acids. Otto's personal contacts with these developments explain his interest in the problem of photosynthesis and the special aspect that preoccupied him.

From the beginning it was obvious that an answer to the problem required new methods of measurement. The physical method for quantitative determination of light, *bolometry,** had been worked out in his father's laboratory, where he had learned how to use it. The method was also used by Max Planck in his experiments on heat radiation, which led to the determination of the quantum of action. The measurements of light intensities had to be correlated with those for the determination of the rate of photosynthesis. Reliable quantitative measurements of this kind were not available. Warburg introduced several decisive innovations. Among them was the modification of manometric techniques, which he had developed for the measurement of respiration (see Kleinzeller, 1965) to the study of photosynthesis. Furthermore, he used suspensions of a unicellular alga—*Chlorella*—as experimental material. The use of single cells, as he found before, was most valuable for studies of cellular reactions. Before him, investigators had used green leaves for their studies. Single cells are far better for quantitative studies than systems formed by many cells, which are much more complex and may contain many additional interfering factors (such as extracellular space and barriers); such factors are eliminated when single cells are used. The introduction of single cells as experimental material has usually initiated significant progress in the field under investigation. In fact, today cellular substructures such as mitochondria, isolated cell membranes, and nuclei are frequently the experimental material, and in many cases are superior to the use of the whole cell. Warburg also introduced a method of intermittent illumination that proved essential for the distinction between reactions taking place in the light and those taking place in the dark.

---

* A bolometer is an electrical instrument for measuring extremely small quantities of heat radiation by changes in resistance of a blackened platinum strip exposed to radiation.

One of Warburg's primary aims was to establish the validity of Einstein's law of photochemical equivalence for photosynthesis, thus reflecting the influence of his early experience in the field. He found that 4 light quanta were necessary to produce one molecule of oxygen (corresponding to the assimilation of one molecule of carbon dioxide).* For red light at 660 nanometers the quantum energy is about 43 kilocalories per mole (a mole is the weight of a compound in grams corresponding to its atomic weight). A requirement of 4 quanta would mean an efficiency of about 65%, an unusually high efficiency for any biological reaction. The figure is based on the requirement of 115 kilocalories for the reduction of 1 mole of carbon dioxide to the level of carbohydrate, assuming that light transforms one molecule of carbon dioxide and one of water (H_2O) to carbohydrate and oxygen.

About half a century has passed since Warburg performed these experiments. They were an extraordinary and outstanding achievement considering the standards of that period. But in this long interval biochemistry, the understanding of the chemical and molecular events in cells, has profoundly changed and has revolutionized our knowledge in every respect. The efficiency of a chemical process is only one aspect and few biochemists will consider it the most pertinent or interesting one. A satisfactory answer must provide insight into the detailed mechanism of a process; that is, how the plant cell using solar energy transforms it step by step with the end result of providing oxygen for sustaining animal life. The problem has preoccupied a number of brilliant investigators and today there exists a vast amount of information about the most essential features of the mechanism, although—as in all fields of science—a number of unanswered questions remain. This book is obviously not the appropriate place to review the field; only a few landmarks may be briefly mentioned before we return to Warburg's reactions to these developments.

First it may be mentioned that the process of photosynthesis, as we know it today, is localized in special organized structures of the plant cell, the *chloroplasts*; they are small organelles similar to the mitochondria in animal cells, in which the processes of oxidation take place as described before. In these chloroplasts are found cylindrical structures, a collection of a multilayered system of lamellae. These lamellae consist of membrane-like sheets, called *quantasomes*. As in the mitochondria, in which respiration (the oxidative and electron transfer chains) take place in the membranes, photosynthesis is effected in the chloroplast membranes. Quantasomes contain quite a few different types of compounds. The energy of light at any wavelength of the visual spectrum, absorbed by the quantasomes, can be transmitted to the specific molecule chlorophyll. The structure of this compound, analyzed by Willstätter and his associates, has been described. It is basically similar to that of the porphyrins

* The energy of a photon is one quantum of electromagnetic radiation.

of the respiratory chain, only the iron is replaced by magnesium. It is universally present in cells with photosynthetic ability.

Among the important steps in the elucidation of the photosynthetic process a few may be mentioned as an illustration of its complexity. In 1939, Robert Hill observed that isolated cholorplasts evolve oxygen when a suitable electron acceptor is present; the acceptor is simultaneously reduced. The oxygen production can occur without the reduction of carbon dioxide. The oxygen comes from water (H_2O) rather than from carbon dioxide (CO_2), since no carbon dioxide is present in the system used. The light-activated transfer of an electron from a substrate against a chemical potential gradient is a primary event in photosynthesis. Another significant step was the evidence offered by Robert Emerson, that there exist two different systems called *photosystem I* and *photosystem II*. The first one takes place in the dark, the second one in the light, that is, at a wavelength smaller than 680 nanometers. System I generates a strong reductant, which reduces NADP to NADPH. System II produces a strong oxidant that forms oxygen from water (H_2O). ATP is generated as electrons flow through an electron transport chain from system II to system I, as was discovered by Daniel Arnon. This latter process is called *photosynthetic phosphorylation*. The two systems can be separated in different particles.* A decisive step in the elucidation of photosynthesis was the investigations of Melvin Calvin and his associates, starting in 1945. It was of crucial importance that radioactive carbon was available at that time, because it permitted the analysis of the different steps involved in the process by which carbon dioxide is taken up and transformed into carbohydrate. Like most cellular reactions, the sequence of steps has a cyclic character referred to as the "Calvin cycle." The carbon dioxide combines with a sugar formed by five carbon atoms and two phosphate groups (called ribulose 1,5-diphosphate); the intermediate form resulting from this association rapidly breaks down into two phosphorylated three-carbon compounds (3-phosphoglycerate). The conversion of the 2 three-carbon compounds to a sugar with six carbon atoms is essentially the same as in the usual glycolytic pathway.† The elucidation of the photosynthetic pathway permitted the estimation of the standard free energy change, $\Delta G°$, required for the formulation of oxygen by sunlight. It turned out to be 8 photons per molecule of oxygen formed, or twice as much as originally estimated by Warburg.

---

* The separation is obtained by a procedure called *chromatography*. It is today widely used in biochemistry in greatly varying forms for the separation of a vast number of compounds present in cells, including proteins (and enzymes).

† On the basis of the Calvin cycle it is possible to write the net balance of the process as follows:

$$6CO_2 + 18ATP + 12NADPH + 12H_2O \rightarrow$$
$$C_6H_{12}O_6 + 18ADP + 18P_i + 12NADP^+ + 6H^+$$

When Warburg spent a year in the United States in 1948–1949, most investigators had already accepted the view that 8 to 10 quanta were required per molecule of oxygen evolved in photosynthesis. He spent the summer of 1949 in Woods Hole on Cape Cod, where there is a large Marine Biological Institute. In the summer it is a great scientific center, the meeting place of many biochemists and biologists (see IV, D). At Woods Hole Warburg met many old friends from the Dahlem days, such as Otto Meyerhof and Carl Neuberg. James Franck and Hans Gaffron were also there. Gaffron had worked in Warburg's laboratory on photosynthesis for many years. He left Germany in the 1930s for the United States. He is not Jewish, but just could not stand the Nazi regime, like so many other scientists. Gaffron continued his work on photosynthesis in the laboratory of James Franck, who had come to the United States in 1934 and whose interest had turned to photochemical reactions even before he left Germany. There were bitter controversies between Franck and Gaffron, on the one hand, who insisted on the higher quanta requirements, and Warburg on the other hand, still rejecting any modification of the 4 quanta that he had estimated in the 1920s. One day Warburg announced a demonstration of his experiments on the quantum requirements in photosynthesis. He was helped in all technical arrangements by Dean Burk from the Cancer Institute in the National Institutes of Health in Bethesda, Maryland. The demonstration was followed by a lecture in a crowded lecture hall. A large section of the scientific community had gathered there, anticipating an interesting and fascinating fight betwen Warburg and Franck. A sharp exchange of contradictory views took place, but both parties did not change their claims. Such controversies are unusual and rare in Anglo-Saxon countries. To those of us who had participated in the Haber colloquia, strong disagreements were nothing extraordinary. However, since the discussants were usually close personal friends, the atmosphere was exciting but friendly, whereas during the discussion in Woods Hole one felt the tenseness of the opponents. Franck was quite distressed about the rather scathing attacks of Warburg. He was a great admirer of Warburg's father, who had been his teacher, and he was most unhappy about the unpleasant situation and the fact that a scientific disagreement should interfere with personal relationships.

The many new results in this rapidly developing field did not shake Warburg's conviction about the unique efficiency of photosynthesis. During the last 20 years of his life he vigorously reaffirmed his earlier findings and together with Dean Burk tried to interpret the new data by introducing a number of assumptions, always maintaining the high efficiency when the experiments were carried out under well-specified optimal conditions. But his arguments were not acceptable to the investigators competent in the field. How could such an extraordinary and brilliant man, who had made pioneering and most ingenious contributions to the problem, as even his strongest critics admit, uncompromisingly uphold a reasoning that was considered a misjudgment? How could he not appreciate the dramatic advances that had taken place after his early observations had been made? This attitude appears even more surprising when we consider that to most investigators the efficiency of a process is not the most pertinent aspect, as compared to the mechanism involved (for an outstanding review of the present information see, e.g., Witt, 1971). The question of Warburg's attitude will be discussed in the context of the features of his personality.

Cancer. One of Warburg's chief aims since his student years was to find an answer to the cancer problem. He was determined to find the fundamental basis, the chemical changes that take place when a normal cell, the growth of which is controlled, becomes a cancer cell, the growth of which is unrestricted. His hope was to find a way, on the basis of this knowledge, to prevent or cure cancer. The question of whether the metabolism of the cancer cell differs from that of the normal one had been asked before, but earlier attempts to find an answer suffered, in Warburg's view, from both conceptual and experimental shortcomings. Since, as he had shown with fertilized sea urchin eggs, growth requires energy, one first had to find the reactions that provide the energy for the growth of cancer.

He started with the determination of the rate of respiration of a certain rat carcinoma, using the technique he had developed for measuring cell respiration. The results were clear-cut. The rate of oxygen consumption of the cancer cell did not differ from that of a variety of normal cells. But cancer cells form lactic acid from glucose to an extent that far exceeds the rates of many tissues, such as kidney, liver, and pancreas. The energy derived from this glycolysis was high enough to make a significant contribution to the energy needs of the tissue. Warburg soon confirmed these results with a variety of types of cancer cells including those of humans. A special feature of the high rate of glycolysis of cancer cells was its occurrence in the presence of oxygen. It has long been known that many tissues, such as muscle, form lactic acid from carbohydrate in the absence of oxygen; the glycolysis greatly increases during muscular contraction, when additional energy is needed.

Warburg then raised the question of whether this glycolysis in the presence of oxygen is specific for cancer cells. He tested, therefore, many tissues as to their capacity to glycolyze in the presence and absence of oxygen. It turned out that in all metabolically active animal tissues there is glycolysis in the absence of oxygen, but in most of them no glycolysis is found in the presence of oxygen. The only exception was the retina of warm-blooded animals, where glycolysis, even in oxygen, was stronger than in cancer tissue. Warburg interpreted this finding as an artifact because the retina is a very delicate tissue.

Some 60 years earlier, Louis Pasteur had found that the rates of fermentation in microorganisms are generally high in the absence of oxygen and low in its presence. He introduced the terms *aerobic* and *anaerobic*. Warburg concluded that cancer cells, in contrast to normal ones, are unable to suppress glycolysis in the presence of oxygen. At that time Meyerhof had established that in muscle lactic acid formation decreases in the presence of oxygen. Only about one-sixth of the lactic acid is oxidized and he assumed that the major part is resynthesized to carbohydrate with the energy derived from the oxidation of the lactic acid. Warburg determined the amounts of lactic acid formed in several tissues under aerobic and anaerobic conditions. He found about the same relations between lactic acid formed and oxidized as Meyerhof and referred to these findings as to the "Pasteur effect." But the real mechanism by which oxygen prevented lactic acid formation remained obscure. The assumption of Meyerhof and that later offered by Warburg turned out to be incorrect. Only several decades later the explanation of the so-called Pasteur effect was found by the work of many investigators (mainly that of F. Lynen, V. A. Engelhardt, Th. Bücher, E. Racker, and A. Sols). The suppression of lactic acid formation is due to a control mechanism. There is an enzyme in the glycolytic pathway, called *phosphofructokinase*, which primarily controls the rate of glycolysis. The enzyme is an *allosteric* one.* Its activity is inhibited by adenosine triphosphate (ATP) and stimulated by the split products adenosine diphosphate and monophosphate and also by citrate, a metabolite that forms part of the Krebs cycle (see Chapter V, A). Respiration causes ATP accumulations and therefore lowers enzyme activity. When the concentration of ATP is low—that is, when the energy charge of the cell is low—the enzyme becomes active. In this way the energy requirements of the cell are regulated. Obviously, all the information required for providing the explanation was unavailable in the 1920s; only because of the rapid advances of enzyme chemistry in the last 30 years has the mechanism been elucidated.

When Warburg started his work on the energy supply of rapidly growing cancer cells, he had anticipated a higher rate of energy supply. The results did not show a high respiration of the cancer cells, but an additional supply of energy by glycolysis. It is evident that in cancer cells the additional energy supply is diverted from the performance of their normal function to growth. This is a feature that is called in genetics *de-differentiation* or *reorientation of gene expression* (genes are the elements that contain the genetic transmitter factors). It thus appears that cancer cells develop due to a genetic defect. This may be the reason

---

* Allosteric enzymes have in addition to their catalytic site another site, which may react with other compounds that may control or regulate the catalytic activity. The term was introduced by Jacques Monod (1910–1976), a brilliant French biochemist; the first enzyme of this type was described, in 1959, by Jean-Pierre Changeux.

that some viruses may cause cancer, although this is not yet a definitely established fact.

Warburg maintained his interest in cancer throughout his life. He still performed a variety of experiments. They seemed to indicate that low oxygen pressure may be a factor in producing cancer. Up to the end of his life he was convinced that the loss of normal respiration is a key factor in carcinogenesis. He insisted that the primary cause of cancer is the replacement of normal respiration by glycolysis. He summarized his views, in 1967 at the age of 83, in a paper entitled "The Prime Cause and Prevention of Cancer." A few significant paragraphs may be quoted, since they indicate in his own lucid style his attitude. Many more paragraphs are quoted by Krebs (1972), from whose article these quotations were taken. Warburg writes,

> There are primary and secondary causes of diseases. For example, the primary cause of the plague is the plague bacillus, but secondary causes of the plague are filth, rats, and the fleas that transfer the plague bacillus from rats to man. By a primary cause of a disease I mean one that is found in every case of the disease.
>
> Cancer, above all other diseases, has countless secondary causes. Almost anything can cause cancer. But, even for cancer, there is only one prime cause. Summarized in a few words, the prime cause of cancer is the replacement of the respiration of oxygen in normal body cells by a fermentation of sugar. All normal body cells meet their energy needs by respiration of oxygen, whereas cancer cells meet their energy needs in great part by fermentation.
>
> Most experts agree that nearly 80% of cancers could be prevented if all contact with the known exogenous carcinogens could be avoided. But how can the remaining 20%, the endogenous or so-called spontaneous cancers, be prevented?
>
> Because no cancer cell exists, the respiration of which is intact, it cannot be disputed that cancer could be prevented if the respiration of the body cells would be kept intact.

After proposing a few very simple methods to influence cell respiration, he forcefully expresses his conviction that these methods could prevent cancer:

> These proposals are in no way utopian. On the contrary, they may be realized by everybody, everywhere, at any hour. Unlike the prevention of many other diseases, the prevention of cancer requires no government help and not much money.

Almost all experts agree that Warburg's judgment is at fault and that his sweeping generalizations spring from gross simplification. Warburg neglected the fundamental biochemical aspect of the cancer problem, that of the mechanisms that are responsible for the controlled growth of normal cells and that are lost or disturbed in the cancer cell. The primary cause is expected to be at the level of gene expression. Although some of the principles are understood, many of the details are unknown.

In addition to the contributions to the four main scientific fields described, Warburg made a number of other interesting and valuable discoveries, some of them even of practical value to clinical biochemistry.

In summary, considering Warburg's achievements in the four major fields of his endeavors, many biochemists will agree with the statement of Ephraim Racker (1972): "Few will challenge the statement that Warburg was one of the great biochemists. His experimental approach was monumental and ingenious. Yet Warburg's views on the two vital areas of his research interest, cancer and photosynthesis, are now almost universally dismissed as erroneous and naive."

It may be mentioned, in the context of this book, that Ephraim Racker came to the United States as a Jewish refugee from Vienna after the *Anschluss* in 1938. He is one of the leading biochemists and has made many outstanding contributions to the mechanism of the oxidative and electron transfer chains in mitochondrial membranes. He is an Albert Einstein Professor of Biochemistry and Molecular Biology at Cornell University, Ithaca. He is a warm friend of Israel, has visited that country frequently, and was most helpful to the Hebrew University in Jerusalem in many respects.

It must be stressed that the criticism refers to Warburg's views. Nobody questions the brilliance of his experiments nor the profound impact of his contributions on the biochemistry of this century. The factors involved in maintaining his views, his passion for his own theories, offer a complex problem that will be discussed later.

Contributions to Methodology. As has been mentioned, Warburg had an extraordinary gift of developing new and original methods or modifying existing ones to suit the specific aims of his research. This ability was based on his broad theoretical background, his rich imagination, his ingenuity and his penetrating mind, combined with great technical skill. Several of these methods have profoundly influenced the progress of biochemistry in general and enzyme chemistry in particular. The use of the spectrophotometric method during his studies on the coenzyme NAD led to startling advances in the techniques of enzyme purification and crystallization. Since then spectrophotometry has been greatly simplified, spectrophotometers are readily available commercially, and their application has been greatly extended by many improvements and refinements. They have become an indispensable tool universally used by biochemists.

Another technique that Warburg developed was the use of the manometric method for measuring respiration. During a visit in Cambridge in 1910, he had seen the Haldane–Barcroft "blood gas manometer," a special device in which the gas space volume was kept constant, so that at constant temperature the pressure was the only variable when a gas was formed or removed. It was originally used for measuring the quantities of oxygen bound to hemoglobin or the carbon dioxide content of blood. Warburg modified the manometric method so that the rates of gas exchanges could be measured. This required quite a few adjustments. The method as developed by Warburg became an extremely important tool in enzyme chemistry; it is far more accurate, convenient, and versatile than the classical methods of gas analysis. It is applicable, under appropriate conditions, not only to processes in which gases directly participate—that is, in cell respiration, photosynthesis, and alcoholic fermentation—but also to a great number of other reactions that can be coupled to gas-producing reactions. Over the decades manometric methods were extended and perfected in a great variety of ways to fit specific problems. They have been used in many laboratories for measuring a vast number of metabolic reactions and proved exceedingly valuable for measuring small amounts of quite a few compounds. An indication of the value and the range of this method, initiated by Warburg, is a 600-page monograph edited by A. Kleinzeller (1965) describing the scope of manometry. It includes over 100 analytic procedures. Although for many reactions quite a few more precise, faster, and still more refined methods have been developed, many brilliant results obtained in biochemistry were performed with this technique.

Finally, Warburg introduced the use of tissue slices as a technique for measuring cell respiration. The tissue slices had to be sufficiently thin to permit diffusion from and to the suspending medium in order to ensure an adequate supply of nutrients and removal of waste products, a function normally effected by the blood circulation. Slices less than 0.5 millimeter thickness are required to satisfy these aims. The slices were cut with a sharp razor, an easy and convenient technique to learn. Warburg developed the mathematical theory necessary for the calculation of the maximum thickness permissible to saturate the tissue with oxygen. Several factors are involved. Slices usually contain 100–150 layers of cells. Since only one cell on each site is cut and directly damaged by the razor, the structure of almost all cells remains intact. This was important because respiration requires the cell structure to be intact, as discussed before.

A decisive advantage of the slicing technique is that it becomes possible to incubate samples of the same organ in parallel under a variety of precisely defined conditions and to compare effects of many different types of compounds on the reactions investigated. Previously, biochemists had used for such studies the perfusion of whole organs, like liver, kidney, or heart. It was a much more cumbersome procedure with several factors involved, the effects of which were difficult to evaluate. Moreover, it did not permit the researcher to carry out a great number of parallel experiments on the same tissue. The tissue slice technique was a great improvement in many respects.

Thus, Warburg influenced and stimulated the advances of biochemistry not only by many original and ingenious ideas, but also by providing many new essential tools, which by their elegance and versatility were invaluable and, for many decades, instrumental for the progress of the field. It is true that in the last 20 to 30 years physics has provided biological sciences with a wealth of highly sophisticated methods and instruments, going far beyond the possibilities of the first part of the century (such as electron microscopy, X-ray diffraction, the methods for measuring reactions taking place in a billionth of a second, optical rotatory dispersion and circular dichroism, spectrofluorometry, and nuclear magnetic resonance). But in evaluating the greatness of the contributions of a scientist, one must keep in mind the limitations imposed by the general knowledge of his era.

Fate during the Hitler Period and after World War II

From the time when Warburg returned to civilian life during World War I up to the end of 1931 he was the head of one of the six sections in the Kaiser Wilhelm Institute of Biology. It was by no means a large department; it provided no more than about 10 places, but comfortable space for all his equipment and his many sensitive instruments. Most of his collaborators were technicians. Usually he had only two to three academic investigators. His technicians were, however, exceptionally well trained and highly skillful in servicing complicated equipment. Several of them had spent some years at the Siemens plant and were thus accustomed to servicing electronic and other delicate precision instruments. For many reasons he preferred technicians to academic investigators. They carried out their assignments without raising questions and had no ambitions to have their own independent research. Nevertheless, over the decades of his scientific activity, many scientists spent one or several years in his laboratory. Among them were Hans A. Krebs, Hugo Theorell, Hans Gaffron, Baird Hastings, Eric Ball, Theodor Bücher, and after World War II, Dean Burk. When a colleague once mentioned the importance of teaching and training scientists for the future, Warburg replied that Meyerhof, Krebs, and Theorell were his pupils. Had he not done enough for the next generation?

During his visit to the United States, in 1929, the Rockefeller Foundation offered to build him a new institute according to his needs and his own plans. This new institute was ready at the end of 1931. The unusually attractive building was modeled in the style of an eighteenth-century rococo mansion house near Potsdam. It provided all the space he needed. He did not extend the size of his group. However, since he always had the best possible equipment for both chemical and physical work, he added more space for equipment and especially for the large-scale preparations required for enzyme isolation and purification, which he started with up to several hundred liters. Although they were available even in the old quarters, they were greatly improved and turned out to be invaluable for the kind of work he did in the 1930s.

Warburg was in his new institute just over a year when Hitler became chancellor. But Warburg continued to work during the whole duration of the Nazi regime. Many colleagues were greatly puzzled. Especially for people in England, Warburg was, as a member of the famous Warburg family, a Jew. Why was he never disturbed or harmed by the Nazis? First of all it may be recalled that his father Emil was baptized and his mother came from an old and distinguished non-Jewish German family. According to Nazi ideology he would be classified as a half-Jew. Like many nonbaptized but fully assimilated Jews, he did not have any connections with the Jewish community nor his Jewish heritage, despite his many distinguished Jewish cousins and relatives. Through his mother's family and his own service in the army during World War I, he had many connections to high-ranking members of the German army. In fact, he respected and admired many qualities of army officers and had many relationships with them. Apparently, influential friends in the *Reichswehr* induced *Reichsmarschall* Göring to arrange a recalculation of Warburg's ancestry with the result that he was considered to be a quarter-Jew. Krebs mentions, and the author heard it from several sources, that some influential scientific friends succeeded in convincing Hitler's entourage that Warburg was the only scientist who offered a serious hope of producing a cure for cancer one day. Hitler apparently suffered from a strong phobia of cancer, and this factor may have played an additional role. Even when Warburg was denounced for having spoken critically of the Nazi regime, he was protected by a high-ranking Nazi official, a *Reichsleiter* in Hitler's chancellery.

Another factor that caused irritation to some, although probably few, people was that the Rockefeller Foundation was willing to build an institute for him in the United States, a highly unusual step in view of the economic situation prevailing at that time, and a sign of the high regard and admiration he enjoyed there. Warburg declined. One of his reasons for refusing the offer was that his colleagues might not be willing to go with him. Moreover, he was apparently afraid that under the best of circumstances his research, by far the most important factor dominating his life, would suffer at least for a considerable period of time. The author recalls several discussions with A. V. Hill—a close friend of Otto Meyerhof —whom he visited frequently in London during the six years he spent in Paris (1933–1939). Hill belonged to the relatively small number of Englishmen who utterly detested the Nazi regime from the beginning. Few have done so much and acted so forcefully as Hill to enable German-Jewish refugee scientists to continue their research. He was one of the very few who foresaw a war with Hitler at a relatively early stage and was active in developing radar for the protection of England against air attacks. Although many Englishmen, and particularly scientists, strongly disapproved of Nazi anti-Semitism, and had some other objections, many people were sympathetic with the aim of fighting against the disastrous effects of the Versailles treaty. They realized the danger of continuously humiliating a great nation such as Germany. One should not forget that even Winston Churchill was until the end of 1934 an admirer of Hitler. Still, in 1938 the author had many discussions during his frequent visits to England, in which he was amazed how few colleagues realized the true character of the Nazi regime. Many still felt that Germany was indeed terribly mistreated by the Versailles peace treaty and that the Nazi movement was a natural reaction. When the author quoted statements and warnings of Churchill, he was frequently advised not to pay any attention to this "old adventurer" whom nobody took seriously in England. Unfortunately for the world, for Europe, including Germany and England, the English government began to see the reality when it was much too late to take the necessary precautions. For Hill, who detested the Nazis, Warburg was a Jew belonging to an old Jewish family. In contrast to the customs on the Continent, in England baptism was a rare exception and poorly understood.*

Hill considered Warburg's refusal to leave Germany as a lack of character and suspected that he actually might have had some sympathy for the Nazis. It took several discussions until the author succeeded, or at least thinks he succeeded, in explaining to Hill Warburg's particular situation and his peculiar character, and that this negative reaction in no way reflected Warburg's attitude or views about the Nazis. In fact, as mentioned before, Warburg endangered himself repeatedly by making critical remarks about the Nazi regime. He kept one of his technicians, Erwin Haas, a Hungarian Jew, until 1938. When the Gestapo indignantly in-

---

* A story that the author heard from Krebs is pertinent here. In 1934 (or 1935) a baptized German-Jewish biochemist, forced to leave Germany, approached Sir Frederick Gowland Hopkins, the most brilliant and outstanding English biochemist of that era and the director of the Sir Wililam Dunn Institute in Cambridge, asking whether he would be able to offer him a place in his institute. Hopkins, known for his extraordinary generosity and his efforts on behalf of German-Jewish colleagues, was anxious to help, but he just had no funds. He asked Krebs, who was at that time in Hopkins' laboratory and for whom Hopkins had a very great admiration, whether he could not go to London and inquire as to the possibility of getting some funds at the Woburn House, where a Jewish committee was trying to help displaced scholars from Germany. When Krebs returned he informed Hopkins that the committee had refused. They were willing to give funds to Jews and non-Jews, but not to baptized Jews. Hopkins' daughter, Mrs. Barbara Holmes, was surprised and asked, "Baptized Jews, what is that?" Hopkins turned to her and explained, "You see, Barbara, these are Jews who are no longer quite orthodox."

quired how he could have dared to commit this outrage against the spirit of the Nazi regime, he innocently claimed to have thought that the anti-Jewish laws applied only to German Jews. He then actively helped Haas find an adequate position in the United States. Warburg had originally intended to participate in the International Congress of Physiology, held in Zurich in 1938, and to present a paper. He was ordered by the Gestapo "to cancel his participation without giving any reasons." He sent a telegram: "Muss Teilnahme absagen ohne Angabe von Gründen." Hill was beaming when he informed the author about this cable. He apparently not only enjoyed this joke, but it also was for him an indication that Warburg had not exactly a deep respect for his *Führer*.

In 1943 air attacks made life in Berlin dangerous and Warburg moved, with his staff and equipment, to an estate 30 miles north of Berlin, which its owner, Prince Eulenburg, placed at his disposal. Here he worked undisturbed until 1945 when the Russians occupied the area and removed all the equipment from the laboratory. It has never been clarified who was responsible for this action. The Russian commander-in-chief, Marshall Zhukov, told Warburg in the name of the Russian government that the dismantling of his laboratory had been an error. Although the marshall ordered the return of the equipment and the books, they could not be traced.

The Dahlem institutes did not suffer any damage during the war, since they were located in a section on the outskirts of Berlin that was not hit by bombs. After the occupation of Berlin by the four Allies, Dahlem belonged to the American sector and the institutes were used, for four years, by the American forces as the seat of their Berlin High Command. Warburg's institute was evacuated in 1949 and with the help of many it was reconverted into the Kaiser Wilhelm Institute of Cell Physiology and reopened officially in May 1950. In 1953 it was renamed the Max Planck Institute of Cell Physiology. Thus for five years Warburg was unable to carry out experimental work. He used the time for writing books, in which he summarized his earlier work and discussed its significance.

He then continued his research much on the lines of his earlier work. He still made quite a few valuable contributions, but they were not of the same outstanding pioneering quality as his previous achievements. Unfortunately he lost contact with the spectacular progress in the biochemical and biological sciences that took place in the last three decades of his life and did not recognize the revolutions in these fields. They were based on new methods and highly refined instruments that were provided by the development of physics. These powerful tools permitted measurements that seemed inconceivable in the first half of the century.

In November 1968 Warburg suffered a fracture of the femur; while looking for books in his library he fell from a ladder. But he recovered and continued to work actively in his laboratory, although he was by then 85 years old. The Max Planck Society had waived in his case the obligatory retirement age. A week before his death on August 1, 1970, he did not feel well and stayed home. He had suffered a deep vein thrombosis and a pulmonary embolus ended his life.

Personality

Warburg's great achievements are due to his creative genius, his exceptional intelligence and originality, his vision and imagination combined with an extraordinary willpower and determination to devote his life to experimental science. His dedication to science was the paramount driving force that dominated his life at the expense of many other values in life and required many sacrifices. Although he was twice in love, he never married. He considered marriage to be incompatible with a total and absolute devotion to research. One is almost reminded of the religious fervor of some historical figures, whose whole life was almost exclusively devoted to their relationship with God.

He was essentially a solitary man, self-contained and never bored with his own company. He had started to play the violin, but gave it up. Listening to his record collection, especially to Beethoven and Chopin, and reading books, particularly biographies, occupied him completely in addition to his work. His deepest satisfaction was his creative work. He came from an unusually brilliant family, to which belonged many outstanding scholars, so genetic endowment was certainly an important element in shaping his personality. But an additional factor was the highly cultured and intellectual environment in which he grew up. The meeting of many illustrious people in the house of his father, many of whom became the teachers who inspired and influenced him most, such as Emil Fischer, Walther Nernst, and Jacobus Henricus van't Hoff, must have had a deep impact on a young boy still in his high-school years, a most receptive age. We have today a much better insight into the strong effects of family and environment than one had previously. His father was also actually one of his teachers, as was described in connection with his early interest in photosynthesis. Krebs mentions a writer who thought that Leonardo da Vinci's saying, "He never turns back who has found his star," was a fitting motto for Otto's cousin Aby Warburg and, Krebs adds, this was equally fitting for Otto.

Warburg was fully aware of his brilliance and his outstanding abilities. He was a great admirer of Louis Pasteur and had his picture in his library. He considered himself to be the Pasteur of the twentieth century. His determination to overcome experimental difficulties expressed itself in the courage to attack great unsolved problems. His supreme mastery of experimental resources and his technical skill enabled him to devise new methods and to use instruments not yet applied at that time by biologists and to adjust them to approach a problem in a novel manner. Humility was not one of his characteristics. When a colleague once mentioned that he had taken part in a discussion in which the question was raised of whether Warburg or Meyerhof was the greater scientist, he was first surprised and asked whether the question really required discussion. But after some reflection he made the comment: "Meyerhof war vielleicht klüger, aber ich konnte mehr." With *Können* ("greater skill") he was correct. Warburg was a genius in devising and handling techniques. Meyerhof's strength was not the use of techniques. In fact, he was not particularly skillful. But as Warburg once himself admitted (see p. 303, Meyerhof's personality was one of an unusual universality, that of a profound thinker who introduced into biochemistry many fundamental notions and concepts. These qualities are hardly expressed by *Klugheit*. The genius of these two men was just different. The author's view regarding the question of which of the two was the greater genius will be discussed later.

Warburg was a fierce, aggressive, and self-righteous fighter, as his fellow scientists know from the literature and from meetings. *Sachlichkeit und Wahrheit* were his ideals. He had vigorous controversies with many leading authorities of his time: Richard Willstätter, Heinrich Wieland, David Keilin, Hans von Euler, Sydney Weinhouse, James Franck, and others. He did not hesitate to use a style that was often scathing, sarcastic, and deliberately ridiculing, sometimes to the point of being insulting. He believed that it was wrong not to contradict erroneous criticism. A scientist had the obligation to fight any unjustified objection. He quoted as a typical example of the danger of avoiding controversy the attitude of Charles A. MacMunn, who, in 1885, had discovered in animal tissues what are now called cytochromes. MacMunn was severely criticized by Hoppe-Seyler, a leading authority in the field, who argued that hemoglobin was the only red substance in animal tissue, and that the "histohematin" of MacMunn was an artifact derived from hemoglobin. MacMunn remained silent. As a result progress was delayed for over 30 years, until Keilin independently rediscovered cytochromes.

It must be emphasized that virtually all the controversies Warburg was involved in concerned views and interpretations. Nobody questioned his experimental data and skill, or the validity and reproducibility of his findings. It was their significance that was the subject of the disagreements.

The fascination of the Haber colloquia (see IV, A) was to a large extent based on the vigorous controversies between the illustrious scientists participating there. Even postdoctoral fellows could raise objections and they were taken seriously. The fights clearly demonstrated the great gaps in scientific knowledge, the difficulty of interpreting striking and revolutionary new data. One may also recall in this context the great dilemma facing Max Planck to break with Newtonian physics in order to accept the corpuscular theory of heat irradiation; his dilemma persisted even after Einstein's observation confirmed the corpuscular nature of light. The most exciting era of atomic physics in the 1920s was characterized by passionate discussions and controversies. The discussions in the Haber colloquia and between atomic physicists were relatively free of emotion; they were exciting intellectual fights among friends trying to find the best answer, the interpretation coming closest to the truth in a set of data of a complex system. Moreover, nobody hesitated to admit that he was wrong. Einstein started a letter to Max Born with the statement that he had made a monumental blunder (Born, 1969). A nice illustration is the incident, reported by Mendelssohn (1973), that took place in a Nernst colloquium held in the Physics Institute of the University of Berlin, where the same free atmosphere prevailed as in the Haber colloquia. These colloquia were limited to physics, but Planck, Einstein, von Laue, and Schrödinger usually sat in the first row. One day Nernst and Einstein had a vigorous fight. Three weeks later, in the discussion following a seminar, Einstein presented the same view that Nernst had taken three weeks before and which he had so vigorously rejected. Nernst was surprised and asked: "But, my dear Einstein, now you take the same view as I three weeks ago and with which you so strongly disagreed?" Einstein's answer was, "My dear Nernst, is it my fault that the good Lord created the world as you recognized correctly three weeks ago and I didn't?"

The fundamental attitude in all these controversies was an awareness of the limitation of scientific knowledge, despite all the spectacular advances, and of the pitfalls of so-called facts, which in the light of new data continuously required reinterpretation. Discussions and controversies were thus essential for clarifying complex problems, as all scientific problems are. It may be appropriate to mention the statement of Einstein quoted by Banesh Hoffman (1972), that although scientific knowledge was primitive, it was the most precious thing people had. A similar statement was made by Isaac Newton when he compared the achievements of his long life with the finding of a precious pebble by a young boy on a beach, but before him was the large unexplored ocean; or one may think about the οἶδα οὐκ εἰδέναι of Socrates ("I know that I do not know anything"). All these statements are expressions of wisdom and humility.

The question then arises as to what made Warburg's controversies so basically different from those just mentioned. It was not the fact that Warburg was a fierce fighter that created a feeling of frustration and futility or even resentment among his colleagues. It was the emotional factor that made the controversies unpleasant, his reaction to the objections as if they were a personal affront; his feeling that his superiority was questioned. He was liable to become angry and to accuse his colleagues of being prejudiced and—in some cases—even dishonest. This kind of reaction must be considered as a feature of his personality and would be difficult to analyze. Personality is an essential factor even in science, just as in all human endeavors, and emotions are bound to prevail over reason. In view of his strong feeling of his own superiority in his field over his contemporaries, he was simply unable to accept any doubt as to his competence. His penetrating intelligence, his critical judgment would hardly permit him to assume that his concepts were the last solution of a problem. When he was not directly challenged, he must have been aware of the many gaps in our knowledge. In fact, in a lecture at New York, arranged by the author in 1949, about the problem of oxidation, his closing remarks were, "One hundred years ago Claude Bernard made the statement that not a single biological process is fully understood. It seems to me, what was correct 100 years ago is still correct today." A few days later, in a round table conference of about 20 people from the New York area, for which the author checked the list of the people to be invited with Warburg, he was remarkably frank in admitting gaps in his knowledge. At that time a book by Linus Pauling (1902–   ), *The Nature of the Chemical Bond*, had produced a considerable stirring among biochemists. When asked about his views on some problems of physical organic chemistry, he turned to Leonor Michaelis with the comment that in this area he was not sufficiently competent and that Professor Michaelis might be able to answer these questions.

Thus it seems justified to conclude that his unusual reactions were due to his emotional inability to accept critical comments, combined with his isolated life, which prevented him from keeping in touch with the breathtaking developments of biological sciences in the postwar period. It is the combination of these two factors that may explain the deep tragedy of this ingenious and extraordinary mind as expressed in Racker's statement; that is, that his views in two vital areas of his research, photosynthesis and cancer, are now universally recognized as naive and erroneous. Certainly it was not lack of intelligence that prevented him from accepting the new information and adjusting his views. Such an attitude would have increased his greatness and his stature in history. After his discussions with Barron, he not only recognized the role of dehydrogenases after rejecting it for a long time, but, as a consequence, he subsequently made some of the most crucial and ingenious contributions to this field. However, he never admitted that he failed or recognized the correctness of Wieland's and Thunberg's views publicly. The same applied to the explanation of the *Atmungsferment* as Keilin's cytochrome oxidase. In the case of the controversy with James Franck as to the efficiency of the quantum yield in photosynthesis, his emotions were so strong that even years after the passing of Franck Warburg continued to denounce his criticism bitterly.

This anguish and bitterness about the critical objections to his views and his hurt pride may perhaps also explain his own assessment of the achievements of his life, written in 1961, at the age of 78. He uses extremely forceful language and describes his contributions as great landmarks seen in historical perspective, emphasizing their outstanding importance. Among them he quotes the discovery of the substance that reacts in the living world with molecular oxygen, thereby providing the general solution to the problem of respiration raised by Lavoisier; the elucidation of the chemical mechanism of enzyme action by separating the protein-bound active groups, the coenzymes, from the enzymes, some of them turning out to be vitamins; the crystallization of a great number of important enzymes; the measurement of the requirements of the quanta of action, calculated by Max Planck, in photosynthesis; and the explanation of the primary cause of cancer. In the discussion of some controversial issues he turns around the difficulties by making some oversimplified assumptions. The remarkable feature of this summary is the uninhibited way in which he praises the greatness of his achievements. It is rare that scientists evaluate and praise the brilliance of their achievements; they usually leave that to the judgment of their colleagues or historians of science.

Warburg's emotional reactions are even more surprising when one considers some additional factors. We know innumerable cases in history where revolutionary notions and concepts were vigorously opposed, and often ridiculed, by a majority of the "authorities" in the field, who chose to embrace currently prevailing and cherished views, accepting them dogmatically. Warburg was a contemporary witness of Max Planck's struggle to get his concepts accepted and the revolutionary character of his contributions recognized. He frequently quoted Planck's statement that scientific opposition never dies except with the death of the opponents. In the free and undogmatic atmosphere at the Kaiser Wilhelm Institutes in Dahlem, these words were frequently quoted. Thus one would expect that Warburg, being fully aware of the brilliance of his contributions, would accept any objections with equanimity. But in fact, in his case the situation was strikingly different. His outstanding qualities and achievements were recognized very early, as indicated in Einstein's letter imploring him to spare his life for the sake of his scientific field or, even more drastically, by the fact that he was elected to full membership in the Kaiser Wilhelm Society at the age of 30. In view of the high standards of excellence used in these elections, it was an unequivocal sign of general recognition. Moreover, it provided him with magnificent facilities for pursuing his creative work with greatest efficiency at an early age. Many other expressions of the great admiration he enjoyed among his contemporaries in his most active years could be cited. Nevertheless, he was extremely sensitive whenever he felt that his greatness was not fully appreciated. For example, in 1938, the botanist and Zionist leader Otto Warburg died. The London *Times* erroneously assumed that it was the famous biochemist Otto Warburg and wrote an obituary full of high praise. But Warburg was dissatisfied with some formulations and omissions, suspecting that one of his opponents was behind them. The author, who worked at that time in Cambridge, heard the story that a few days later the *Times* received an indignant letter from Warburg complaining in essence that the obituary did not do justice to him at all.

Scientists have always been aware of the progressive and accumulative character of scientific knowledge. As Michael Faraday (1791–1867), a brilliant experimenter, once said: "I may be wrong, I am free to admit it, but who can be right altogether in natural science, which in its nature is progressive and corrective?" In our century the accumulation of knowledge has been much more rapid and dramatic. It appears all the more difficult to understand Warburg's insistence on the correctness of his early interpretations and his anguish about any doubts based on new knowledge long after the acme of his most creative period. Any student reading his textbook on biochemistry about the extraordinarily complex electron transfer or oxidative chain, structurally organized in the mitochondrial membranes and formed by a great number of enzymes and coenzymes, may easily fail to understand the greatness of Warburg's contribution. The same is true for the understanding of an enzyme mechanism, since the tridimensional structures of many proteins, including enzymes, have been elucidated with the use of X-ray diffraction analysis in the last 15 years. This applies to many fields, not just to biology. Every student of physics knows about the many elementary particles in the atom. It may be difficult to grasp the excitement caused by Niels Bohr's atom model, which is primitive compared to our present concepts, and difficult to understand the admiration that Planck expressed for Bohr's model in his Nobel lecture in 1920. A correct evaluation of an accomplishment is only possible within the context of a period. Thus a scientist must know that no matter how great and revolutionary his contribution is, it will be the beginning of new developments, increasing and enlarging knowledge, but it will never be the final answer.

As mentioned before, Einstein and Planck revolutionized epistemology by their recognition that scientists arrive at their theories only by speculative means. Quantum theory, in the 1920s, broke even further with the classical philosophical attitude prevailing since Newton. Although accepting Newton's physics as an indispensable basis for further developments, quantum theory emphasizes its limited applicability; it introduced the subjective element (see, e.g., Heisenberg, 1958). When we consider also that Warburg's father was an illustrious physicist, who lived until 1932, it seems difficult to believe that Warburg was completely unaware of these profound changes in basic philosophical attitudes. In contrast to physics, and even to chemistry, the biological sciences have joined the exact sciences only in this century. The unknown areas are much wider than in physics, the complexities infinitely greater. We are just beginning to explain some biological mechanisms in terms of physics and chemistry. Large sections and areas of biology are still mainly descriptive phenomenology. In view of Warburg's penetrating intellect it is difficult to assume that he was not aware of this situation. The most likely explanation for his fierce rejection of any critical comment seems to be that the emotional factors were so strong that they clouded his lucidity and his usually superb and critical judgment.

When we leave aside this special weakness, no one who had personal contacts with Warburg could fail to be impressed and fascinated by his personality, whether or not one agreed with his views. What made such a deep impression on everyone was his extraordinary intelligence, his remarkable insight and understanding of scientific problems. One could not help but admire his personal charm. He looked at one with his penetrating eyes and listened carefully. In personal discussions, whether they were concerned with scientific problems, history, or public affairs, his questions were to the point, his comments and answers always pertinent and stimulating. His intellectual honesty, his straightforwardness, his wit and humor were outstanding characteristics of his personality.

For Meyerhof's associates, Warburg was relatively easily available in view of his personal association with Meyerhof in the Heidelberg years and the high respect he had for Meyerhof. When the author worked in the Kaiser Wilhelm Institute of Biology, the simplest way to see Warburg was to go up two flights of stairs between 5 and 6 P.M., when Warburg had usually finished his experiments and was therefore available for a discussion. Since he had no secretary, it was unnecessary to make special appointments. When Warburg was in the United States, in 1949, the author had quite a few discussions with him, in New York, in Washington, and in Woods Hole. When he visited Berlin after World War II, which after 1957 was an almost annual event, a visit with Warburg was obligatory. It was usually arranged by Hans Herken, professor of pharmacology at the Free University, who was on excellent terms with Warburg, making a telephone call to Jacob Heiss. These visits were always pleasant and stimulating. It should be added that the author took pains to avoid discussing cancer or photosynthesis. When he mentioned this to Hugo Theorell in a discussion about Warburg's personality, Theorell smiled and said that he did the same thing. Warburg was by no means indifferent to the work of others. When he was about 80 years old, he asked the author how his work on the proteins and enzymes in excitable membranes was progressing. He became quite obviously very interested and raised a great number of precise and detailed questions. The discussion continued over one and a half hours, although Jacob Heiss had announced that lunch was ready. A sign that he had followed the discussion with real interest was his call to Professor Herken, saying, "Der tut ja noch etwas" ("he is still doing something"), which for Warburg meant high praise.

Lifestyle

Warburg lived in an extremely regular way. He paid much attention to his health and fitness. For many years he would get up at 5:30 A.M. From 7:00 to 800 A.M., when there was daylight, he would ride on horseback and then go to his laboratory. Except for a lunch break he remained at the bench until late in the afternoon. He expected his associates to keep equally long and regular working hours. In August and September he had his vacations, which he spent, until 1944, in his country home on the island of Rügen. There he walked and rode, but also spent much time reading, and writing his papers. After the war and the partition of Germany, he spent his vacations in the Hunsrück Hills, at the country estate of Jacob Heiss.

Horseback riding was his main physical recreation. In 1924 he fractured his hip when falling from his horse and for some time lay helplessly in the woods. After that he never rode without the company of Heiss. When in 1955 Heiss had to give up riding after an operation, Warburg built a riding arena in his own garden. He regularly rode until he was 85, when he fractured his femur. In 1952 he started a new sport, sailing almost every Sunday on the river Havel. He continued this sport too until the femur fracture. He was also fond of walking on the outskirts of Berlin, especially on winter Sundays. He was very much attached to the landscape around Berlin, with its many lakes and woods.

In his personal outlook he was very aristocratic in the Greek sense of aristocracy (i.e., the rule of the best). From this attitude it is easy to guess how much he must have despised the Nazi mobs, the vulgarity of their mass meetings, the low stature of the men with whom Hitler surrounded himself. It was certainly with not a trace of sympathy for the Nazis that he decided to stay in Germany and to decline a move to the United States. Being disinterested in politics and sometimes naive in his political views, he certainly did not foresee the disaster that cost him five years interruption of his scientific research.

During the years he spent in the army in World War I he became attracted by what he considered to be the best features of the army. He respected the discipline in manners, the straightforwardness in thought and argument, and the elimination of inefficiency. In his view a military leader was dismissed if incompetent, whereas in academic life many people got away with incompetence and with neglecting their primary duties. Obviously, one may question these—as all—generalizations. In fact, the majority of academic people may have a quite different opinion. But one consequence of his attitude was that he had more friends among military men than among university professors. However, none of his friendships amounted to a close relationship. The only exception was Jacob Heiss. When Warburg looked, in 1919, for someone to keep his house, Heiss was recommended by his military friends. Born in 1899, he served in

World War I in a Prussian infantry regiment of the Guards. Slowly there developed a close personal relationship between him and Warburg. Heiss became a faithful companion. He was an understanding, devoted, and self-denying friend, who did everything he could to support and protect Warburg and to make his life as congenial as possible. He also did much in calming and appeasing Warburg's emotions and resentments when they were raised by controversies. The two men spent not only the daily routine together, but also their vacations and travels. When they were in the United States, in 1949, Warburg did not accept invitations unless he could take Heiss with him on his travels. Later Heiss became unofficially his secretary and manager of the institute.

In the laboratory Warburg worked with great intensity. He disliked being disturbed during his work and wasting his time with unimportant activities. Once he opened the door and a reporter asked for an interview with Professor Warburg. Warburg said, "Professor Warburg cannot be interviewed. He just died. Goodbye," and shut the door. He did not like the visits of professors who came to meet the famous Warburg out of curiosity, without really having any serious problem to discuss with him. He could be in such cases quite rude and impolite. Similarly, he considered *Antrittsvisiten* as a complete waste of time. Such visits were customary when somebody was appointed to the faculty of the University of Berlin as *Ordinarius* so that he might meet colleagues in his field. The author recalls one occasion when he was with Meyerhof and a newly appointed *Ordinarius* entered on his way back from Warburg to visit Meyerhof. He was still trembling and indignant about the way Warburg had treated him. In contrast, when colleagues, old or young, visited him for serious discussions of a scientific problem, he was usually very friendly and the meetings could be very pleasant and stimulating. His comments were always original and valuable.

Style and Outlook on Scientific Writing. Warburg's scientific papers are distinguished by an exceptional clarity, simplicity, brevity, and rigid scientific logic. Occasionally striking and colorful passages enliven the text. When he discussed, for instance, the results of his spectrophotometric measurements of the monoxide compound of the oxygen-transferring enzyme in yeast cells, he wrote, "If one uses reagents which react only with the enzyme and not with any other cell constituents, then these cell constituents do not interfere any more than the wall of a test tube interferes with a chemical reaction. Hence enzymes can be studied like pure substances and from their reactions conclusions about their compositions can be made." This was actually his answer to the criticism of Richard Willstätter, who had questioned the results on the basis that spectrophotometric measurements on living cells do not provide precise information since too many other cell constituents may interfere. The forceful formulation and the logic are typical of Warburg's style.

His aim was to formulate conclusions reached from his experiments without too many qualifications, which might detract from the value of any statement. In his tendency to be clear and straightforward, he occasionally tended to oversimplify his conclusions deliberately. Krebs wrote that Warburg subscribed to the view of the English physiologist W. M. Bayliss (1860–1924):

> Truth is more likely to come out of error if it is clear and definite than out of confusion, and my experience teaches me that it is better to hold a well understood and intelligible opinion, even if it should turn out to be wrong, than to be content with a muddle-headed mixture of conflicting views sometimes called impartiality, and often no better than no opinion at all. [Bayliss, 1920]

The endorsement by Warburg of Bayliss' statement, written more than half a century ago—especially the last words, that not having an opinion at all was just as bad as presenting a muddle-headed picture—is a timely and welcome reminder dealing with a pertinent problem of science. Today many investigators present "facts," but are afraid to offer any explanation of their significance, any concept, assumption, or working hypothesis, to use whatever expression one may choose. As was more fully discussed in the chapter on atomic physics and briefly mentioned here before, all our knowledge is extremely limited and new facts present just a tiny bit of additional information about the whole problem. Experimental facts collected by an investigator always present only a very small fraction of the total problem in which we are interested (see II, E). The facts must therefore be supplemented by ideas, imagination, intuition, or a working hypothesis proposed by the scientist, if they are to become of real value for the progress of science. Warburg's attitude just underlines his qualities of leadership. This should, of course, not be interpreted as an encouragement to cheap speculations, poorly supported by facts. Science is based on facts. The facts must form the basis of any theory or hypothesis. But in view of our limited knowledge, even the apparently best hypothesis must be modified, changed, or abandoned when new facts are discovered that contradict it and offer a more satisfactory explanation. The value of a hypothesis should be judged according to its ability to discover new pertinent facts and to enrich our knowledge and understanding of a problem.

Love for England. Warburg had a great love for everything English. Ever since he first visited Cambridge before World War I, he was an Anglophile. He loved England because, as he once said, "the English tolerate headstrong and eccentric [*eigensinnige*] people—such as me." He also loved the adherence to tradition and the dignied pomp and circumstance of the ceremonial that he experienced when he was given an honorary degree at Oxford in 1965. "England is the last bastion of old Europe," he said. Earlier in his life he traveled twice a year to England for brief periods to purchase the finest suits. Hill once told the author a typical story. He met Warburg just leaving the Carlton, the hotel in which he was staying. He welcomed him and said, "Professor Warburg, I am happy to see you; you probably came to attend the meeting of our Physiological Society." Warburg replied, "No, sorry, I just came over to buy a pair of riding gloves." Warburg was also very fond of antique English furniture and surrounded himself with beautiful pieces that he had bought in England. He held that when something was English it could not be wrong. He was a long-standing subscriber to the *Times*.

Eccentric Food Habits. Warburg was convinced that most cancers arose from chemicals when applied for long periods of time. He became increasingly obsessed with avoiding food that had been treated with special chemicals ("additives"). He had heard that bakers tend to get eczema and he thought that this had something to do with the bleaching chemicals added to te bread. Because of this he would not, when he could avoid it, eat bread from a baker's shop. For the last 15 years or so of his life, he insisted on eating only bread baked by Heiss at home. He was also afraid of artificial fertilizers and insecticides. He enlarged his garden by buying adjacent land and increasing the garden to 4500 square meters in order to produce food. He kept hens, ducks, geese, turkeys, and rabbits and employed a full-time gardener to look after the animals and the garden. He grew most of the fruit and vegetables he needed. He would not permit the use of artificial fertilizers and pesticides. One can easily picture how these eccentricities, his whims, fancies, and anxieties concerning food, must have at times been quite exasperating to Heiss.

On Warburg's seventieth and eightieth birthdays, in 1953 and 1963, the scientific community of the world paid tribute to his genius in two *Festschriften*. On his eightieth birthday 80 eminent biochemists contributed to the *Festschrift*. The latter was edited by Feodor Lynen (1911–   ).

Summary

Even if he was a half-Jew, like Adolf von Baeyer, Wolfgang Pauli, Hans Bethe, and others, one cannot deny that Otto Warburg was a product of the association of the two ethnic groups. With all his weaknesses, emotions, and eccentricities, he belongs to the illustrious galaxy of geniuses who were the product of this association and among whom were an extraordinarily large number of the founders of the science of the twentieth century.

D. OTTO MEYERHOF (1884–1951)

Otto Meyerhof stands out as a giant in biology and biochemistry, two fields in which this century has seen spectacular achievements by many scientists of great ingenuity and extraordinary stature. The revolutionary character of Meyerhof's thinking, the originality of his approach, and the great variety and number of his superb experimental contributions had a profound impact on the whole field of biological science. His monumental and epoch-making pioneer work cannot be appreciated without the realization of the breadth and richness of his rather unique personality, the remarkably wide range of his knowledge and interests. These features may explain his extraordinary ability to integrate many seemingly unrelated facts.

Meyerhof's fame in the history of science is inseparably associated with several fundamental concepts and notions that today form an integral part of the biological sciences. The three laws of classical thermodynamics were well established when Meyerhof started his work. But he was the great pioneer in recognizing the importance of applying thermodynamics and energetics to the analysis of the chemical processes taking place in the living cell. This approach, today generally referred to as *bioenergetics*, has become an indispensable element of biological thinking, although—as we know today—biological systems require, in addition to classical, the use of non-equilibrium thermodynamics. Meyerhof's work represents the first attempt at explaining the function of a cell in terms of physics *and* chemistry. This aim offered a particular challenge in view of his basic philosophy, which vigorously rejected vitalistic and neovitalistic theories of cellular function. In the course of these investigations Meyerhof and his associates discovered two types of phosphate derivatives; one releases a large amount of energy during its splitting, and the other releases only a small amount. This distinction was a monumental discovery. Its revolutionary importance for the progress made in exploring and understanding chemical reactions in living cells can hardly be exaggerated. In view of the paramount role of phosphate derivatives in almost all cellular processes and metabolic pathways, this discovery was destined to become one of the most valuable clues in the analyses of the sequence of energy transformation in all metabolic cell processes. His work opened the way to the first complete understanding of the complex and intricate steps of a metabolic pathway that takes place in cells, the so-called glycolytic pathway—the degradation of carbohydrate (glycogen in muscle) through a dozen steps catalyzed by enzymes to its end products, alcohol in yeast and lactic acid in muscle.

Career

Otto Meyerhof was born in Hannover in 1884. He was the son of Felix Meyerhof, born in 1849 in Hildesheim, and Bettina, *née* May, born in 1862 in Hamburg. The family moved to Berlin and it was there that Otto attended school. When he was 16 years old he developed kidney trouble, which necessitated a long period of rest in bed. This period of forced seclusion was apparently responsible for a great mental and artistic development. He read constantly and matured perceptibly. In the autumn of 1900 he was sent to Egypt for recuperation. He was accompanied by a second cousin of his and a personal friend, Max Meyerhof (1874–1945). Max was 10 years older than Otto and was a highly competent ophthalmologist. But in addition he had a broad general education and a deep interest in Egyptology, influenced by a cousin who was a professor of Egyptology in Strasbourg. Otto's interest in archaeology throughout his life was certainly greatly stimulated by this friendship. The trip to Egypt, which lasted four months, apparently greatly enhanced his intellectual development; it was an invaluable experience for a boy at this early and receptive age.

During his school years Otto became passionately interested in philosophy. He became a close friend of Leonard Nelson, the son of a prominent Jewish lawyer in Berlin, who was active in the field of literature and art and who entertained a great number of interesting people at his home. Leonard Nelson was a forceful and dynamic personality. Meyerhof was strongly influenced by him and expressed his admiration on several occasions. One of Nelson's major interests was philosophy. He devoted many studies to the relationship between psychology and philosophy and his views had a great effect on Meyerhof. Nelson later became professor of philosophy in Göttingen and Otto maintained close contact with him and his associates until Nelson's death in 1927 (Hieronimus, 1964). There exists a large collection of letters that were exchanged between Meyerhof and Nelson. Their philosophy closely followed the teachings of Kant and Fries. Otto's active interest in philosophy is best demonstrated by the fact that for several years he was editor of *Abhandlungen der Friesschen Schule*, where Nelson and his associates published their writings.

But philosophy and archaeology were only two aspects of his wide range of interests. He was an avid reader of history, including the history of art and literature. Among his favorite poets were Goethe, and in his later life Rilke. These interests greatly affected Meyerhof's scientific attitude and the direction of his lifework.

After Meyerhof passed his *Abiturientenexamen* in 1903, he studied medicine in Freiburg, Berlin, Strasbourg, and Heidelberg. For his promotion in Heidelberg, in 1909, he chose a topic in psychiatry for his doctoral dissertation. It was during this period that he started to write *Beiträge zur psychologischen Theorie der Geistesstörungen*, which was published in 1912. The book shows the influence of Nelson's ideas, but also that of other friends in Heidelberg, in particular Arthur Kronfeld, an expert in the field of psychology and psychiatry.

But then came a turning point. The director of the Department of Medicine was L. Krehl, one of the pioneers in stressing the paramount importance of basic science for the progress of medicine. In Krehl's clinic Meyerhof met Otto Warburg, one year his senior. Warburg's brilliant mind, his strong personality and willpower, his unflinching determination to explore the chemical reactions taking place in a living cell, made a deep impression upon Meyerhof. With his many interests and his remarkable abilities Meyerhof could have easily dissipated his energy and his talents. When he started to work with Warburg on cell respiration he soon realized that creative work requires deep devotion, the concentration of one's efforts on a special field to which one must dedicate one's enthusiasm, imagination, and energy. Endeavor in any field of human creativity, be it music, art, or science, requires a person's total devotion. To express it with Goethe:

> Und solang Du dies nicht hast
> Dieses Stirb und Werde
> Bleibst Du nur ein trüber Gast
> Auf der dunklen Erde.

It is primarily in this sense that Meyerhof became a "pupil" of Warburg's. Warburg introduced him to the fascination and the challenge that exploring the mysteries of the living cell then offered to the biochemist. The great advances of organic and physical chemistry, mentioned before, had opened the doors to exploring living cells on an entirely new level. In this large and rapidly developing field, Meyerhof soon found his own line of research, one tailored to his main interests and abilities. In 1912, after spending a couple of years with Warburg and some additional time studying physical chemistry with Georg Bredig (1868–1944) in Heidelberg, he joined the Department of Physiology at the University of Kiel. At that time the department was under the direction of Albrecht Bethe, the father of the famous physicist Hans Bethe; after 1915 it was under the direction of Rudolf Hoeber.

Scientific Contributions

Energetics of the Cells (Bioenergetics). The first developments in bioenergetics took place in the field of energy conservation when Robert Mayer formulated the concept of chemical forces being transformed into mechanical forces during muscular contraction. In 1844, in a letter to his friend Friesinger, Mayer wrote, "For some time a logical instinct has led physiologists to the axiomatic proposition: No action without metabolism" (quoted from Florkin, 1975). Mayer realized that this generality was inadequate and required the explanation of the exact chemical reactions taking place and their transformation into function, problems that could not be attacked with the methods and the knowledge available at that time. Mayer's intuitive views were supported by the experimental observations of Helmholtz, in 1847, that the heat combustion of nutrients is equivalent to the heat given off by the animal.

The specific problem of how to approach the investigation of the use of chemical energy in living cells by applying thermodynamics and energetics was discussed by Meyerhof (1913) in his lecture, "Zur Energetik der Zellvorgänge." This lecture, which he gave on July 31, 1913, at the Philosophical Society in Kiel as a young *Privatdozent*, marks a milestone in the development of modern biochemistry. The notions and concepts discussed in this lecture and his subsequent great experimental contributions form the beginning of the use of thermodynamics and energetics in the analysis of chemical cell reactions. To remedy a misunderstanding about the difference between thermodynamics and energetics, Meyerhof (1931) stressed that *energetics* includes all chemical energy transformations, whereas *thermodynamics*, for example of the muscle, refers only to the correlation between heat exchanges and mechanical processes.*

In his 1913 lecture Meyerhof began with the law of energy conservation as formulated by Mayer, Joule, and Helmholtz in the first law of thermodynamics for the maintenance of life processes, which he recognized as being a dynamic, rather than a static, equilibrium: The organism requires a continuous supply of energy, which is provided in the form of food. But the real problem starts with the fundamental question of how this energy is eventually used for the performance of a function. This process must require a whole series of steps between the initial energy input, its use for the function, and its final dissipation as heat. For the exploration of these steps, it is essential to know the whole sequence of energy transformations. A prerequisite for this type of analysis is the knowledge of the energy of the compounds involved in the long series of reactions. Willard Gibbs, in 1878 in the United States, and Helmholtz in 1882 had introduced the distinction between total energy (expressed by the symbol ΔH, where Δ stands for the change and H for heat) resulting from the heat produced by the combustion of a chemical compound, and the free energy that is available for work. This free energy was referred to for a long time as ΔF and later as ΔG, in recognition of

---

* The term *bioenergetics* was introduced in the 1950s. It is a convenient and appropriate term, but obviously its meaning does not differ from "energetics of cell processes," the title of Meyerhof's lecture in 1913, and does not represent the introduction of a new notion.

the importance of the contributions of Gibbs, never recognized during his lifetime. Meyerhof emphasized in his lecture that for the evaluation of a reaction the total energy is not as important as the free energy of Helmholtz. Meyerhof also referred to Nernst's heat theorem, later called the third law of thermodynamics, which makes it possible to establish a quantitative relationship between total and free energy. Florkin (1975), in his masterfully written *History of Biochemistry*, stresses the remarkable insight shown by Meyerhof in recognizing the importance of free energy at that time. As he points out, "In the U.S.A. biochemists were little aware of the distinction of ΔH and ΔG until the publication in 1923, ten years later, of the book of Lewis and Randall, *Thermodynamics and the Free Energy of Chemical Substances*." The free energy, under standard conditions (at a given temperature and concentration and in the absence of other interfering reactions), can be calculated when the equilibrium constant (K) of a reversible reaction is known (by the equation $\Delta G^0 = -RT \ln K$, where G^0 stands for free energy under standard conditions). Furthermore, Meyerhof also recognized the importance of another thermodynamic parameter, entropy, which indicates the degree of order of a system. This parameter was introduced by Clausius, who together with Carnot had established the second law of thermodynamics. The free energy is actually the function of the two other parameters. (The equation is $\Delta G = \Delta H - T\,\Delta S$, where S stands for entropy and T for absolute temperature.) Meyerhof also discussed some of Boltzmann's views on entropy. This parameter, as has become increasingly apparent in the last few decades, is a particularly important factor in biological systems.

It appears unnecessary, in the context of this book, to describe the many details of this extraordinary lecture. The few data given appear to be an adequate illustration of the statement that this lecture marked the opening of a new era in biochemistry; it was the first profound discussion of the role of thermodynamics and energetics in the analysis of cellular reactions and their relation to function, and it gave the first indications of the way to approach this aim.

It appears appropriate that this lecture was presented at the Philosophical Society. Fundamental philosophical questions were raised and ideas discussed that would hardly have fit into the frame of physiological journals of that period. The lecture reveals the lucidity of Meyerhof's thinking and his broad knowledge. The publication made a profound impression in the scientific world. Jacques Loeb, at that time professor at the Rockefeller Institute in New York, invited Meyerhof to write an enlarged version of his paper for Loeb's series, Monographs on Experimental Biology. Because of the outbreak of the war, the monograph did not appear in the United States until 1925, when it was published under the title, *Chemical Dynamics of Life Phenomena*. The ideas expressed in this essay had such an influence on biochemistry and biology that as late as the 1950s it was still recommended to biology students at Harvard University as a useful introduction to the physicochemical approach to the study of life processes.

Studies on the Biochemical Basis of Muscular Contraction. In order to apply thermodynamics and energetics and to achieve his aim of explaining a cellular function in terms of physics and chemistry, Meyerhof chose to study the processes taking place during muscular contraction. The reason for this choice was not a particular interest in this process, but a realization that in this particular case chemical and mechanical changes were on a sufficiently large scale to offer a relatively favorable material for testing their relationship using the rather primitive and cumbersome methods available at that time. Moreover, in 1910, A. V. Hill had started his research on the heat production of muscle and other biophysical phenomena during activity and was trying to integrate them with the observations of Fletcher and Hopkins on lactic acid: They had found, in the years 1905 to 1907, that lactic acid appeared during activity under anaerobic conditions (i.e., in the absence of oxygen). In the presence of oxygen lactic acid disappeared. Heat production measurements do not explain the mechanism, but the precision of physical data offers an important guide to the magnitude of the chemical reactions to be expected. Hill's work had been mentioned in Meyerhof's lecture in 1913. The outbreak of the war stopped or reduced most scientific activities. Meyerhof did not start his investigations until 1918. Within a few years he achieved a number of important and impressive results, published in a series of papers that earned him the Nobel Prize of 1922 (awarded in 1923), which he shared with A. V. Hill. Hill had found that the extra heat produced during activity was about half of the total extra heat; the other half was produced during recovery. Meyerhof demonstrated that the lactic acid formed was proportional to the tension developed (i.e., to the mechanical work performed). He also showed unequivocally that the source of the lactic acid was glycogen, as had been suspected by Justus von Liebig and Claude Bernard. Parnas had tested this possibility but his data were not unequivocal. Lactic acid is a very strong acid. A cell must immediately neutralize it. There are many "buffers" in a cell that permit it to keep its neutrality, which is vital for its function. When Meyerhof determined the heat produced by the lactic acid formed, and added to this value the heat produced by the neutralization required, both values together accounted for only about 50% of the extra heat measured by Hill during activity; about 50% remained unaccounted for. Only one-fifth to one-sixth of the lactic acid formed was oxidized. Meyerhof assumed that the energy of oxidation was used for re-forming the glycogen in muscle. Some observations seemed to support this idea. He suggested a cyclic reaction: carbohydrate breakdown, via some intermediary phosphorylated compounds, to lactic acid; resynthesis of the lactic acid, again via intermediary phosphorylated compounds, to carbohydrate using the energy provided by the lactic acid oxidized. These experiments were, however, performed under inadequate conditions. Many years later, in 1940, Carl F. and Gerty Cori in St. Louis established the actual cycle (referred to as the Cori cycle). The unoxidized lactic acid diffuses out of the muscle into the blood and is transported to the liver, in which it forms glycogen. In the liver glucose is formed from glycogen and is transported to the muscle, where it forms glycogen.

It took about 10–12 more years for the total amount of heat produced during activity to be accounted for by chemical reactions. This became possible only after the discovery of two new phosphate derivatives that were associated with muscular contraction, discussed subsequently.

The data of Hill and Meyerhof mark the first attempt in the history of biology to correlate chemical, physical, and thermodynamic data, even though the aim was only partially achieved. But what accomplishment that tries to explain a mechanism of any physical or biological phenomenon has given the final answer? Were the atom models of Rutherford or of Niels Bohr or even later models the final answer? No—but they were momentous contributions and initiated one of the most exciting developments in the history of science. Some biologists were under the impression that Meyerhof received the Nobel Prize for the demonstration that the lactic acid formation not only provided a part of the energy of muscular contraction, but was directly responsible for the elementary process of contraction. Since there was no evidence for such an assumption, it was at best a speculation. The Nobel Prize is not given for assumptions. The author had an opportunity to discuss the reason the Nobel Prize was given to Meyerhof with Professor Gören Liljestrand, the long-time secretary of the Nobel Prize Committee. He emphatically stated that this assumption was not even considered. The prize was given for the brilliant attempt by Hill and Meyerhof to correlate biochemical and biophysical data. Moreover, this assumption was never seriously proposed. On the contrary, in a paper by Meyerhof and Karl Lohmann submitted in November 1925 and published in 1926 (Meyerhof and Lohmann 1926), they specifically rejected the notion that the proportionality between the lactic acid formed and the mechanical work performed was an indication of a direct action; they stated specifically: "Nun ist es ja niemals in Vorschlag gebracht worden, den chemischen Vorgang der Milchsäurebildung direkt die mechanische Arbeit leisten zu lassen."

When Meyerhof received the Nobel Prize in 1923 he had just been through an unpleasant and humiliating experience. Rudolf Hoeber, the chairman of physiology, had established a separate division of physiological chemistry and had proposed Meyerhof as *Abteilungsleiter* (head of the division). The faculty declined—an unusual affront against a chairman—and selected another member, by no means of equal stature and distinction. Anti-Semitic feelings were apparently an important factor.

One has to remember the strong rise of anti-Semitic feelings in the Weimar Republic after the lost war and the continuous humiliation of Germany during the years following the Versailles treaty. Hoeber himself was a Jew. He left when the Nazis came to power and joined the University of Pennsylvania in Philadelphia, where he spent the rest of his life. There he and Meyerhof met again in 1940. Meyerhof had declined an attractive offer from Yale University. But now Warburg and other members of the Kaiser Wilhelm Institute of Biology, among them the director C. E. Correns, R. Goldschmidt, and M. Hartman approached the Kaiser Wilhelm Society and urged the creation of a new department for Meyerhof in the institute. Each of them offered to give up a room in order to facilitate and expedite the decision. Meyerhof joined the institute in 1924. Although the rooms were rather primitive (even for the standards of that time) and quite scattered, Meyerhof was able for the first time to build up a group, to have technical help, and to accept scientific collaborators. In Kiel he had no technical help whatsoever. The brilliant series of observations for which he received the Nobel Prize were all performed single-handedly. Only H. H. Weber had spent six months with him in 1922.

The first and most important person to join Meyerhof was Karl Lohmann (1898–1978), a brilliant organic chemist, who stayed with Meyerhof from 1924 to 1937, when he accepted an offer from the University of Berlin to become chairman of the Department of Physiological Chemistry. There is no question that the association of these two men had a deep impact on the advances in Meyerhof's laboratory, as was emphasized by Meyerhof on many occasions. Although their personalities were quite different, scientifically they supplemented each other in a most fortunate way. It was a cordial, happy, and invaluable association. Of the many important contributions Lohmann made, the most outstanding was the discovery of ATP (discussed subsequently), the elucidation of its structure, and the interpretation of its function, jointly with Meyerhof. But soon Meyerhof was joined by a number of other young collaborators, among them Hermann Blaschko, Severo Ochoa, Fritz Lipmann, Ken Iwasaki, Paul Rothschild, and the author. But at any given time there were very few scientists. This was an extremely important factor that promoted the development of close ties between them and Meyerhof and also among themselves. Meyerhof's personality attracted a special type of person. This may have been an important element in the formation of the close friendships that developed among his collaborators and lasted for a lifetime. Several of them became eminent leaders in various fields during the rapid progress made in biochemistry in the postwar period.

The Exploration of the First Metabolic Pathway. Hoppe-Seyler (1825–1895) had suspected the presence of enzymes *inside* cells. In 1897 Eduard Buchner, a pupil of Adolf von Baeyer, published his famous discovery that the juice obtained, under certain conditions, from pressed yeast cells was capable of alcoholic fermentation; he had obtained the enzyme responsible for alcoholic fermentation in cell-free solution. Buchner referred to this intracellular enzyme as *zymase*. As it was later discovered, Buchner's zymase is a mixture of a great number of enzymes. But the observation was nevertheless a landmark in the history of biochemistry. It was the basis and starting point of a new concept. The observation showed unequivocally that reactions taking place in the living cell can be performed in a solution outside the cell. No "vital" force had to be invoked. As happens frequently in the history of science when new concepts emerge, this view was immediately vigorously opposed from many sides, all citing the different ideas of many authorities, including Pasteur. Buchner vigorously rejected all objections. One of his compelling statements was that even Pasteur's authority was no argument against experimental facts, a statement that should be kept in mind when scientists adhere to dogmatic views, no matter how many facts contradict them. Nevertheless, the importance of Buchner's discovery and its enormous implications soon found wide recognition and initiated many important investigations. For example, it was enthusiastically endorsed by Émile Duclaux (1840–1904), a close associate of Pasteur. Franz Hofmeister (1850–1922) became one of the foremost supporters of the enzymic theory of intracellular metabolism, of the view that enzymes (or ferments), were *inside* the cells and that sooner or later a specific enzyme would be found for every vital reaction, an almost prophetic idea when one looks at our present knowledge. Three brilliant pupils of Hofmeister, all of them Jews, became eminent leaders in the field of enzyme chemistry and its role in intermediary metabolism: Gustav Embden (a half Jew) in Frankfort on the Main, Carl Neuberg in Dahlem, and Jacob K. Parnas in Lvov, Poland. Each developed a large school in enzyme chemistry.

In the two decades following Buchner's discovery, many significant advances were made in the understanding of several steps of alcoholic fermentation as well as the intermediary steps of glycolysis in muscle. But in general, progress was slow, primarily because of the cumbersome methods available at that time. Again a major breakthrough was the success of Meyerhof, in 1925, in extracting the glycolytic enzymes of muscle, which were capable of carrying out all the steps of glycolysis in solution when glycogen was added. This achievement is comparable to that of Eduard Buchner: It opened the way to isolate the enzymes and to study the individual steps involved in the complex glycolytic pathway of muscle from glycogen to lactic acid. It took about 15 years to unravel the complete pathway. Although the contributions of Meyerhof and his associates played the largest role in this development, many brilliant discoveries were made by Gustav Embden, Jacob Parnas, Carl F. and Gerty Corti, Otto Warburg, Dorothy Needham, and their collaborators, and many others; several of the enzymes were, in the 1930s, crystallized by Otto Warburg (see IV, C). Embden, with admirable ingenuity and intuition, proposed the main skeleton of the sequence of reactions forming the pathway in 1932. He died tragically in 1933. It remained to Meyerhof and his associates to verify most of the essential steps. The pathway is today generally referred to as the Embden–Meyerhof pathway. But whereas the glycolytic pathway was the first metabolic pathway to be completely analyzed, the stimulus derived from this achievement initiated an intensive period of investigations of a great number of metabolic pathways in the following decades, particularly in the United States. Today most pathways are well established, and the main interest of biochemists has long since shifted to a variety of other subjects.

One of the by-products of the unraveling of the pathways of glycolysis in yeast and muscle cells is of general significance. It turned out that the many single steps of glycolysis in yeast, a rather primitive cell, and in muscle are the same except for the last one, in which the action of one single enzyme leads to alcohol as end product in the former, and that of another enzyme leads to lactic acid in the latter. This remarkable similarity in the mechanism of glycolysis in the course of evolution gave strong support to the idea of the biochemical unity of life, a notion greatly cherished by Pasteur, and a basic element in the thinking of Warburg and Meyerhof. Obviously, this notion requires a considerable amount of flexibility. It would be naive to assume that in the course of evolution no changes took place, but these modifications are usually quite minor. As a guiding principle the idea of the biochemical unity of life played, and is still playing, an essential role in the advances of biochemistry.

The Discovery of Energy-Rich Phosphate Derivatives. One of the most fundamental contributions of Meyerhof to the field of biochemistry was, as mentioned before, the discovery of two types of phosphate derivatives; one of them is rich in energy and the other is poor. When the first is split with the formation of inorganic phosphate, a large amount of heat is produced, as measured in the calorimeter. The change of total energy (ΔH) is of the order of magnitude of 10,000 to 12,000 calories per mole. The second type—to which most phosphate derivatives belong—releases relatively small amounts of heat, of the order of 1500–3000 calories per mole.

The experiments were prompted by the discovery, in 1926, of a new phosphorylated compound in muscle, first referred to as "phosphagen," by C. H. Fiske and Y. SubbaRow in the United States, and shortly afterward and independently by P. and G. P. Eggleton in England. Fiske and SubbaRow established that the new compound was phosphocreatine. Since there were some indications of a breakdown of this compound during muscular contraction, Meyerhof, always deeply concerned with bioenergetics and looking for the possibility of a source of chemical energy, in addition to lactic acid, immediately tested the heat produced by this new compound. Since only about 50% of the extra heat produced during muscular contraction was so far accounted for, it was essential to learn whether and how much chemical energy the new compound might contribute by its breakdown during contraction. The first paper about the surprisingly high ΔH of phosphocreatine breakdown, coauthored by Lohmann, appeared in 1927 in *Naturwissenschaften*; a second more detailed paper, coauthored by Suranyi, appeared in the same year in *Biochemische Zeitschrift*. The excitement in the laboratory about this finding was great (the author worked at that time in Meyerhof's laboratory). It appeared obvious that this factor might become the starting point for new, exciting developments in correlating chemical, physical, and thermodynamic events in muscular contraction.

In his lecture in 1913, Meyerhof had pointed out that the important thermodynamic parameter for evaluating a chemical reaction was not the change in total energy (ΔH), but the change in free energy (ΔG). One may therefore ask why he did not calculate the free energy, since as he had pointed out in his lecture this had become possible by the heat theorem of Nernst. Polanyi, a close friend of Meyerhof in Dahlem, had calculated the standard free energy of some reactions in solutions in 1914. Hill, discussing the problem in a paper in 1912, had pointed out how difficult it was to make estimates for reactions in which the equilibrium constants were not known, as was the case for most reactions during glycolysis. Moreover, the standard free energy (ΔG^0) is not the really important factor in reactions taking place in muscular contraction, an extremely complex system very poorly understood in the 1920s. For Meyerhof the most essential value was the total energy (ΔH), since for correlating the heat measured during contraction with the chemical reactions taking place the total energy released during the breakdown of a compound must be used, including the wasted heat. In the 1920s, when he determined ΔH with the calorimeter, Meyerhof assumed that ΔG was not too much different from ΔH and that the high total energy a compound might release could be used as an approximate indication of its ability to provide a rather large amount of usable free energy. This assumption proved to be correct many years later, in the 1950s, when such estimates became possible. Actually, when the single steps of glycolysis were finally elucidated, calculations of ΔG^0 (not yet of ΔG) became easy and Meyerhof himself made a number of such calculations. But even the free energy, ΔG, does not yet provide the complete picture of the bioenergetics of a biochemical process, since we also have to know the change in entropy S, as pointed out before. Even in a simple system, formed essentially by two reactants only, such as hemoglobin and oxygen, it is extremely difficult to separate the three parameters, as was discussed in the 1960s by E. Antonini, A. Rossi-Fanelli, and J. Wyman, Jr. In any event, there is no question that Meyerhof was right to use the total energy (ΔH) for his special aim, to correlate the chemical reactions with the heat data of Hill.

In the following years, when more information accumulated about the intermediary steps of glycolysis, Meyerhof tested several of the phosphate derivatives found. The heat of hydrolysis (ΔH) of two other phosphate derivatives, 1,3-diphosphoglycerate and phosphoenolpyruvate, was found to be high. The most important energy-rich phosphate derivative turned out to be adenosine triphosphate (generally abbreviated ATP). This compound was discovered by Lohmann in Meyerhof's laboratory in 1929.* In the following years Lohmann also established its exact structure. When adenosine triphosphate was split so two inorganic phosphates were released and adenylic acid, which contains one phosphate, remained, the ΔH was found to be about 25,000 calories per mole; that is, 12,500 calories per mole of phosphate, about the same heat as per phosphate of phosphocreatine. Meyerhof and Lohmann (1931) described as early as 1931 the utilization of the high energy of ATP in biosynthesis, at that time during the glycolytic process. They recognized that the uptake of phosphate during carbohydrate breakdown to lactic acid is associated with a simultaneous splitting of ATP. In the subsequent reactions ATP is resynthesized. This cycle of ATP maintains lactic acid formation.

Meyerhof's discovery of the two types of phosphate derivatives grew out of his efforts to correlate the chemical reactions occurring during muscular contraction with the heat produced. Although the strong heat released by phosphocreatine breakdown was strong evidence for its contribution to the heat measured by Hill, it did not answer the question of how its action was involved in the mechanism of contraction. Several observations suggested that phosphocreatine was closely associated in some way with the speed of contraction (Nachmansohn, 1928, 1929, III, 1929). Soon afterward a most important series of observations was performed by Einar Lundsgaard in Copenhagen. He found that muscles poisoned by iodoacetate did not produce lactic acid, but were still able to contract as long as phosphocreatine was present. In his elegant experiments he measured the tension developed (work performed) and correlated this work with the heat released by the phosphocreatine breakdown, using the ΔH data of Meyerhof. It was this combination of chemical, mechanical, and thermodynamic data that greatly impressed Meyerhof, since the approach was so similar to his own. He therefore took Lundsgaard's evidence quite seriously.† Lundsgaard sent Meyerhof a preprint of the still unpublished

---

* It is sometimes written that Fiske and SubbaRow discovered ATP simultaneously. Florkin (1975) has documented in his *History of Biochemistry* that this statement is incorrect and that the priority belongs to Lohmann. But it is possible that they discovered ATP independently shortly afterward.

† It has been said that Lundsgaard's observations were a rediscovery. As stated by Sir Rudolph Peters in his biographical note on Meyerhof, this is a half-truth, more false than true. Several years earlier A. Schwartz and S. Oschmann had described how contracture occurring in the presence of bromoacetic acid took place without increase of lactic acid. But Lundsgaard established that regular muscular contraction (in contrast to poorly defined contracture) occurred without lactic acid, and he recognized its importance in combination with the other data.

manuscript with a letter asking him for permission to test the correctness
of his findings under Meyerhof's guidance, since it seemed to be in contradiction to his theory. This letter shows Lundsgaard's scientific attitude,
is search for truth under the strictest possible control. This spirit is unfortunately not the rule; how beneficial it would be for science if this kind
of attitude prevailed.

Meyerhof's reactions to Lundsgaard's observations and letter reveals
his admirable flexibility of mind, as so vividly described by Hermann
Blaschko in Meyerhof's biography prepared by Sir Rudolph Peters
(1954). For some years Meyerhof had held certain views that conflicted
with those of Embden, who had proposed that the phosphate of "lactacidogen" (a phosphate derivative of glucose formed during glycolysis)
might be the substance directly responsible for contraction. It was more
the reasoning of Embden with which Meyerhof disagreed than the hypothesis of the direct action of phosphate. As mentioned before, he and
Lohmann had explicitly stated in 1926 that lactic acid had never been
claimed to be directly responsible for the elementary process of contraction. Although he obviously had some reservations and felt that
Lundsgaard's experiments raised some questions that had to be answered,
it was remarkable how little prejudice or rigidity he showed in evaluating
the new observations. When the author had a long discussion with him
about the new development, he was deeply impressed by the detachment
of Meyerhof and his willingness to follow up the new findings. It must
be emphasized that the new data were actually not as disturbing to him
or to us, since there was no real conviction as to what compound was
directly responsible for contraction. Meyerhof's reaction will be further
discussed later in the context of his general personality features.

In the following years Lundsgaard's experiments seemed indeed to sug-
gest that phosphocreatine might be directly associated with the process of
contraction. But very soon both Lundsgaard and Meyerhof became
skeptical. A decisive turning point was Lohmann's (1934) observations
that two phosphates of two molecules of phosphocreatine could form
ATP from adenylic acid without heat production. Lohmann was thus
the first to recognize that high-energy compounds could accomplish a
biochemical reaction without heat production. The synthesis of ATP
from phosphocreatine and adenylic acid is thus performed with almost
100% efficiency. These experiments, and investigations carried out independently and simultaneously by Parnas and his associates, soon indicated
that phosphocreatine acts as a "storehouse" for the energy for phos-
phorylating ATP. Thus they explained why phosphocreatine is associated
with the speed of muscular contraction. In rapidly contracting muscles
ATP must be restored quickly and with a minimum of extra energy.
These muscles require, therefore, the presence of phosphocreatine. The
suggestion made by the author in 1928, before Lundsgaard's findings, had
finally been confirmed.

In 1934 Meyerhof and Lohmann proposed that ATP might be the compound providing the energy for the yet unknown elementary process. Meyerhof referred to ATP as "the specific operative substance." This view was accepted by many biochemists and physiologists, especially after a discovery made in 1939 by V. A. Engelhardt and M. N. Ljubimova (discussed subsequently). However, after the article of A. V. Hill entitled "A Challenge to Biochemists" (Hill, 1950) in the *Festschrift* on the occasion of Meyerhof's sixty-fifth birthday, quite a few investigators began to question Meyerhof's assumption that ATP hydrolysis was the primary event during contraction in the living muscle. Attempts to measure ATP or phosphocreatine breakdown during a single contraction failed. This led to several controversies. It was not until 1963, after the demonstration that even in a single contraction a breakdown of phosphocreatine takes place, that Meyerhof's view was generally accepted. One of the chief aims of Meyerhof's lifework was achieved: The heat measured by A. V. Hill during muscular contraction was virtually completely explained in terms of chemical reactions (Meyerhof, 1937). A major breakthrough in the exploration of the molecular mechanism of contraction was made in 1939, when Englehardt and Ljubimova found that myosin, one of the muscle proteins assumed to be involved in the contractile process, changed its shape in the reaction with ATP and was able to split ATP. When the splitting was inhibited, no change of shape was produced by ATP. These observations were confirmed and extended by Dorothy M. Needham (1971). They supported Meyerhof's assumption of a direct involvement of ATP in muscular contraction. But Meyerhof did not live to see the exploration of the detailed mechanism of muscular contraction. These developments were initiated by H. E. Huxley and his associates using electron microscopy analysis, X-ray diffraction, and other refined methods and instruments that became available after the 1950s. Today, owing to the work of several outstanding investigators, we have a remarkably good picture of the molecular mechanism of muscular contraction and the proteins involved, but the description of these advances does not belong here.

When the detailed pathway of glycolysis had been established by the end of the 1930s, it turned out that its main function was to provide ATP. The yield is not very high; for each molecule of glucose, 2–3 molecules of ATP are produced. At was learned later, oxidation has the same main function, but it is a much more efficient and powerful source of ATP. For each molecule of oxidized glucose 36 molecules of ATP are formed, as first calculated by Severo Ochoa (see V, B).

The discovery of Meyerhof and his associates of the two types of phosphate derivatives, just 50 years ago, was one of the most fundamental contributions to biochemistry of this century. The fact that, in the decades following the discovery, the energy released by phosphate derivatives, in particular ATP, turned out to have a much more general significance than originally suspected only emphasizes its importance. In fact, it is a criterion of the greatness of a discovery when it initiates entirely new aspects by subsequent investigations. In the 1940s and 1950s many investigators showed that ATP is the commonly used source of energy in most metabolic reactions of the human body; it is frequently referred to as the universal currency of free energy in biological systems. The first reaction for which the free energy of ATP was shown to be used in biosynthesis outside the glycolytic cycle was the process of acetylation; the author demonstrated in 1942 that ATP energy may acetylate a compound in solution (for a review see the Prefatory Chapter of Nachmansohn, 1972). Acetylation is a reaction widely used in many important intermediary steps, and this discovery later catalyzed a whole series of investigations on the role of ATP as energy source outside glycolysis.

In the spring of 1951, William McElroy and Bentley Glass organized a symposium at Johns Hopkins University, "Phosphorus Metabolism." Many of the most eminent leaders in the field attended the meeting. As was pointed out in the summary by Bentley Glass, the meeting dealt with virtually all aspects of the chemistry of life, in view of the central role of phosphates. Meyerhof, as the uncontested master and great pioneer of the field, was asked to give the opening address, in which he pointed out the paramount role of phosphate derivatives in view of their role in energy transfer. He closed his address with the following words: "Just as the role of iron in biological oxidation is now made completely understandable by the work of Otto Warburg as being necessary for the catalysis of oxygen transfer, so the role of phosphate compounds is made understandable by their importance for energy transfer." It must be added that it was the work of Otto Meyerhof that made this understanding possible. Meyerhof did not live to see the publication of the symposium; he died in October 1951. But as was stressed by many colleagues, the volume is a living monument to his trailblazing work in this field.

In an article entitled "The Roots of Bioenergetics," Fritz Lipmann (1975) made the statement that "when two new phosphate-containing compounds, phosphocreatine and ATP, were discovered, *although far from realizing their bioenergetic importance*, Meyerhof and Lohmann determined their heat of hydrolysis [italics added]." This statement appears to be due to a misunderstanding. An analysis of Lipmann's remarks requires scientific comments of importance for biochemists mainly and is not pertinent for most readers. The statements of Lipmann are, therefore, discussed in an Annex at the end of this section.

Life during the Nazi Period

At the end of 1929 Meyerhof moved from Berlin–Dahlem to the newly established Kaiser Wilhelm Institute in Heidelberg, built on the initiative of

Krehl, in whose clinic he had met Warburg and had worked for two years. The institute was made up of four independent divisions: physics, chemistry, physiology, and basic medicine. Krehl was the director. There were close relationships between the various departments. For the first time in his life Meyerhof had magnificent facilities. The laboratories were built according to his plans. Everything was arranged to facilitate his work which required a great variety of refined methods and instruments. Today they would be considered rather primitive. He was able to accept more scientific collaborators than before. His fame attracted many investigators from all over the world. He was able to take with him his very able technical assistants, W. Schultz and W. Möhle; they were essential for carrying out his experiments. The Kaiser Wilhelm Society built a magnificent house for him, with a lovely garden. One can readily imagine his happiness. He had finally achieved the dream of a true scientist: not glory, but the opportunity to carry out his creative work under the most comfortable, pleasant, and efficient conditions. He was 46 years old and in the midst of an active and dynamic phase of his investigations. Most of the essential steps in the metabolic pathway of glycolysis were established when he was in Heidelberg; also many relationships were found between physical and chemical events, always an important aim in Meyerhof's thinking.

But the happy situation did not last very long. Two years after Meyerhof's move to Heidelberg Hitler became chancellor. The whole atmosphere changed drastically, as has been described earlier. Meyerhof was extremely worried, but decided to stay on for the time being. Like so many other Germans and Jews he was convinced that such a regime could not last very long. He was greatly encouraged in this decision by Krehl, Max Planck, the president of the Kaiser Wilhelm Society since the death of Harnack in 1930, and later Karl Bosch, Planck's successor as president. Not only was his scientific work just then proceeding in a most exciting way, but many highly qualified investigators joined him in this period, among them M. Dubuisson, André Lwoff, Severo Ochoa, Paul Ohlmeyer, W. Kiessling, and many others. But the political situation deteriorated rapidly and became quite ominous. Around 1936 Meyerhof realized that his position was untenable and that he would have to leave sooner or later.

He and his wife Hedwig visited the United States, hoping to find a position there. But the situation for scientists at that time, with the country just coming out of the depression, was still quite tight. The only offer he got was most unsatisfactory: a small laboratory in a commercial enterprise with a salary of $5,000 per annum. The author, who worked at that time at the *Faculté des Sciences* in Paris, spent several weeks in the United States at about the same time as the Meyerhofs, and met them quite frequently. Meyerhof was depressed. After discussing the situation very thoroughly with Meyerhof, the author, knowing the great respect and admiration the French had for Meyerhof, asked him whether he would be interested in the author's investigating the possibility of a suitable place in Paris. Meyerhof's response was most enthusiastic. He had always been a great admirer of French civilization, its art and poetry, its science, its great cultural achievements. A special code was agreed upon for correspondence in view of possible censorship. Although the Nazis wanted to dismiss all Jews, they were beginning to prevent them from going abroad. On the author's return to Paris he immediately approached René Wurmser, Henri Laugier, and Jean Perrin. All three promised their enthusiastic and strong support. The French acknowledged Meyerhof's brilliance as a scientist, but they were also attracted by his extraordinary background in the humanities. His personality had made a great impression on them during a visit to Paris in 1934, when he gave two lectures there. During the dinner parties and receptions on that occasion, his French colleagues gave strong expression to their admiration, as the author vividly recalls. René Wurmser was at that time the head of a subdivision of the *Institut de Biologie Physico-Chimique*. As a brilliant biochemist and physicochemist he was well familiar with Meyerhof's work and delighted with the prospect of having him in his division. He arranged for Meyerhof most satisfactory working facilities and a position as *directeur de recherches*, equivalent to a research professorship. All the negotiations were carried out by the author with the aid of the code agreed upon in New York. The problem was now to get him and the family out of Germany. With the help of some friends Meyerhof got permission in September 1938 to go to Switzerland for a few weeks with his wife and his youngest son Walter for reasons of health. His daughter Bettina had already left for Paris and went in November to the United States; she had been accepted by Swarthmore College near Philadelphia. His eldest son Geoffrey was already living in England. The Meyerhofs never returned to Heidelberg. Of course, since they were leaving the country on the pretext of taking a few weeks of vacation, they were unable to take anything with them except the bare necessities. A few years ago, Heidelberg honored its great citizen by naming a street after him: *Otto Meyerhof Strasse*.

The reception in Paris was extremely warm. Meyerhof soon formed many friendships with his French colleagues. The rue Pierre Curie, with several institutes and many famous scientists, was a great intellectual center and Meyerhof greatly enjoyed the atmosphere; it reminded him in some respects of that of Dahlem. For the author, the admired teacher became a personal friend, a friendship getting closer with the years and lasting until Meyerhof's death. There were many occasions to discuss not only scientific problems, but a wide range of common interests in literature, philosophy, and many other fields. The two couples met frequently and made several excursions together. Having visited places like Chartres or the chateaux of the Loire before, the author was deeply impressed with how much more he appreciated the monuments of art under the guidance of Meyerhof. The visits became a great and delightful experience. They showed Meyerhof's impressive detailed knowledge and understanding of art and history, more than a discussion would reveal.

The happy time in Paris, which the Meyerhofs deeply loved and to which they became genuinely attached, was unfortunately destined to be a short episode. Just one year after their arrival the war broke out. For a few months it was thought to be a "phony" war. But in May 1940 the Nazis invaded France and when they threatened Paris, the Meyerhofs had to flee to southern France with their son Walter. Although the Meyerhofs were most cordially received everywhere and greatly helped by their colleagues, it was obvious that they had to escape from France as soon as possible. The author had in the meantime accepted an invitation from Yale University, where he had arrived a few days before the outbreak of the war. He contacted a few friends. It was A. V. Hill, Meyerhof's longtime scientific colleague and personal friend, who made his escape possible. He contacted A. N. Richards of the University of Pennsylvania, at that time the president of the National Academy of Sciences. Richards managed to create a professorship for Meyerhof at the University of Pennsylvania in the Department of Physiological Chemistry directed by Wright Wilson. But before coming to the United States, the Meyerhofs passed through a painful and difficult period. With the help of Varian M. Fry of the Emergency Relief Committee, a predecessor of the International Rescue Committee, they finally escaped to Spain, crossing the border over the Pyrenees, a very exhausting effort. They reached the United States via Lisbon in October 1940.

Scientific Contributions after Heidelberg

Tragedy, sufferings, and hardships did not break Meyerhof's spirit. After his arrival in Paris he immediately built up his laboratory with great energy. Several significant papers appeared during the relatively short period there. He continued his experiments on the application of radioactive phosphorus, which he had begun in Heidelberg in association with Niels Bohr's laboratory in Copenhagen. Among his other contributions may be mentioned his demonstration that glycolysis in embryonic tissue requires phosphorylation, just as it does in adult tissue. This was in contrast to some reports that were not acceptable to him, since he considered it almost inconceivable that there should be such a fundamental difference in the mechanism of glycolysis between the early phase of development and a later one. On his arrival in Philadelphia he started to build up his laboratory immediately. He had a few gifted young collaborators and trained them in using his methods and making them familiar with his basic concepts. Many of these collaborators and several younger colleagues whom he met in the department always speak of Meyerhof with great admiration and affection and emphasize how much they were inspired and influenced by his ideas and his personality. The scientific output soon began to increase again.

Meyerhof participated in quite a few meetings and symposia. The glory and fame of his work were by now great and he was always a center of attraction at these meetings, and a source of inspiration to many colleagues who had known him only from the literature. At a meeting of the New York Academy of Science in 1944 he gave a brilliant lecture about his favorite topic, the importance of thermodynamics and energetics for biochemistry, and he summarized his findings on the energetic aspects of muscle chemistry.

During the last decade of his life he made an impressive number of valuable contributions, although they were not of the same pioneering character and fundamental importance as those made in the preceding decades. For example, he discovered the mechanism by which an intermediate compound (hexose diphosphate) accumulates during fermentation by yeast extracts (the so-called Harden–Young reaction). This fact had puzzled biochemists. Meyerhof demonstrated that this accumulation was due to the absence of an enzyme (adenosine triphosphatase) that dephosphorylates ATP. On addition of the enzyme the yeast extracts fermented glucose to alcohol and carbon dioxide in the same way as living yeast cells. He also succeeded in separating this enzyme in muscle from myosin, one of the important structural proteins to which the enzyme is attached. His former associate in Heidelberg, Paul Ohlmeyer, joined him for a year in Philadelphia and they both purified the enzyme from yeast. Working on this enzyme led to the discovery that there are two types of adenosine triphosphatase; one was activated by magnesium ions, but the other was not. Several years later it turned out that one type of these adenosine triphosphate enzymes played an important role in the "active" transport (i.e., requiring energy) across membranes, a very important function of all cells. He demonstrated with a variety of compounds that transphosphorylation (transfer of phosphate from one compound to another) is catalyzed by enzymes termed *phosphatases* in the absence of ATP. Such transphosphorylations seem to go preferably from higher to lower energy phosphates. Still intensely interested in the energy content of some key metabolic intermediates—so important for the understanding of the mechanisms underlying the biological utilization of metabolic energy—he measured the heat produced on hydrolysis of various compounds (ΔH), rechecking some of his old measurements. He determined the equilibrium constants of the hydrolysis and synthesis of several phosphate derivatives, for instance, the conversion of phosphopyruvate and adenosine diphosphate to pyruvate and ATP, and calculated the standard free energy of ATP (ΔG^0) from these measurements. Just as an indication of the intensity of his activities, it may be mentioned that during this decade about

50 papers appeared from his laboratory, bringing the total number of his publications to about 440. This activity is all the more remarkable since he suffered a severe heart attack in the summer of 1944 and was hospitalized for about 10 months. He continued his work until a second heart attack led to his death, which came suddenly during sleep and without suffering, in the midst of creative work and the preparation of various projects for the future.

Woods Hole. During the 10-year period that Meyerhof spent in the United States, a special role was played by Woods Hole, a tiny village on Cape Cod in Massachusetts. In Woods Hole there is a very large Marine Biological Institute. Founded near the end of the last century, it has slowly grown to become one of the largest institutions of this kind. Jacques Loeb had greatly contributed to the establishment of a modern center of biology there. The laboratory facilities have steadily grown, thanks to generous private and government gifts. There are now several large and well-equipped buildings. With the help of the Rockefeller Foundation a library was established that is almost unique in its kind. It has an extraordinarily large and complete collection of journals, some going back to the seventeenth, eighteenth, and nineteenth centuries, and all starting with volume 1, such as the *Proceedings of the Royal Society*, the *Annales de Physique et Chimie*, and *Poggendorf's Annalen*. The library is open 24 hours a day, 365 days a year. It is superbly organized and offers many special facilities for people who want to work there (writing a book, reviews, papers, and the like). In the summer Woods Hole is an extraordinary meeting place where many biochemists, biophysicists, and biologists from all over the country meet; distinguished visitors come there from all over the world. Many interesting seminars are given, offering excellent opportunities for exchange of views and information. Beaches nearby facilitate personal meetings of investigators in a relaxed and pleasant atmosphere.

During World War II, when travels to Europe were completely cut off, and several years after the war, Woods Hole was an ideal place for scientists to spend their summers. It offered facilities for a combination of vacation and work, either in the laboratory or in the library, a chance to catch up with the literature or to write papers. The place was very quiet, since there were no tourists. The scenery is lovely and the climate is mild and pleasant, especially when compared with the mainland. The accommodations, although primitive, were very moderately priced, an important consideration for most scientists in view of the low salaries prevailing at that time. Meyerhof and his wife Hedwig spent two months there every summer, from 1941 to 1951, and they met many other European friends. The intellectual and relaxed atmosphere and the many friends were a great attraction. They loved the place and it became a second home for them. The author and his associates did research work at the laboratories during the summer months for over two decades, since there was a special material available that was useful for his investigations. The friendship with the Meyerhofs that had begun in Paris became intensified. Another former associate of Meyerhof, Severo Ochoa, and his wife Carmen, also spent the summers there. Ochoa had spent several years with Meyerhof. The Meyerhofs were extremely fond of the couple, and Meyerhof considered Severo Ochoa the most brilliant and outstanding scientist he had ever had in his laboratory. He followed, very closely and with great pride, the rapid rise of Ochoa as he became one of the leading figures in biochemistry (see V, B). The three couples spent much time together. The close friendship with Meyerhof intensified the friendship between the Ochoas and the author that had started in 1929 in Berlin–Dahlem.

In addition, there were many other European scientists and friends. Leonor Michaelis, a pupil of Paul Ehrlich, who was a member of the Rockefeller Institute since 1929, spent his summers there. Among others may be mentioned Carl Neuberg, who had come from Israel to the United States to join his daughters; Otto Loewi from Graz and his family; and Hans Gaffron, who had spent many years in Warburg's laboratory in Dahlem and, although not Jewish, had left Germany in 1936 because he could not stand the Nazi regime. He was working at that time with James Franck in Chicago on photosynthesis, the effects of sunlight on plants, a problem he had already studied in Warburg's laboratory (see IV, C). James Franck spent many summers on Cape Cod, although not in Woods Hole, but he visited Woods Hole frequently. The two had become engaged in a vigorous controversy with Otto Warburg about a special problem of photosynthesis. Another frequent visitor was Max Delbrück, who also had left Germany in 1936. In 1949, Otto Warburg spent a whole summer in Woods Hole and there were many pleasant meetings, despite his controversy with Franck and Gaffron.

Although it was quite natural that there were special ties among the European scientists because of many common interests, similar experiences, previous friendships, and common heritage, this in no way means that the Europeans formed a kind of separate community. On the contrary, it was in Woods Hole that Meyerhof met many American colleagues for the first time and the relaxed atmosphere favored long and serious discussions. The great scientific stature of Meyerhof made a deep impression on his colleagues, especially on the biochemists interested in his line of work, and influenced many investigators. They admired his dignity, his wisdom and, last but not least, his familiarity with the contemporary developments in science.

Personality

Meyerhof was one of the greatest thinkers among the biologists of our time. His intimate knowledge of philosophy greatly influenced his biological views and judgments, his concepts and his imagination. The personality of a scientist is affected by many factors that play an essential role in his or her creative work, just as in the work of artists and in most other human endeavors. It may be difficult to trace such factors, but their role appears undeniable. Meyerhof's continuous preoccupation with philosophy is reflected in his discussions about his concepts of relationships between the phenomena of life and physics and chemistry. He vigorously supported the view that the laws of physics and chemistry must be applicable to the forces acting in the living cell, and he emphatically rejected the vitalistic and "neovitalistic" views of Hans Driesch, with whom he had several letter exchanges. He was convinced that many manifestations of life will become understandable in physicochemical terms. But deeply influenced by the transcendental idealism of Kant and Fries he was constantly aware of aspects belonging to other categories in the meaning of Kant, aspects that cannot be analyzed by physical methods: consciousness, ethical values, freedom of will, the creative forces of man, emotions and psychological phenomena in relations between men, the beauty of a painting or of a symphony. He felt that in the last analysis the whole of scientific truth, the understanding of physical reality, becomes relative to other values that do not refer to processes that may be recognized by our senses, or analyzed by science.

In view of the crucial role that philosophy played in Meyerhof's life and work, it seems appropriate to discuss, at least briefly, some of his basic ideas in the context of the conceptual changes resulting from the revolutionary discoveries in modern physics that were discussed in Chapter II. Problems resulting from the manifestations of the human mind were discussed on a remarkably high level by Greek philosophers and have continued to preoccupy all thoughts about the nature of man throughout the ages. When modern science began to emerge in the seventeenth century, Descartes vigorously supported a mechanistic approach to all biological phenomena, but strongly defended a dualistic concept of mind and body, as has been discussed before. Basically this view was dominant for a long time among scientists and philosophers, as just mentioned in reference to Kant. But even in this century, the dichotomy is a controversial problem, although with some pertinent modifications resulting from the new philosophical notions that were outgrowths of physics. Schrödinger, one of the founders of wave mechanics (see II, C), in a lecture series entitled "What Is Life?," given in Cambridge in 1944, raised the question of whether it would one day be possible to explain what happens in a living organism, at a given time in a given space, in terms of physics and chemistry. Although it is not possible as yet, he was convinced that it eventually would be. Few biologists will disagree with this judgment, provided one limits the answer to the specific question raised. The spectacular progress of biology in this century was the result of applying physics and chemistry to the study of living organisms. But in respect to other aspects of life, such as the phenomena of the human mind and ethical values, Schrödinger (1951), like Meyerhof, flatly rejected the possibility of a physical explanation.

Basically similar is the attitude of Heisenberg, who was not only one of the founders of quantum mechanics but also one of the great philosophers of our time. In *Physics and Philosophy* he discussed the limitations of a physical approach (Heisenberg, 1958). As a scientist he obviously favored the attempts to explain biological phenomena, the function of living cells, including those of the brain, as far as possible on the basis of physicochemical laws. But in respect to the functions of the human brain, specifically those that include psychology, ethical attitudes, and man's creative forces, Heisenberg seriously questioned the possibility of physicochemical analysis:

> If we go beyond biology and include psychology in the discussion, then there can scarcely be any doubt but that the concepts of physics, chemistry, and evolution together will not be sufficient to describe the facts. On this point the existence of quantum theory has changed our attitude from what was believed in the nineteenth century. During that period some scientists were inclined to think that psychological phenomena could ultimately be explained on the basis of physics and chemistry of the brain. From the quantum-theoretical point of view there is no reason for such an assumption. We would, in spite of the fact that the physical events in the brain belong to the psychic phenomena, not expect that these could be sufficient to explain them. We would never doubt that the brain acts as a physicochemical mechanism if treated as such; but for an understanding of psychic phenomena we would start from the fact that the human mind enters as object and subject into the process of psychology. [Heisenberg, 1958]

Heisenberg discussed the different sets of concepts in the development of science that led to the increasing part played by the subjective element. He distinguished four sets of concepts in the development of physics. The first is that of Newtonian mechanics; the second introduces concepts of probability, heat, thermodynamics, free energy, entropy, and the like; the third is the specific theory of relativity and electrodynamics; and the fourth is essentially the quantum theory. The "a priori" notion of Kant's philosophy was based on the first set. The complexity of nature became increasingly apparent from the further developments. The quantum theory does not allow a completely objective description of nature; man as the subject of science is brought in through the questions that are put to nature in the a priori terms of human science.

Heisenberg's teacher and friend Niels Bohr repeatedly discussed the limitations of human knowledge by physicochemical analysis, probably more than other atomic physicists, perhaps because of his passionate interest in biology from his early youth as the son of a distinguished biologist (Bohr, 1958). Like so many other scientists, he seriously questioned the possibility of explaining intelligence, imagination, consciousness, and freedom of will in terms of physics and chemistry. Since even in the much simpler systems of physics, in the analysis of the atom, different answers may result from different observational situations, much more serious difficulties should be expected in the infinitely more complex problems of psychology. The notion of complementarity as an extension of the law of causality may be useful in the dichotomy of psychological phenomena (e.g., feeling and thinking), but in its extreme applications may lead to difficulties. In the light of conceptual concepts evolved from quantum theory we must keep our minds open to the possibility that a better understanding of many manifestations of the human mind may result from the introduction of revolutionary conceptual notions and new dimensions in our thinking. But the question of the limits of physicochemical analysis of the mind remains wide open.

Meyerhof followed with great interest the developments of atomic physics, particularly its philosophical implications. His views on the revolutionary effects of quantum mechanics are well expressed in his lecture on the philosophical basis of physiology ("Betrachtungen über die naturphilosophischen Grundlagen der Physiologie"), given in January 1933 in Berlin and published in the same year in *Abhandlungen der Friesschen Schule* (volume 6, pp. 33–65, 1933) and summarized in *Naturwissenschaften* (Meyerhof, 1934). In this lecture he examined the relationship between the physical and the philosophical approach to the analysis of the universe and stressed the limitations of the former and the importance of the latter.

Meyerhof specifically disagreed with the opinions of Niels Bohr and Pascual Jordan on the limitations to all experimental analyses of biological processes. Bohr saw an analogy between the causal explanation of elementary biological processes and that encountered in the experiments on the atom, in which the measuring instruments modify the process under investigation in such an uncontrollable way that it is no better than without the use of the instruments. Bohr applied this difficulty to biological macroevents: They cannot take place and be observed simultaneously. Since the methods used would first destroy the life, they would prevent the possibility of obtaining a causal insight into the process under investigation. Meyerhof fully agreed that the atoms of the living organism are the same as all other atoms of nature, and that inside the atoms the same quantum jumps take place. Thus it is possible that in some undefined elementary events of the living organism such atomic events may play a role. But these are pure speculations. They do not offer any difficulty to a causal explanation of those biological processes that present the most significant areas of biological research. Biology tries to keep the process intact while working at the same time on its analysis. This is quite obvious in the experiments on the respiration of isolated yeast cells under conditions that did not at all interfere with the cellular processes. Even when we test some reversible inhibitory effects on certain cell constituents— for example, in the observations on the role of heme in oxidation—we are sure that the living organism has not been damaged in an uncontrollable way. But even using isolated surviving organs of higher organisms, such as muscle or nerve, we can ascertain in many ways that the processes analyzed take place in the same way as in the intact organism. The concern expressed by the physicists may apply to some methods used in experimental psychology, but this is a different problem. In any event, in the investigations of the macroevents of the elementary biological processes those taking place in a single atom are irrelevant.

However, although we are able to achieve a causal analysis providing an understanding of individual processes in terms of physics and chemistry, we are unable at present to explain some other phenomena meaningful and specific for the functional and structural maintenance of the total organism. Mechanisms found in inorganic nature do not offer an adequate explanation. In the best case we can point to some simple analogies of self-maintaining systems. One such analogy—but just an analogy—is the unusual dynamic stability of the atom. An atomic nucleus deprived of its electrons is capable of a complete regeneration of its electron shell. We cannot derive the properties of the complete atom from the properties of the isolated elementary particles, such as protons and electrons. The dynamic totality of the atom has additional forces and abilities. In this respect Meyerhof agreed with Bohr's statement: "The inadequacy of the mechanistic analysis of the stability of the atom forms a close analogy with the impossibility of a physical and chemical explanation of the characteristic functions of life." Meyerhof believed, however, that it is not, as Bohr thought, the same experimental paradox of atomic physics that forms the obstacle. Bohr considered that the smallest freedom that we are forced to permit to the organism under investigation is just sufficiently important that it hides its last secrets. In Meyerhof's view we encounter in the organism, on a higher level, the problem of the organization of the atom: Outside the organization there are no phenomena from which we are able to derive inductively natural laws capable of informing us about the properties of the total entity. Certainly we can say that the formative and functional forces of the living matter are already latently present. In this respect one may, for instance, refer to the specific ability of carbon atoms to form complexes. But for many special processes of the living organism there is no possibility of finding them in inorganic nature.

The corresponding difficulty of a reduction is found in the transformation of the atom into its elementary particles. All matter has the organization of the atom. Physics has found a new basis in quantum mechanics for accepting the prerequisites on which the binding of all elementary particles in the atom may depend. However, such an analogy would be inadequate for biology. Living organisms form only a small part of matter. Obviously it is impossible to put physics on a new basis in order to interpret the great number of various vital functions as physically perceivable consequences of the theory of matter.

One more topic among those discussed in Meyerhof's lecture may be mentioned, namely E. D. Adrian's discovery of the rhythmic nature and frequency relationship in nerve conduction in the stimulation of peripheral sense organs. These observations show that the intensity of a sensory perception is the result of a certain frequency of the action currents, but that these currents have always the same amplitude. The data show another aspect of biological research: the psychophysical nature of the organism. Thus the biologist has to include consciousness in his considerations, especially in sensory physiology. Just as physics taught us that the effects of objects and their diffusion in space and time are based on discontinuous processes, the events in the central nervous system, which transmit to our consciousness a continuum in space and time, are discontinuous. A great number of objects in a space will excite a greater or smaller number of nerve elements, whereas the intensity of a sensation will depend on a higher or lower number of impulses in a unit of time in the same element.

These few points may suffice to illustrate Meyerhof's views on the differences between contemporary physics and biology. However, a brief summary of the conclusions and the basic philosophical attitude that he expressed in this lecture appears pertinent. Meyerhof emphasized the limitations of the physical approach when it comes to the problems of psychology and other manifestations of the human mind that physics is unable to explain. The physical concepts and notions of the universe disregard the events of consciousness. Thus physics and chemistry are unable to include and explain *all* of biology. They cannot provide an explanation of the meaning of the universe or that of life. But there exist different ways of analyzing reality that are only apparently in contradiction to each other, such as the concepts presented in the philosophy of Kant's transcendental idealism. In its most general formulation it states that there are different ways of analyzing reality and of conceiving things; the contradictions are the result of different perspectives. Meyerhof recalled a statement of Fries, who spoke of a law of division of truth. According to the notion of Fries the separation of the truth of different approaches leads to concepts that are relative to each other and form part of a higher truth. This higher truth is not based on sense experience or perception that is open to analysis, but looks for the meaning of the world behind the manifestations. The assignment of the philosophy of science is to let the apparent contradictions between the world pictures disappear by rendering them relative to each other. Kant's transcendental idealism provides the fundamental insights into such a relativity. This fact is not changed by the statement that some of Kant's conclusions were erroneous and conditioned by the period in which he lived. Meyerhof believed that the words of Arthur S. Eddington about Newton apply equally to Kant: "To imagine that Newton's scientific fame increases or decreases with the revolutions of one time means to confound science with omniscience."

The lecture gives a vivid impression of Meyerhof's passionate interest and insight into the philosophical developments that took place in the first decades of this century, stimulated particularly by quantum mechanics and the theory of relativity. In spite of his deep involvement in his experimental work, his philosophical mind was continuously driving him to follow closely the impact of the new knowledge on the foundations of biology and philosophy. The revolutionary advances in biology in the half century since this lecture was given, the resulting modifications of many concepts, would not affect the fundamental attitude of Meyerhof. It is essentially similar to the views of Heisenberg mentioned previously, which were expressed more than two decades later, and to those of Born, Schrödinger, and other leading physicists who fully recognized the limitations of physics in explaining manifestations of the human mind in terms of physics and chemistry.

In contrast to the views just discussed, Pascual Jordan, an eminent atomic physicist and coauthor with Born and Heisenberg on their important contribution to quantum mechanics in the 1920s, proposed some interesting theories pointing in a different direction. In *Anschauliche Quantentheorie* (Jordan, 1936) he considered the possibility that seemingly complex organisms may actually not be macroscopic systems in the meaning of quantum theory, since microscopic structures may play a decisive and central role and may be made up of a few molecules. He quoted as illustration genetic processes or the few light quanta required to induce a response in the retina. Jordan recognized the difference between the initial cause and the sequence of events leading to the final function, but he emphasized the possibility that processes involving a few molecules may determine and control macroscopic effects. An evaluation of the examples given reveals the pitfalls of the argument. It is correct that Selig Hecht established that 3–4 quanta of light may induce a response in the retina. But the reaction in the retina involves many molecules, some of them macromolecules, in an extremely complex, well-organized structure. Forces that are still poorly understood send messages to brain cells before the reaction can take place. Equally open to question is the reference to genetic transmission. The molecules that have this function, the nucleic acids, are giant molecules, formed by millions of atoms, with a large number of possibilities for action. But in addition a great number of additional cell components control and regulate the actions of the nucleic acids. Proteins, also macromolecules, although much smaller ones, and many other small molecules are among these components. Compared with these complexities, an atom as we know it today, with its many elementary particles, appears simple, and even here we are not yet at the end of the road. The problems facing a complete understanding of the simplest biological process, seemingly involving only a few molecules, are many orders of magnitude greater than those in atomic physics. Thus Jordan's premise is built on weak ground. It is true that in 1936, when his theory was proposed, the extraordinary complexity of even seemingly simple biological processes was not yet fully appreciated. In fact, the rapid advances and brilliant achievements of genetics and molecular biology in the last four decades have greatly increased our insight and awareness of the complexity of biological processes. In this connection it may be useful to remind the reader that the human brain is formed by more than 10 billion neurons. Each neuron has many connections with other neurons, ranging from 300 to more than 3000. If we assume an average of only 1000 connections per neuron, 10 billion neurons would have 10 trillion connections. The cell bodies of neurons form only a small fraction of the mass of the brain (probably about 1%, although reliable and precise estimates are not available). In discussing the mental process of the brain in the context of the complexity of the atom, it appears useful to keep these figures in mind, although they are by no means the decisive factor in the problem of bridging the gap between physical and mental processes.

Many scientists rejected Jordan's theory, among them Max Born, with whom Jordan had been associated. In *Natural Philosophy of Cause and Chance* (Born, 1949), Born discussed the inadequacy of the solution offered in Jordan's theory, using the indeterminacy of quantum mechanics for the interpretation of the autonomy of the human mind by applying the laws of physics.

But let us assume that the development of science will one day permit us to establish, with the help of computers and a host of highly sophisticated instruments, the number of light quanta of a Van Gogh picture hitting our eyes, or the intensity and number of sound waves produced by a Beethoven symphony reaching the receptors in our ears. Let us further assume that one day molecular biology will succeed in analyzing the number and type of molecules activated in the brain, their function and the molecular modifications induced. Will the information obtained from these analyses provide the answer required for the understanding of our feelings of beauty, our various emotions of elation, joy, or sadness? A remark of Einstein's quoted by Gustav Born at a symposium entitled "The Creative Process in Science and Medicine" (Krebs and Shelley, 1975) seems quite illuminating. Asked whether he thought that everything could ultimately be expressed in scientific terms, Einstein replied, "Yes, that is conceivable, but it would make no sense. It would be as if one were to reproduce Beethoven's Ninth Symphony in the form of an air pressure curve." The pertinent question is that of whether we will be able to bridge the gap between the physical and mental processes. There for many scientists the *ignorabimus* of Du Bois-Reymond still holds.

It must be mentioned that there are biologists who, in view of the rapid advances of genetics and of molecular biology in general, are convinced that it is only a question of time before the understanding of the workings of the human mind in terms of physics and chemistry will become a reality. They do not accept the opposite views discussed. Francis Crick, for instance, one of the most eminent founders of molecular genetics, is convinced that all life can be ultimately accounted for by the laws of inanimate nature; he has expressed his view in his book *Of Molecules and Men* (Crick, 1966). His views have been criticized by several scientists, among them Michael Polanyi, an outstanding physicochemist, whose interests shifted to economics and philosophy after he left Germany in April 1933 and settled in Manchester. During the years he spent in the Kaiser Wilhelm Institutes in Berlin as head of the physicochemistry division in Haber's institute, he and Meyerhof were for many years close friends and had similar philosophical views and interests. In an article entitled "Life Transcending Physics and Chemistry," Polanyi (1967) rejected Crick's views; in particular he specifically stressed that genetics would not offer, for instance, a possible physical explanation of human consciousness.

The negative attitude of Meyerhof and the other physicists mentioned as to the possibility of understanding psychological and other manifestations of the mind in terms of physics and chemistry did not apply to other functions of the brain. Like every other organ in the human body, the brain is subject to the laws of nature and therefore open to physical and chemical investigations in spite of its complexity. Most of our information, valuable as it is, still belongs in the category of descriptive phenomenology; this includes behavioral studies, whether they are performed with highly sophisticated electronic equipment or by means of the fascinating observations of Konrad Lorenz and others on animals. They have brought us much valuable and interesting information about many aspects of behavior. Many chemicals and drugs have been found to have remarkable and strong effects on psychological behavior and have revolutionized psychotherapy. As to the other functions of the brain, many studies have tried to find clues to the underlying mechanisms, but so far no major breakthroughs have been achieved.

Amazing progress has been made in several areas of biological sciences completely obscure three or four decades ago. The insights gained had a deep impact, not only on our knowledge of the living organism but on our philosophical thinking and our understanding of the universe. For example, the advances made in genetics have strongly influenced another field with which scientists have been preoccupied for many decades: the origin of life and evolution (i.e., how the different cell constituents originated and eventually succeeded in forming organized structures). Manfred Eigen (1927–   ), one of the most eminent scientists of our time, a man with a thorough knowledge of modern physics, and a brilliant mathematician, has been for many years deeply engaged in research on this problem of biology and has raised many challenging biological and philosophical questions. In his paper, "Self-Organization of Matter and the Evolution of Biological Macromolecules" (Eigen, 1971), he offered a penetrating analysis of the prerequisites of self-organization and the phenomenological theory of selection; he proposed several reaction models that might have led to the formation of precursors of living cells. His stochastic* theory of evolution is the extension of the theory of probability to dynamic problems. He discussed in detail how far physical factors are already capable of analyzing and explaining biological processes, and he stated that the limitations derive not so much from principle but from the complexity of certain biological phenomena. Eigen's analysis introduces new, revolutionary dimensions into biology, applying fundamental concepts of physics to the understanding of essential characteristics of the living cell. The theory provides a quantitative basis for the evaluation of laboratory experiments on evolution. However, in view of the role of randomness of the process in its origin, the beginning can, at least at present, not yet be explained. This raises the question of the role of chance in evolution. This problem is discussed in two books: *Le Hasard et la Nécessité* by Jacques Monod (1970) and *Das Spiel: Naturgesetze steuern den Zufall* by Manfred Eigen and Ruthild Winkler (1975). Whereas Monod emphasized the predominant role of chance in the origin of life and in the course of evolution, Eigen and Winkler have documented in a penetrating and convincing analysis the decisive role of physical laws that control chance. They present precise and unequivocal calculations that, although chance may have played a certain role in the initial phase, the whole duration of the universe would not be an adequate period of time to generate life in its present form on the basis of chance, without the decisive role of physical laws that control the processes. The book offers a wealth of fascinating new aspects of interest to physics, philosophy, biology, and to the problems of science and society.

The developments in many areas of biology are truly breathtaking. They have been decisively promoted by the revolutionary advances of physics in this century. Nevertheless, to return to the starting point of the discussion of Meyerhof's philosophical credo, there is no indication that the gap between the physical analysis of the living organism and the manifestations of the mind can be bridged.

One of the problems in which Meyerhof was greatly interested was that of freedom of will. In view of its obvious importance and its implications it has been intensively discussed by philosophers, scientists, and humanists throughout history. The Greek philosophers of the Hellenistic

---

* στοχάζομαι, meaning "aim," "hit," or "guess."

period were probably the first to consider it. An excellent history and analysis of this problem starting with the views of Greek philosophers and including those of leading atomic physicists may be found in the lecture entitled "Die Willensfreiheit in Wandel des physikalischen Weltbilds," given by Sambursky at an Eranos meeting in 1971 (Sambursky, 1971; 1977).

Of the atomic physicists, Planck was particularly interested in the problem of freedom of will, especially in connection with the law of causality. In a lecture at the Prussian Academy of Sciences in 1923, Planck took a firm deterministic attitude. The law of causality was for him an axiom valid in all fields of nature, science, and humanities. He recognized the necessity of the distinction between its validity and its applicability. Physics has come close to the goal with Newton's and Einstein's gravitational theories and Maxwell's electrodynamics. Although there are also statistical laws, such as those in heat processes, that seem to have only the character of probability, the underlying phenomena actually obey the laws of causality. Although in biology the processes are much more complex, Planck assumed that in time the law of causality would be found to be valid there also. He believed that just as force causes movement in nature, motive directs will and actions. This motive is based on the nature of the personality, even if it may be difficult to trace it. This attitude seems rather close to the determinism of Joseph Priestley, proposed at the end of the eighteenth century. For Priestley the motive was the decisive element in determining will. He even went so far as to assume that the motive was based on the specific structure of the brain. Actually, however, there is a fundamental difference between the two views. The advances made in physics in the nineteenth century led Planck to accept the reality of the existence of the physical universe, in contrast to the philosophical views expressed in the positivistic philosophy of the nineteenth century. In fact, this was for Planck not only a reality, but a necessary postulate of all scientific thinking, an axiom that could not be proved or disproved by logic. Since the reality of the universe was not only a legitimate but an indispensable element of scientific thinking, Planck could not ignore the paramount importance of consciousness, which must include the outer as well as the inner world. It follows that we should have at any given moment the possibility of acting according to our will; the action may be wise or foolish, good or evil. This obviously is in contradiction with a deterministic attitude. Planck tried to overcome this obvious contradiction with some not very convincing logical assumptions.

However, in a series of later lectures Planck clearly shifted away from his original determinism, although he was still convinced that all physical and mental processes were causally conditioned. Like Einstein, Schrödinger, and other physicists of this generation, Plank was never fully convinced of the philosophical implications of quantum mechanics, and he believed that the gap between the two different views would somehow be bridged. Nevertheless, it is apparent in his later lectures that his attitude toward the autonomy of the mind had markedly changed, perhaps influenced by the new developments. In particular, the importance of psychological factors was increasingly stressed. The motives that determine the will may be strongly influenced and changed by thinking, and this dichotomy leads to a new situation. The perceiving and recognizing ego as observer and the will of the ego as observed object almost preclude an insight into the motives and thus a causal understanding. Self-recognition has its limits and thus, from the subjective point of view, the freedom of will is a reality. The interaction between the perceiving ego and its will thus becomes increasingly an essential point in the indeterminacy of the inner will. Here the new philosophical attitude resulting from quantum mechanics may have affected Planck's views.

Other physicists, such as Born and Schrödinger, strongly rejected any deterministic attitude as to the autonomy of will; the latter vigorously objected to the implications of such a view especially as to ethical behavior. Niels Bohr, who extended the notion of complementarity to many fields of physics, considered it to be of universal importance for all fields of science and humanities. Applied to the problem of freedom of will, it permits a more precise definition than that proposed by Planck. According to Bohr feeling and thinking are two complementary properties of the human mind. The combination of their analysis is a prerequisite for the understanding of the will. Accepting only the feeling as the determining factor of the will, we will be free, whereas the other extreme, thinking alone, would undermine our decision. Thus both aspects, freedom and limitation of the will, must be considered as complementary factors; it would be difficult to separate them (Bohr, 1934, 1958).

Meyerhof's philosophy was strongly influenced by the role of psychology, especially in view of his friendship with Leonard Nelson, as mentioned before (Hieronimus, 1964). He maintained his strong interest in psychology throughout his life. He was familiar with the modern concepts of psychoanalysis as initiated by Freud, and with the important influence of such factors as the unconsciousness and repression of earlier experiences. But in addition he had the advantage of being a biologist and realizing that many additional factors affected the mind, such as hereditary factors, genetic endowment, family tradition, education, and other qualities of the personality. They are all essential elements in the determination of the will. They add to the complexity of the freedom of will and to the difficulty of analyzing it in physical terms. As in other manifestations of the mind, analogies taken from physical laws of nature are not applicable to the analysis of this problem.

A few short remarks about the problem posed by Manfred Eigen in a lecture (Eigen, 1977) may be mentioned in view of the rare and profound insight that his analysis of genetics and evolution has given to many aspects of biology. As in all systems, the freedom of will is, according to

Eigen, based on a combination of chance and laws of nature. One's will is controlled by a mechanism that strictly follows laws as exemplified in the process of evolution. The triggering factor may depend on insignificant variations, for instance the electrical discharge of one or a few nerve cells. There are many alternatives—even for a single individual—to strengthen or suppress such variations. The alternative that prevails is determined by individual value criteria based on the preceding history of the individual, but is not unequivocally predetermined. It is only in this context that we must consider freedom—or, more appropriately, autonomy—of our will. This concise formulation of Eigen of the fundamental laws underlying the problem can hardly be questioned. Like all functions of the human body, those of the brain, including the mind, will be controlled by laws. The understanding of the laws controlling many brain functions are open to physicochemical analysis. Their knowledge will sooner or later bring us pertinent insight into many mechanisms and laws that are at present unknown. This does not contradict the view that some laws controlling the mind, of which freedom of will forms a part, are of a different nature. The factors mentioned before, such as unconsciousness and repression, put the problem in a category not open to physicochemical analysis. Thus a full understanding of the freedom of will remains, like other manifestations of the mind, an open question.

Meyerhof, as mentioned before, was one of the staunchest supporters of the view that many of the mysteries of life still existing will be eventually explained by physics and chemistry. In the early 1940s, in a discussion with the author, he mentioned that the chemistry of the transmission of genetic information was for him one of the greatest mysteries. But he forcefully expressed his conviction that this problem would be explained in the foreseeable future in view of the rapid advances of biological sciences. A few years later, in 1944, O. T. Avery, C. M. MacLeod, and M. McCarthy discovered that nucleic acids are the transmitting material for genetic information. Meyerhof did not live long enough to see the spectacular development of molecular genetics, one of the most fascinating and admirable achievements of the human mind in the history of science. One of his pupils and a personal friend, Severo Ochoa, received the Nobel Prize for his outstanding contributions to the field (see V, B).

Meyerhof was greatly influenced by Goethe. He loved his poetry and frequently quoted his philosophical views. He admired Goethe's many interesting observations in the field of descriptive science. However, he objected to Goethe's criticism of Newton's methods applied to the analysis of color vision and discussed this problem in several lectures, the last time in a lecture presented at the Goethe bicentennial celebration of the Rudolf Virchow Society in New York in 1949. Newton's famous experiments, in which he used a prism to obtain the color spectrum produced by sunlight, were vigorously criticized by Goethe; he rejected this purely physical approach. For Goethe the psychological and artistic views were central, and the sensation of color was an artistic experience. Plato in his *Timaeus* had already distinguished between two types of approach: the scientific and the "divine," or, as we would say today, the "artistic." Meyerhof did not accept Goethe's criticism of Newton's method of analyzing colors. Goethe's views are in conflict with the laws of physics, with the mechanistic approach on which the progress of science was based in the nineteenth and twentieth centuries. But Meyerhof realized that scientific analysis was not Goethe's real aim. It was the deeper meaning of creation—"die Ahnung des Ewigen in Endlichen," to use the words of Fries.

Van't Hoff, one of the founders of physical chemistry and the recipient of the first Nobel Prize in chemistry in 1901, spoke in his 1878 inaugural lecture in Amsterdam on the topic, "Imagination in Science" (van't Hoff, 1967). He had studied the life histories and personalities of the founders of modern science, starting with Copernicus, Kepler, Galileo, and Newton and continuing with the leaders in science up to and including the nineteenth century. He found an amazingly large number of them to be great and passionate lovers of art, poetry, and music. He raised the question about the relationship between these two seemingly different types of expression of creative forces. He suspected that this relationship is not an accident, but an indication of some basic similarity of the working of the mind that accounts for the profound and genuine love of art among the truly great architects of science. If this study were to be extended to the great architects of science in this century, the results would be strikingly similar and would add further support to van't Hoff's view. Planck, Einstein, Emil Warburg, Max Born, Werner Heisenberg, Manfred Eigen, Ernst Chain, Hugo Theorell, and Jacques Monod—to give just a few examples—were not only music lovers, but accomplished players, some of them of the highest quality. Many other leading scientists, even when not active, are great music lovers; many of them are passionate lovers of art and poetry. At international scientific meetings one frequently meets friends at art galleries, museums, cathedrals, and other monuments, viewing the expression of the creative forces of man, documents of the glory of the past. However one interprets the close association of the devotion to art and to science, it seems to be a significant fact. Meyerhof's genuine enthusiasm for art, poetry, and archaeology and his extraordinarily broad knowledge in these fields have been mentioned. He himself liked to write poetry. His poems are of exquisite beauty and reflect a keenly sensitive spirit. Since in poems some features of a personality find an expression that cannot be easily described, one of his poems is reproduced on p. 302. It appeared already in print in the book of Hieronimus (1964), who received the poem from the author. It reveals a rare spiritual attitude, his admiration for the divine nature of the universe, his deep feelings for the beauty of the manifestations of the human soul.

Ostergedicht für Hedwig

1951

Der einige Weltgeist, der das All ersann
Und Leben rief aus Millionen Sternen,
Dunkel umkreisend fremder Sonnen Bahn
Endlos in Zeit, in ungemessenen Fernen

Hat *uns* gesetzt in dieser Erde Raum
Der gross—doch klein—uns hält in seiner Runde,
Schenkt uns des Lebens flüchtig-bunten Traum
Und mass uns zu des Hierseins Ort und Stunde.

Sagt nicht, wir wären nur lebendiger Staub,
Den Motten gleichend die im Lichtstrahl flimmern,
Verweht im Winde wie verwelkend Laub
Halme von Gras, die in der Wiese schimmern.

Nein, Wunderbar, hoch über Erdensinn
Kam uns die Seele und der Seele Reifen
Dass langsam von des Menschseins Anbeginn
Wir lernten unsern Auftrag zu begreifen.

Wir lernten Gott, und dass Er uns erschuf,
Das Heilige Werk der Deutung zu erfüllen,
Dass er uns gab am einzigen Beruf
Voll Staunen Sein Geheimnis zu enthüllen.

Wars mir vergönnt, auch einen Faden nur
Klar zu erschaun an Seines Mantels Saume,
So bleibt ein Hauch von meiner Erdenspur
Unlösbar ausgestreut im Sternenraume.

Und Alle Liebe, die wir uns geschenkt
In Stunden voller Angst, in Glück und Sehnen
Ward in des Weltalls Tiefe eingesenkt
und glänzt vor Gott, wie eines Engels Tränen.

So sei getrost, dass nimmer wir vergehn,
Auch wenn wir lösen uns vom Erdenstaube.
Die Seele sucht ein *reines* Auferstehn.
Des Wissens Stückwerk überwölbt der Glaube!

It may be mentioned that Meyerhof's notions of "Weltgeist" and "God" are probably not those of the traditional Jewish or Christian religions, but are closer to the views of Einstein or Spinoza, who envisioned a God that reveals itself in the harmony of all being. This is, however, the interpretation of the author. Spiritual notions and religious attitudes cover a wide range of different concepts.

Meyerhof was deeply devoted to experimental work. From the time he came to Dahlem until he left Heidelberg he had in Walter Schulz a capable, competent, skillful, and devoted technical assistant. As soon as Schulz had made the necessary preparations for starting an experiment, Meyerhof was at his bench. Experimental work was an integral part of his life. Nevertheless, he was always available for discussions with his collaborators. The group was of course small. But in these discussions one could raise any problem, and occasionally the talks would go on for quite a while. More significant, there was no difficulty in contradicting his views, in contrast to the attitude of Otto Warburg. He would listen carefully, raise questions, and take objections very seriously. Sometimes he would be reluctant to accept new ideas or proposals for new experiments; but a few days later he might come back to the new proposals. He had given more thought to them and perhaps there was more merit to them than he had at first realized. These discussions were a precious part of the formation of his pupils' minds. They had a deeper influence than even a perfect lecture.

The flexibility of his mind was truly remarkable. Up to the end of his life he followed the new developments in biochemistry very closely. His memory was extraordinary. In Woods Hole, where many investigators were anxious to meet the "great old man," whom they knew only from the literature, he would surprise his visitors by asking them whether they were the authors of an article that appeared a few years before, and he would discuss with them their observations. They all were impressed by his dignity, his wisdom, and his personality. Meyerhof was actually shy. He was at his best when only a few people were present. When Warburg spent the summer of 1949 in Woods Hole, he and the author visited Meyerhof one evening in his garden. It was a lovely summer night. Meyerhof was in high spirits. He was relaxed and delighted to spend an evening with his old friend, whom he always admired as one of the most resourceful, imaginative, and original experimenters in biochemistry. Meyerhof discussed various trends in modern science. He made most stimulating comments, including some philosophical implications; they revealed his remarkable ability to keep up with recent advances. Warburg listened very carefully. He was quite obviously fascinated. When we finally left, he was silent for a few minutes, then he said: "Wissen Sie, er ist doch die grösste Persönlichkeit von uns allen." This was quite a remarkable statement from a man who certainly did not suffer from inferiority complexes. He considered himself, as mentioned by Hans

Krebs in Warburg's biography, as the Pasteur of this century. The question has sometimes been raised of whether Meyerhof or Warburg was the greater genius. To the author this seems a moot question. Both had a profound revolutionary impact on the development of biochemistry in this century, Warburg by his monumental experimental contributions, the originality and ingenuity of introducing and applying highly refined methods, and Meyerhof by his revolutionary new concepts and notions and the depth of his thinking. The greatness of a genius is hard to measure. When we consider the founders of modern atomic physics—Niels Bohr, Werner Heisenberg, Wolfgang Pauli, Paul Dirac, and some others —they all were great geniuses of about equal stature. It seems difficult and in fact irrelevant to discuss differences in the greatness of their genius.

A flexible mind is a characteristic of a truly great spirit. We have seen the remarkable attitude of Meyerhof in the face of Lundsgaard's observations. Dogmatic views and a lack of flexibility block the progress of science, as history has shown time and again. The lack of flexibility may become a disastrous and tragic factor in the work of even a great scientist, as the section on Warburg has shown. It was one of the greatest and most decisive experiences in the scientific formation of the author to listen to the vigorous controversies in the Haber colloquia, to hear how the great scientists, some of the giants in their time, participating in these discussions had not the slightest hesitation to admit that they were wrong.

Meyerhof's personality and fame attracted a great number of investigators from all over the world to join his laboratory. Many of them have not only further developed his ideas, but have opened new fields of investigation and created in their turn schools that have been referred to as "the second Meyerhof generation." Many of his collaborators became friends for life, bound by the extraordinary effect that the time spent in Meyerhof's laboratory had on their thinking and their work. For many, another element may have been the motives that led them to join Meyerhof. His personality attracted a special type of scientist, one who had an affinity to Meyerhof's intellectual approach and way of thinking. There were, of course, a few exceptions. Moreover, Meyerhof's laboratory offered little reward in terms of salary or career. In the Dahlem days people were happy if they received a very small fellowship from the *Notgemeinschaft der Deutschen Wissenschaft*; some even worked without any payment. It was the devotion to science, the enthusiasm for exciting work for which they were willing to sacrifice material advantage. There were no chances for advancement. Since the quality of Meyerhof's collaborators was high, many of them could easily have obtained comfortable positions with a chance of making a career. These factors led to a kind of special selection and may also have been important in forging the close ties that lasted a lifetime.

The admiration and affection of his collaborators found a vivid expression in a *Festschrift*, "Metabolism and Function," prepared and edited by the author on the occasion of Meyerhof's sixty-fifth birthday in 1949 (Nachmansohn, 1950). The number of outstanding associates who contributed to the *Festschrift* was truly impressive; they were joined by quite a few friends and colleagues, among them A. V. Hill, Otto Warburg, Dorothy Needham, Carl Neuberg, René Wurmser, Mme. Sabine Filliti-Wurmser, Claude Fromageot, Hans A. Krebs, Leonor Michaelis, Carl F. Cori and G. Cori. A. N. Richards, on reading it, wrote an enthusiastic letter to the author, calling it "a truly glorious *Festschrift*" in the old sense of the name. The author received innumerable letters from all over the world expressing admiration for this impressive collection as a monument to Meyerhof's scientific impact. The *Festschrift* was presented to Meyerhof after a dinner in New York at the Faculty Club of Columbia University in which a small but highly selected group (about 50 people) participated. As Hedwig Meyerhof told the author, it was one of the happiest moments in Meyerhof's life, to receive this testimony of deep affection and admiration from pupils and friends.

A similarly remarkable expression of these feelings for Meyerhof's personality and his impact on biochemistry took place at the second International Congress of Biochemistry in Paris, in 1952, a year after he had passed away. A special meeting was called to pay tribute to his genius. Three speakers described his personality and his work: André Mayer of the Collège de France, Alexander von Muralt from the University of Berne, and the author. To the best knowledge of the author, it was the only time that this kind of tribute was paid and a scientist honored by an International Congress of Biochemistry.

Meyerhof's philosophical background, the extraordinary range of his knowledge and interests, the originality of his concepts, his human qualities—all these factors contributed to make his work outstanding and may account for the wide scope of his achievements. The combination of a great scientist and a great human being made him a real leader and one of the most distinguished representatives of modern biological science.

Views on Science and Society

Although Meyerhof was never active in any field of applied science, he was deeply concerned with the effects of science on society. As an illustration may be mentioned his reaction to the use of nuclear energy for the atom bomb, which he expressed in a lecture in New York in 1946. He was greatly shocked and distressed by the misuse of science for destructive purposes, as most scientists were, and profoundly upset about the catastrophe of Hiroshima and Nagasaki. In his lecture he mentioned a letter of Winston Churchill's, written as war leader in 1943 to A. V. Hill, in which Churchill recognized the monstrous distortion of the meaning of science—that it served for wholesale destruction instead of being employed for the benefit and blessing of man. Meyerhof quoted an American poet who pictured the release of atomic energy as a sort of rebirth of Doctor Faustus, who sold his soul to the devil for the secret of conquering nature by black magic. This contract with the Prince of Shame brought about utter mischief and destruction. The nuclear physicists were well aware of this awful covenant and its terrible implications; one needs only to read their prophetic memorandum, written some weeks before the first bomb test in the New Mexico desert, to realize how conscientiously and truthfully they visualized the consequences of a thoughtless and improper use of the terrible weapon. Their moral conflicts, the anxiety that Hitler might develop the weapon, which was their essential motive, have been frequently described and are well known today.

But Meyerhof was disturbed that this perversion of science, its misuse by politicians and administrators, might have very dangerous consequences for science itself; he feared it might lead to a restriction of the freedom for which scientists have fought since ancient times, freedom from dogma or any other imposed authority. He was worried that this successful use of science might lead to planned science and to organized research. We can see in the efforts prevailing at present how farsighted and justified his fears were. Many dangerous side effects have developed due to the rapid progress of technology—environmental pollution and the population explosion are just two—some of which were predictable and could have been avoided, others that nobody could have foreseen. These side effects have encouraged many people, public administrators, and even some scientists, to call for planned and organized science at the expense of free and unrestricted research. Nobody will question that governments can and should make good use of planned scientific efforts in the fight against certain dangers and problems. The fight against cancer and the need to develop new sources of energy offer good examples. As was pointed out in an excellent document, *Science and the Challenges Ahead* (1974) much can be done by competently planned and organized efforts. But the solution of the real problems requires free and basic research, to uncover the still obscure cause of cancer and its origin, or to develop new sources of energy such as fusion, which will require great scientific efforts in the field of atomic physics and other fundamental scientifically unexplored areas. As Meyerhof pointed out so forcefully in his lecture, "The true life of science does not consist of applications and exploitations; they are only an endproduct or even a byproduct. It consists in revolutionary thought, in the concept of new theories and in the basic discoveries which are made by single prepared minds in a creative act like a piece of art." Few leaders of science will disagree with this statement. It recalls the views of Einstein and Planck on epistomology (see II, E).

Ancestry and Jewish Attitude

Meyerhof was born in Hannover, but the family came from Hildesheim. The records of the family there go back to about 1720 to Otto's great-great-great-grandfather. It was a well-established and large family, and included two artists: Agnes Meyerhof, a painter, and Leonie Meyerhof, a writer publishing under the pseudonym of Leo Hildeck. Isak Meyerhof, in 1753, established a foundation to honor the memory of his father and thanks to this legacy there exists a register of the large family tree, printed by another Otto Meyerhof in 1932. Hans Krebs (1972) has studied this register and has compiled a considerable amount of information about the family background. Krebs was born in Hildesheim and his maternal family had intermarried with the Meyerhofs. Their families were also closely connected through intermarriage with that of Carl Neuberg (see IV, E). These interconnections between the families were due to the smallness of the community (about 1000–2000 Jews in a radius of 25 kilometers), which favored some measure of inbreeding. In 1854 Israel Meyerhof and his wife Therese, *née* Gumpel, moved to Hannover. He played a prominent role in the Jewish community there. His son Felix married Bettina, *née* May, from Hamburg. Otto was their second child. Both Israel and Felix were merchants. The cultured atmosphere in Otto's family strongly encouraged learning and interest in literature, music, and art. The cultural atmosphere was typical for the higher-middle-class Jewish families of the nineteenth century and greatly influenced the members from their childhood on in their feelings, their thinking, their actions, and their attitude. After the Hitler Holocaust not a single member of this large family survived; all vanished, including the other Otto Meyerhof, who had been responsible for the printing of the family tree.

Otto Meyerhof met his wife Hedwig, *née* Schallenberg, in Heidelberg and they married in 1914. She had studied mathematics and was a painter. They had three children. The elder son Godfrey is now professor and chairman of civil engineering at Nova Scotia Technical College in Halifax, Canada. The daughter, Bettina, is a physician and is married to Donald Emerson, professor of history at the University of Washington in Seattle. The younger son Walter is a distinguished professor of physics at Stanford University. He was chairman of the department from 1971 to 1977.

Meyerhof's life was completely absorbed by philosophy, art, and science. He had only limited interest in politics, in contrast to his friend Nelson, and tried to stay away from political activity. He was progressively minded and as a young student he volunteered to teach courses to workers to improve their education. As might have been expected from his family background, he was completely assimilated. Like many other upper-middle-class Jews of his generation, he had a rather limited knowledge and interest in his Jewish heritage. He was never baptized and was always fully aware of his Jewish origin, but it had little meaning to him. After World War I, when the anti-Semitism in Germany became violent, he was quite concerned about the bad effects these anti-Semitic movements might have in general, not only for the Jews but for the efforts to establish a liberal, progressive, and democratic German Republic, which he strongly endorsed. He was encouraged by Gustav Stresemann's efforts to establish good relations with France and Great Britain and quite upset when Stresemann died.

But even the tragedy of the Hitler period, the disaster and the personal sufferings of Nazi persecution, did not change his basic attitude about the Jewish problem. These violent explosions of anti-Semitism were for him a regression into the Middle Ages, an expression of sick minds, an evil force bound to disappear with the progress of human society. When after the war the news of the Holocaust became known, it was for him a terrible shock. He was profoundly upset by the unimaginable horror of the events. He greatly resented the pro-Nazi attitude of some German scientists, although he knew that the overwhelming majority of the members of the Kaiser Wilhelm Society abhorred the Nazis. While still in Germany, he had witnessed the rapid decline of German science which, before the Nazis, had been at the height of its greatness. Deeply rooted in European civilization, he felt like a citizen of the world, as did many other Jews of his generation. His attitude led him to consider nationalistic forces as being based essentially on irrational emotions that could lead to an unnecessary restriction of free creative development and a permanent threat to peace. He failed to see that whatever the roots of these emotional and irrational forces were, national aspirations and a national heritage were powerful and legitimate factors that could not and should not be ignored, and that these forces could be very constructive and inspire highly creative aspirations, provided they were channeled in the proper direction. However, his attitude toward Israel was by no means indifferent. He greatly admired the creative achievements of Israel in building up a homeland in the desert. He fully endorsed Weizmann's efforts to create a great scientific center. He welcomed every encouragement of creative achievements and followed with genuine interest these developments. He did not live long enough to see the great accomplishments of Israel in the arts, humanities, and sciences of the last decades after his death in 1951.

ANNEX

The great pioneer work of Meyerhof in the introduction of thermodynamics and energetics into biochemistry and the profound impact of his fundamental contributions, which opened the field of bioenergetics, have been described. A much more detailed, extended, and well-documented description of Meyerhof's monumental role is found in Florkin's *History of Biochemistry*, the thirty-first volume of *Comprehensive Biochemistry*, edited by Florkin and Elmer Stotz. Therefore only a few comments on the statement made by Lipmann (1975) may suffice.

For Meyerhof, the explanation of Hill's heat production data on the basis of chemical reactions—knowledge of the total energy change ΔH, measured as the total irreversibly produced heat—was his primary concern. When phosphocreatine was discovered and its breakdown seemed to coincide with muscular contraction, Meyerhof immediately measured the heat change, reflecting ΔH, of the hydrolysis of phosphocreatine. The high value found for ΔH was startling and made it immediately apparent that there was a significant chemical reaction *in addition* to lactic acid formation that would account for part of the missing 50% of chemical energy. Meyerhof, extremely competent in thermodynamics, knew that the free energy change under standard conditions, ΔG^0, was of limited value, although he and his associates made several such calculations. What was physiologically relevant was ΔG, which is the useful convertible energy under the conditions and actual concentrations of the reaction partners in the cell.

At that time most emphasis was placed on equilibrium thermodynamics. Biological structures such as the muscle require, as is now well known, nonequilibrium thermodynamics. The distribution of substrate and products within a cell is basically a nonequilibrium distribution characterized by

$$\Delta G = RT \ln \frac{C_{Products}}{C_{Reactants}}$$

Gibbs' standard free energy, $\Delta G^0 = \Delta H^0 - T\,\Delta S^0$ (where $T\,\Delta S^0$ is the heat exchanged with the environment), is only applicable to equilibrium systems, because $\Delta G^0 = -RT \ln K$ (where K is the equilibrium constant). Since in biological systems $T\,\Delta S$ is dissipated, ΔG is equal to ΔH of the system. In equilibrium systems to which living cells do *not* belong, part of the internal energy is converted to heat and there ΔG differs from ΔH. In the light of the knowledge of thermodynamics of the 1920s, when Meyerhof established the difference between phosphate derivatives rich in energy release on hydrolysis and those that were poor, he was completely right in using ΔH as the parameter. The developments of nonequilibrium thermodynamics have explained much more correctly and precisely the significance of the parameters in biological systems and have provided new support for the correctness of Meyerhof's measurements and his assumptions. Thus, Lipmann's statement that Meyerhof, "far from realizing their bioenergetic importance, . . . determined their heat of hydrolysis," has no justification.

As has been pointed out, Meyerhof's assumption that ΔH was a reasonably fair indicator of the maximum value of the energy change of a reaction turned out to be correct. In fact, ΔG^0 of phosphocreatine hydrolysis is now estimated to be -10.5 kilocalories/mole, about the same as the ΔH value found in 1927. When, in the 1950s, estimates of ΔG of ATP in muscle were made, they turned out to be quite consistent with the original ΔH values of Meyerhof and Lohmann, although ΔG^0 is slightly lower (-7.5). Lipmann wrote in his article that he was at that time not much interested in thermodynamics and that his real interest started in 1939. This may explain his misunderstandings, since many events look different in retrospect than at the time they take place.

For Meyerhof and those familiar with thermodynamics, the discoveries of the very high ΔH values of phosphocreatine hydrolysis, and later those of the hydrolysis of ATP and other phosphate derivatives, were terribly exciting events. They motivated the author to investigate the quantitative relationship between phosphocreatine breakdown and muscular contraction (Nachmansohn, 1928/29, I). The direct relationship was unequivocal. There was never any doubt in Meyerhof's or the author's mind that, in view of the data obtained and the high ΔH values, the breakdown of phosphocreatine was an essential factor in muscular contraction and contributed to heat production. In addition, on the basis of some data, the author had the idea, quite intuitively, that phosphocreatine was somehow connected with a special parameter of contraction, namely its speed. Meyerhof was at first reluctant to accept this idea, but soon quite a few experimental data supported this view. It was therefore proposed that phosphocreatine action was somehow associated with the speed of contraction (Meyerhof and Nachmansohn, 1928; Nachmansohn, 1929) (or possibly with chronaxy, but since the two parameters are usually affected in a similar way, a distinction was rather difficult); the author always felt that it was the speed of contraction with which phosphocreatine breakdown was associated (Nachmansohn, 1928/29 III and 1929). The molecular mechanism (i.e., the interaction of these compounds with the muscle proteins) was at that time completely unknown. After Lundsgaard's observations, Lundsgaard and Meyerhof for a short while considered the possibility that phosphocreatine might be directly associated with the mechanical process of contraction. Both soon discarded this view when in 1933 and 1934 the role of ATP was clarified and appeared to be, as described earlier, the substance directly providing the energy for the mechanical process. At the same time, Lohmann's studies (1934) on ATP in Meyerhof's laboratory and similar work in that of Parnas indicated that phosphocreatine was a "storehouse" used for rapidly supplying phosphate for the resynthesis of ATP when it was hydrolyzed. Thus it became clear why phosphocreatine was required in rapidly contracting muscles. The assumption of the author that it was associated with the speed of contraction was fully confirmed and explained.

Meyerhof and Lohmann wrote in 1925 that lactic acid had *never* been claimed to be directly associated with the mechanical process. This fact was also emphasized by Florkin. Thus, Lipmann's statement that "in view of the convincing link established with lactic acid, a connection with phosphocreatine was not seriously considered in Meyerhof's laboratory when I joined it in 1927 [cf. Meyerhof and Nachmansohn, 1928]" again is obviously due to a misunderstanding.

One other statement in Lipmann's paper may be mentioned. The first demonstration that the free energy of ATP hydrolysis is used for biosynthesis outside glycolysis was achieved by the author in 1942. Interested in the transformation of chemical energy into electrical energy, the author studied this problem with electric organs of electric fish, particularly suitable for this analysis. Using Meyerhof's basic approach and notions, he established the sequence of energy transformations during electrical activity. From the data, he concluded that ATP provides the energy for acetylation (in this particular case of choline). The first acetylation achieved with ATP, and thus the first biosynthesis, in cell-free solution was achieved by the author in September 1942. In Lipmann's view at that time, acetylphosphate and not ATP was the compound that performed acetylation, since there was no explanation of the mechanism by which ATP was able to acetylate a compound. At the end of 1944, Lipmann spent two weeks in the author's laboratory and became convinced that it was not acetylphosphate but ATP that effected acetylation. The details of the story are detailed in the Prefatory Chapter of the *Annual Reviews of Biochemistry* (Nachmansohn, 1972).

E. CARL NEUBERG (1877–1956)

The rise of modern dynamic biochemistry after the turn of the century is closely associated with Carl Neuberg's name. The range of his contributions to a great variety of problems is stupendous; he stimulated many pertinent developments by his dynamism, his enthusiasm, his encyclopedic knowledge, and the ingenuity of his concepts. He was widely referred to as one of the "big three" in biochemistry at the Kaiser Wilhelm Institutes in Berlin–Dahlem (the two others were Warburg and Meyerhof).

His contributions to the process of alcoholic fermentation will always be remembered as one of his most magnificent achievements. His discovery of the enzyme carboxylase was one of the important milestones in the elucidation of alcoholic fermentation, one of the great problems that had preoccupied scientists in the nineteenth century. Neuberg's discovery showed that the zymase of Buchner was actually not a single enzyme, as Buchner had assumed, but a complex system of several enzymes. The demonstration of this important step led Neuberg to propose his ingenious schemes of fermentation. It was a turning point in the history of enzyme chemistry, since for the first time alcoholic fermentation was envisaged as a process formed by a series of successive enzymic steps. The schemes had a deep impact on the thinking of enzyme chemists; they made apparent the necessity of considering that a variety of steps might be involved. They created the pattern of inquiry into the mechanism of metabolic pathways, thereby making a brilliant contribution that played a key role in the further study of the chemistry of cell reactions.

There are, however, many other important contributions by which Neuberg initiated and stimulated various developments in biochemistry. Moreover, he was a passionate and inspiring teacher and as the director of a very large institute he attracted a large number of pupils from all over the world, including the United States, and created one of the largest schools of biochemistry of that period. Many of the scientists who were trained in his institute became leaders in their native countries. Thus Carl Neuberg holds a foremost place in the early period of dynamic biochemistry.

Family Background and Career

Neuberg was born in Hannover. His parents belonged to the relatively small group of Jewish families in the region that included Hannover, Pyrmont, and Hildesheim. Since the early eighteenth century, they had enjoyed a relatively high degree of freedom and security compared to that found in other parts of Germany. As mentioned before, there were many intermarriages between these families. Neuberg's family was related to that of Otto Meyerhof and Hans A. Krebs (on his mother's side). His father was Julius Neuberg, a textile merchant, married to Alma, *née* Niemann. Carl Neuberg married Hela, daughter of Sigmund Lewinski, a prominent lawyer and high official in Posen, and Johanna, *née* Poznansky, from Lodz, Poland. It was an upper-middle-class family.

Like many scientists of his generation in Germany, he attended a humanistic gymnasium. After the *Abiturientenexamen*, he started studying astronomy—his father had wanted him to become a "brew master"—but he soon turned to chemistry. He studied at the universities of Würzburg and Berlin, where he received his Ph.D. in chemistry in 1900. The teachers who had the greatest influence on Neuberg were Emil Fischer, Alfred Wohl (one of the last assistants of Rudolf Virchow, the founder of cellular pathology), and Ernst Salkowski. Wohl directed Neuberg's doctoral thesis on the chemistry of glyceraldehyde. Salkowski, at the *Landwirtschaftliche Hochschule* in Berlin, introduced him to physiological chemistry.

In 1903 he became *Privatdozent* and in 1906 professor at the University of Berlin. When the Kaiser Wilhelm Institute of Experimental Therapy was opened in 1913, with A. von Wassermann as director, Neuberg became the head of the biochemistry division. After the death of Wassermann, in 1925, he became director of the whole institute, now transformed into a biochemical institute. He was forced to resign his position in 1934 and left Germany a couple of years later. He arrived in Israel in 1938. Although he loved the country, he felt that he was not young enough—he was in his sixties—to start a career in a country that needed young and vigorous pioneers for the use of the extremely limited facilities and funds available at that time. Since his two daughters, Dr. Irene S. Forrest, a biochemist, and Mrs. Marianne Lederer, a health physicist, had settled in the United States, he decided to join them and arrived in New York in 1940. Here he was associated for varying periods of time with New York University, with the Polytechnic Institute of Brooklyn, and at the time of his death with the New York Medical College. Although he worked under rather difficult conditions, his productivity and the number of his publications continued to be remarkably high, his age notwithstanding.

Scientific Achievements

Fermentation. Neuberg's most brilliant contributions were concerned with alcoholic fermentation. The full appreciation of his accomplishments requires a brief description of the historical background. Fermentation was one of the central problems of chemists and biologists, starting with Lavoisier and continuing throughout the nineteenth century. In 1815 Joseph Gay-Lussac, the teacher of Justus von Liebig, demonstrated that in alcoholic fermentation sugar is reduced to equal portions of alcohol and carbonic acid.* The experimental evidence that the agents that effect this fermentation were living organisms, namely yeast cells, was offered, independently and simultaneously in 1837 by Charles Cagniard de la Tour (1777–1859); Theodor Schwann (1810–1882), a pupil of the famous physiologist Johannes P. Müller (1801–1858); and Friedrich T. Kützing (1807–1893). The view that alcoholic fermentation and other reactions referred to as fermentation were catalyzed by living organisms was soon widely accepted; however, the nature of the process was a subject of widely different views and bitter controversies between such leading figures as Berzelius, Liebig, and Pasteur, owing to the limited knowledge of the actual processes involved. The term *fermentation* was applied to many different cell reactions with vaguely defined and frequently contradictory interpretations, but it was a key problem in biology. The fas-

---

* The Gay-Lussac equation is written $C_6H_{12}O_6 = 2CH_3CH_2OH + 2CO_2$; that is, one molecule of glucose breaks down into two molecules of alcohol and two molecules of carbon dioxide.

cinating history of the developments and the concepts cannot be described in the frame of this book. It may only be mentioned that Marcelin Berthelot vigorously supported the idea, in 1860, that soluble "ferments" (today we call them enzymes) *within* the cells, similar to those that had already been isolated, were responsible for fermentation. He flatly rejected any notion of vital forces. One of the strongest supporters of this view in Germany was Moritz Traube (1826–1894). With remarkable vision he insisted that most cellular reactions are catalyzed by soluble ferments, which "are the causes of the most important vital chemical processes, and not only in lower organisms, but in higher organisms as well" (Traube, 1878). This view was later endorsed by Felix Hoppe-Seyler, who fully accepted the view that ferments act inside living cells. The word *enzyme* was coined by Friedrich W. Kühne in 1878, meaning "in yeast" (*zyme* in Greek), but for decades the terms *ferment* and *enzyme* were used interchangeably, thereby creating considerable confusion.

It was during this period that Louis Pasteur entered the field of alcoholic fermentation. He is one of the greatest figures in the biological sciences of the last two centuries. His genius laid the groundwork for many fields—biology and biochemistry, as well as microbiology and immunology. Meyerhof and Warburg, as mentioned before, were great admirers of Pasteur and his concepts and his work were an inexhaustible source of inspiration for both of them. Even today, the author must confess that when he reads Pasteur's original papers he cannot help but be struck by the brilliance and the extraordinary perspicacity of his mind. (For biographies of Pasteur see Dubos, 1950; Duclaux, 1896; Vallery-Radot, 1900). One of Pasteur's first concepts to be introduced into the process of fermentation was that of the fundamental similarity of the fermentation of sugar to lactic acid or to alcohol, a concept that guided Meyerhof decisively in his work. His almost prophetic vision was experimentally verified only in this century. His many classic contributions to alcoholic fermentation, like so many of his other achievements, not only clarified basic scientific problems but also decisively helped the brewing industry; like Haber, Pasteur always contributed to the solution of fundamental scientific problems as well as to their practical applications. Without going into details, as tempting as it would be for the author, it may only be mentioned, in connection with the specific problem discussed, that Pasteur, in the beginning of his studies on alcoholic fermentation in 1857, found several additional products such as glycerol and succinic acid. He discussed the possibility that the process might be too complex to be attributed to a ferment, that it might require additional factors, such as organization and other characteristics of the cell. Two decades later Pasteur (1878) stressed the possibility of isolating soluble ferments (enzymes) from living cells capable of forming alcohol from sugar.

In view of the many controversial views on the nature of fermentation it is not difficult to understand that Buchner's isolation of zymase from yeast produced great excitement. Although it was vigorously rejected by some scientists, others enthusiastically endorsed this achievement, particularly Pasteur's school, including Émile Duclaux (1840–1904) and Pierre P. E. Roux (1853–1933). The discovery almost put an end to the notion of vital forces, so vigorously fought by Berthelot and Traube. One of the strongest fighters for the physicochemical explanation of cell processes, Jacques Loeb, expressed his deep satisfaction about the decisive blow to the notion of vitalism.

It must be recalled that quite a few enzymes had been isolated during the nineteenth century, before Buchner's discovery. Theodor Schwann had discovered pepsin in 1836, and F. Tiedemann (1781–1861) and L. Gmelin (1788–1863) had found "pancreatin" in pancreatic juice (later named trypsin by Wilhelm Kühne). A trypsin-like enzyme was later found in plant extracts. Liebig and Wöhler had observed, in 1837, that amygdalin, a crystalline substance obtained from almonds, was broken up by emulsin; this was actually the first enzyme described to have a chemical action on a well-defined crystalline substance. In 1848 Claude Bernard (1813–1878) had identified lipase in pancreatic juice. Lipase, which hydrolyzes fats, also occurs in plant seeds. In 1869, Berthelot had obtained from yeast a soluble ferment (enzyme) that split cane sugar (sucrose) into a mixture of two other sugars (dextrose and levulose); he called it "ferment glycosique"; later the enzyme was named *invertase*. As the last example may be mentioned the discovery, by Frederic Musculus in 1872, of urease, the enzyme that splits urea. By the end of the century about two dozen enzymes had been isolated in soluble form. Thus the many vigorous objections to Buchner's discovery of an enzyme effecting alcoholic fermentation from sugar are rather surprising. Buchner received the Nobel Prize in 1907 for his work.

In addition to the discovery of zymase, the elucidation of the chemical structure of many sugars by Emil Fischer, Neuberg's teacher, in the last two decades of the nineteenth century set the stage for new progress in the exploration and the understanding of alcoholic fermentation and glycolysis. It may be mentioned that Harden and Young, in 1906, found that inorganic phosphate disappeared rapidly when added to yeast juice; they suggested that a phosphoric ester of glucose might be formed. Young soon isolated the ester and identified it as a hexose diphosphate; two decades later it was established to be a fructose diphosphate. Harden and Young also found that alcoholic fermentation required a heat-stable, dialyzable coenzyme. Meyerhof, in 1918, found that this same coenzyme was required for glycolysis, a new forceful support for the fundamental similarity of the mechanism of the two processes. The coenzyme was later found to consist of several components. However, the meaning of the phosphorylation of the sugar remained obscure for almost two decades.

This brief, sketchy, and incomplete outline of the history of alcoholic fermentation was necessary to help the reader realize the outstanding importance of Neuberg's great achievements. In early 1911 he had found that pyruvic acid ($CH_3COCOOH$) is fermented by yeast. This was an unexpected finding, since it was in contradiction to the prevailing assumption that the compound was not fermented by yeast. Shortly afterward Neuberg discovered the enzyme carboxylase, which catalyzes the decarboxylation of pyruvic acid into acetaldehyde and carbon dioxide:

$$CH_3COCOOH \longrightarrow CH_3C\overset{\displaystyle \nearrow O}{\underset{\displaystyle \searrow H}{}} + CO_2$$

pyruvic acid   acetaldehyde   carbon dioxide

Acetaldehyde accepts a hydrogen molecule, H_2, and is transformed into alcohol:

$$CH_3C\overset{\displaystyle \nearrow O}{\underset{\displaystyle \searrow H}{}} + H_2 = CH_3CH_2OH$$

By 1913 Neuberg had worked out a complete metabolic sequence in the fermentation of sugar to alcohol (Neuberg and Kerb, 1913, 1914). He suggested that in the first step sugar is cleaved into two molecules of methylglyoxal:

$$C_6H_{12}O_6 + 2H_2O = 2CH_3COCHO$$

sugar   water   methylglyoxal

One molecule of methylglyoxal is reduced to glycerol and the second molecule is oxidized to pyruvic acid:

$$2CH_3COCHO + 2H_2O = CH_2OHCHOHCH_2OH + CH_3COCOOH$$
methylglyoxal   water   glycerol   pyruvic acid

Pyruvic acid is then decarboxylated by carboxylase into acetaldehyde and carbon dioxide, as mentioned before. A molecule of methylglyoxal and a molecule of acetaldehyde yield alcohol by the reduction of the acetaldehyde and pyruvic acid by the oxidation of methylglyoxal:

$$\begin{array}{ccc} CH_3COCHO & O & CH_3COCOOH \\ + & | & = & + \\ CH_3CHO & H_2 & CH_3CH_2OH \end{array}$$

Pyruvic acid is again decarboxylated. The sequence suggested is thus

sugar $\rightarrow$ methylglyoxal $\rightarrow$ pyruvic acid $\rightarrow$ acetaldehyde + carbon dioxide
methylglyoxal + acetaldehyde $\rightarrow$ pyruvic acid + alcohol.

The assumption of methylglyoxal as an intermediate of alcoholic fermentation was based on several observations. The existence of an enzyme, methylglyoxalase, in animal tissue was found independently and simultaneously by Neuberg (1913) and Dakin and Dudley (1913) in the United States. The Russian biochemist Alexander N. Lebedev (1879–1949) had observed methylglyoxal among the products of yeast fermentation, and this was confirmed by other investigators.

Another important step in the analysis of alcoholic fermentation was made when Neuberg introduced a technique involving what is referred to as a "trapping mechanism." By the addition of sodium sulfite (Na_2SO_3) to a fermenting mixture of yeast and sugar, acetaldehyde is trapped by reacting with the sulfite and equivalent amounts of acetaldehyde and glycerol are formed. The production of these two compounds is increased, but the yield of alcohol and carbon dioxide is decreased. This step was frequently called the second scheme of fermentation. The finding became of great practical importance for the mass production of glycerol in Germany during World War I. Neuberg was awarded the Delbrück Medal of the *Landwirtschaftliche Hochschule* in Berlin for this discovery.

Soon Neuberg suggested two other schemes of fermentation. The first one was the decomposition of two molecules of sugar into one molecule of alcohol and acetic acid and two molecules of glycerol and carbon dioxide (third scheme of fermentation) and the second one was, under special conditions, the breakdown of sugar into equal quantities of pyruvic acid and glycerol (fourth scheme of fermentation).

In the course of his investigations on alcohol fermentation, Neuberg discovered, in 1918, a hexose monophosphate with properties different from the hexose diphosphate of Harden and Young, mentioned before. The ester is referred to as the "Neuberg ester." Still another hexose monophosphate was discovered by Harden and Robert Robinson in 1914; it was later identified as a glucose monophosphate (the "Robinson ester"). All three esters were fermented by yeast. Because of new chemical methods developed in the 1920s, the exact structures of the three esters with the position of the phosphates were finally established:

fructose-1,6-diphosphate	glucose-6-phosphate	fructose-6-phosphate
Harden–Young ester	Robinson ester	Neuberg ester

Neuberg's discovery of carboxylase and his brilliantly conceived four schemes of fermentation were milestones in the understanding of the mechanism of alcoholic fermentation. It became apparent that zymase was not a single enzyme, but a group of enzymes capable of catalyzing a whole series of chemical reactions. Thus, the new facts and concepts presented biochemists with a great challenge: to attempt to separate the enzymes responsible for the various steps in the fermentation schemes and thereby elucidate the mechanism. Studies of this kind might provide the answer to the question that was so puzzling to Pasteur, namely why several products appeared during fermentation. But beyond the special problem of alcoholic fermentation the schemes opened new perspectives; they were a significant step in the realization of biochemists that seemingly simple reactions frequently consist of a complex series of reactions. They were an important factor in attracting to the study of metabolic pathways many investigators who might have devoted their chief efforts to the isolation and identification of cell constituents. It is this feature of Neuberg's achievements for which he must be considered one of the founders and leaders of dynamic biochemistry. His schemes were widely accepted for more than two decades. After Meyerhof succeeded in obtaining the glycolytic enzymes of muscle in solution, as described in the preceding section, and after the various enzymes catalyzing the different steps had been isolated and purified—by Meyerhof and his associates, Gustav Embden and his school, Warburg, Carl and Gerty Cori, Dorothy M. Needham, and others—a complete picture emerged of the complex pathways of glycolysis and alcoholic fermentation in the late 1930s. Only then was the idea of methylglyoxal discarded. It is almost needless to add that these rapid new advances do not in the least decrease the historical importance of Neuberg's pioneer work and concepts. We have seen in the relatively few cases described before that almost all pioneer work in science, which opens new territories and leads to new information, invariably requires modifications and great changes of the original concept, no matter how crucial and decisive the step that advanced the field. Neuberg's discovery was widely recognized as a major breakthrough. He received in recognition of his achievements, in 1912, the prestigious Emil Fischer Medal of the Society of German Chemists.

Neuberg's Other Scientific Achievements. Although his contributions to the problem of fermentation were Neuberg's most brilliant and pioneering accomplishments, his drive and dynamism and his imaginative mind are apparent in the great number of pertinent achievements covering a large scope in many areas of biochemistry. His strong personality, his enthusiasm, and his profound and broad knowledge of biochemistry attracted a great number of excellent co-workers. His large institute and its excellent facilities were used for work on a great diversity of problems.

At the turn of the century there was much discussion about the physiological role of catalase, an enzyme that catalyzes hydrogen peroxide (H_2O_2) to water and oxygen ($2 H_2O_2 \rightarrow 2 H_2O + O_2$). The enzyme was discovered in 1901 by Oscar Loew (1884–1941). It was assumed by Frederico Battelli (1867–1941) and Lina S. Stern (1878–1968), in 1910, that hydrogen peroxide was a normal product in biological oxidation and, since it was known to be toxic when accumulating in tissue, that the main function of catalase might be the removal of hydrogen peroxide. The puzzling problem of the role of catalase led to chemical studies using model systems. Ferrous salts (divalent, i.e., reduced, iron compounds) are able to catalyze the oxidation of organic substances by hydrogen peroxide. In this process the divalent ferrous iron was considered to act as an oxygen carrier through oxidation by hydrogen peroxide to the trivalent (oxidized) ferric iron, from which the divalent ferrous form was regenerated by organic substances. Neuberg, in the first decade of this century, and independently and about simultaneously with Henry Dakin (1880–1952) in the United States, carried out a number of investigations on the oxidation of fatty acids and amino acids, using hydrogen peroxide in the presence of ferrous salts. By testing the products formed in the chemical oxidations in the model systems and comparing them with those that appear in the urine or are formed in animal tissue after administration of amino and fatty acids to experimental animals, a considerable amount of information was accumulated about the intermediates of the oxidative breakdown of cell constituents. These investigations were the forerunners of Otto Warburg's work on the role of iron in biological oxidations, described before. Catalase, as is known today, belongs to the group of enzymes involved in biological oxidations and in which heme is the active group (see IV, C).

In the last decade of his life Neuberg was greatly interested in the mechanism by which insoluble matter occurring in nature may become solubilized. This problem has several practical aspects, many of them still poorly investigated and understood. Many polyphosphates form soluble complexes with metal ions. In view of this ability, metaphosphate may be used as a water softener, for example. Inorganic pyrophosphates occur in almost all tissues in the course of certain enzyme reactions and are very actively hydrolyzed into inorganic phosphate. Many inorganic polyphosphates are of biological interest, since they exist in a great variety of living structures, especially in bacteria. They occur over a wide range of polymerization, the lowest form having only two phosphorus atoms (i.e., pyrophosphate). They occur either as straight-chain polyphosphoric acids or in a cyclic form in which they are called metaphosphoric acids:*

Several enzymes splitting poly- or metaphosphoric acids (phosphatases) are known. Pyrophosphatases are ubiquitous. Neuberg's interest in phosphatases goes back far. In 1928 T. Kitasato found, in Neuberg's labora-

---

*

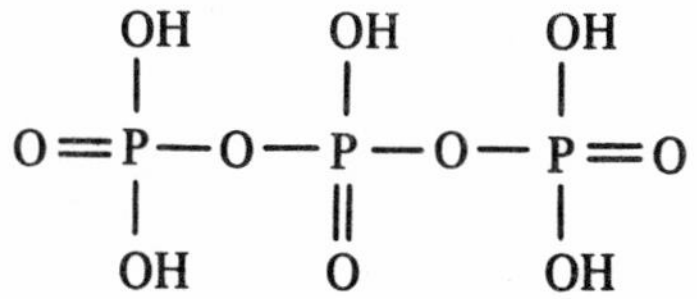

tripolyphosphoric acid 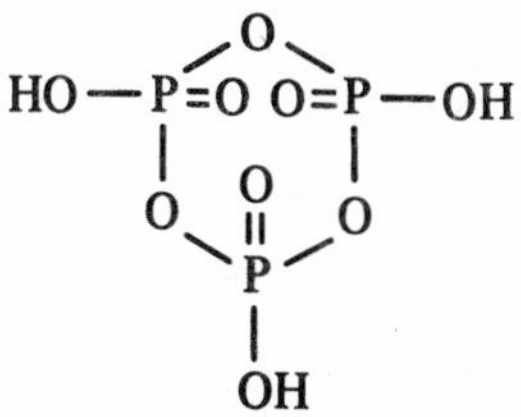   trimetaphosphoric acid tory, that several tissues (e.g., yeast and rabbit kidney or liver) may transform tri- as well as hexametaphosphates into orthophosphate (which has only one phosphorus atom). No accumulation of pyrophosphate was observed during the action of metaphosphatases. Several metaphosphatases were obtained from yeast by Neuberg and H. A. Fischer, in 1937 and 1938, and their properties as well as those of triphosphatase were investigated. In 1950 Neuberg found, with A. Grauer and I. Mandl, a water-soluble tripolyphosphatase in *Aspergillus oryzae*.

Many investigators have devoted great efforts to elucidating the role of polyphosphatases; they have been associated with a variety of cellular functions. Polyphosphates do not accumulate in tissues and several investigators considered the possibility that one of their functions might be the prevention of the accumulation of their substrates in tissues, since they form complexes with metal ions and may thus dangerously interfere with vital cellular functions. Neuberg and Mandl found that this complex-forming ability is not limited to inorganic polyphosphatases but is also a property of ATP (adenosine triphosphate). In view of the great importance of ATP, such complex formation may seriously damage many cellular functions that depend on the energy of ATP hydrolysis. They suggested, therefore, that one important function of polyphosphatases and particularly of pyrophosphatase might be the protection of the cell against the accumulation of polyphosphates and the prevention of complex formation.

Neuberg's scientific activities covered a period of more than half a century. Since it is not the intention of this book to bring a complete survey of the contributions of the scientists described but to convey to the reader some of the most pertinent accomplishments in order to indicate his scientific stature, only a few additional achievements may be briefly mentioned. In addition to carboxylase and several phosphatases already mentioned, he discovered quite a number of other enzymes. Among them was the enzyme that reversibly transforms the Harden–Young ester (fructose-1,6-diphosphate) into the Neuberg ester (fructose-6-phosphate). He elucidated, in 1908, the enzyme transformation of adrenaline into adrenochrome. He established the structures of a great number of natural products. He obtained the first evidence of the relation of inositol (a six-carbon ring structure) to common sugars by its transformation into furfural in acid solution, a reaction that is characteristic for sugars. Inositol occurs in many animal tissues—for example, in muscle, heart, lung, and liver. In phosphorylated form it is an important constituent of lipids. Neuberg's studies of inositol led him to the elucidation of phytin, a compound occurring in plants. There inositol is found in phosphorylated form as phytic acid or a mixed magnesium calcium salt, phytin. As a result of this work it became apparent that substances not directly involved in fermentation by yeast or other microorganisms may act as hydrogen acceptors and that thereby a vast number of aliphatic or aromatic aldehydes (compounds containing a $C{<}^{O}_{H}$ group) may be reduced to corresponding alcohols. Thus Neuberg's work led to a broad application of the concept of phytochemical reduction. These experiments became significant in later developments in the investigation of the biological oxidation and reduction of steroids. Finally it may be mentioned that Neuberg was the first to describe the conversion of α-ketoglutaric acid into succinic acid, a step that forms part of the citric acid cycle (see V, A).

The few examples cited convey only a very sketchy and insufficient idea of the large amount of Neuberg's contributions and especially the profound influence of Neuberg on the whole field of biochemistry. Many of his observations were the initial steps of important new developments, as were his contributions to alcohol fermentation. Moreover, biochemistry also owes to Neuberg a great number of methods that he used in his work, which were developed by him and his associates. For decades these techniques were important tools for biochemists. An indication of the wide range of his interests is the huge number of publications—about 900—that appeared from his laboratory. In addition he published six books, three of which became famous and found a wide distribution (Neuberg 1911a, 1911b, 1913).

Neuberg was unquestionably one of the most illustrious biochemists of his generation. The admiration for his great achievements found an eloquent expression in the great number of distinctions he received. He was elected to membership in many academies as a foreign member. He received many of the most coveted medals of that era. In the 1920s he was repeatedly proposed for the Nobel Prize. In 1954 he received the Great Cross of Merit from West Germany's President Heuss.

Personality

Neuberg was a typical representative of the upper middle class of imperial Germany. Few people realize today the high standard of culture by which this class was distinguished, the progressive and liberal spirit most of them had, its high rank among the groups representing the finest European tradition. Unfortunately, because of the kind of government established by Bismarck, as mentioned in Chapter I, this class had virtually no influence and no political power. It has been greatly reduced by the various tragic events in the history of this century.

Neuberg was a forceful although complex personality, a gentleman with all the charm and chivalry typical of his class and that era. Because of his broad educational background obtained in the humanistic gymnasium, he read Greek, Latin, and Hebrew fluently, and was a scholar well versed in classical literature and history. He was always full of youthful enthusiasm, full of humor and wit, which could be rather caustic at times. He maintained these characteristic features up to the end of his life. He was always frank and outspoken, never hesitating to express his views in unequivocal terms. It is thus not surprising that many colleagues admired but feared him; however, quite a few had a warm affection for him and appreciated his strong and genuine personality and character.

Neuberg loved Germany as he had known it in his younger days. He felt deeply rooted in German civilization and strongly attached to the great era of imperial Germany at the turn of the century. He had met many leading personalities in art and science, many important political figures. It was always fascinating to listen to him when he recalled events of that era and evaluated, in a critical but fair spirit, the personalities involved, divulging much inside information otherwise not available. Since he had lived through the glorious period of the Kaiser Wilhelm Institutes in Berlin–Dahlem—he became division head in 1913—very few people were as familiar as Neuberg was with all the illustrious scientists there. He knew many amusing stories about them, including much gossip. He also had many friendly relationships with colleagues at the University of Berlin and at the *Technische Hochschule*. In his characterization of the personalities he was always fair and tried to be as objective as possible. The author frequently implored him to describe the exciting era, especially the personalities involved, since he was in an extraordinary position in view of his familiarity with the whole atmosphere; such a book would have been an invaluable contribution to the history of that period of science and a rich source of many aspects usually not found in history books. It would have been a presentation seen with the eyes of a competent contemporary, sometimes a severe critic, but always just and candid. Unfortunately, the author was unable to prevail. Neuberg remained active and devoted to his laboratory research. This was more important to him than writing memoirs.

Two weeks before his death he gave a lecture before a distinguished group of biochemists of the New York area. It was well attended by his many friends and admirers. He described some fundamental enzyme reactions, which were also demonstrated by simple test tube experiments. He was already quite ill and too weak to stand. Part of his lecture was read by his daughter, Dr. Irene Forrest; the experiments were performed by his assistant, Dr. Amelie Grauer. The audience was deeply moved by his spirit of devotion to biochemistry and the alertness and vigor of his mind in spite of the physical weakness. All knew that it was his farewell lecture.

A simplicity in technique was typical of Neuberg's research work. His strength was in his ingenuity, his keen sagacity and imagination, and his superb judgment, all of which made him a leading figure in the early period of biochemistry.

Nazi persecution, personal hardships, and sufferings did not change his deep attachment to Germany. His strong character did not permit him to burn today what he had adored yesterday. In the living room of his New York apartment was a photograph of Kaiser Wilhelm taken at the opening ceremony of the first Kaiser Wilhelm Institute in 1912. It was not a sign of admiration for the kaiser; on the contrary, Neuberg was most critical of his personality and his actions. The photograph was for him a symbol of the glorious era of imperial Germany. The true greatness of German culture had been, as he hoped, only temporarily overpowered by a group of gangsters and criminals made possible by a movement grown out of despair and the continuous humiliation of a great nation.

Despite the difficult and adverse conditions under which Neuberg had to work in New York, he continued to be productive, unaffected by age and change of environment, as documented by the continuous flow of his publications. He still trained quite a few young biochemists. One of his last pupils was Dr. Ines Mandl, at present a professor at Columbia University. On the occasion of his seventieth birthday, in 1947, the American Society of European Chemists arranged a festive ceremony attended by many outstanding biochemists. One of his oldest pupils, F. F. Nord, at that time professor of biochemistry at Fordham University, was the main promoter. A Neuberg Medal was established in his honor and Neuberg was the first recipient. It became a coveted prize, given once a year, to some of the most distinguished biochemists. The medal was distributed for many years; however, because of some unfortunate circumstances, the society ceased its activities in the 1960s.

Neuberg was a passionate teacher; he loved, and always found time, to discuss problems with young people. His extraordinary memory and his knowledge of literature made these discussions stimulating and an invaluable source of information. It was very easy to get an appointment to see him, even in the Dahlem days, in spite of his many obligations. In one hour of discussion with Neuberg his students frequently learned more than they could have learned in days spent at the library. For example, at a dinner in the home of the author in New York, Neuberg asked him whether he had tried to use hydroxylamine (H_2NOH) for determining acetylcholine by a colorimetric method. Neuberg knew that the author was working on an enzyme that he had discovered in 1942, an enzyme that forms acetylcholine. To make real progress in studying the formation or splitting of a compound by an enzyme, it is essential to have a rapid and simple method to determine the compound. At that time, 1948, there was no convenient and rapid chemical method available for the determination of acetylcholine. The use of existing pharmacological bioassays in enzyme work is cumbersome and for many reasons unsatisfactory. Fritz Lipmann had shortly before described the determination of acetylphosphate. The author had indeed tried to apply this method for the determination of acetylcholine, but without success. Lipmann also had tried to use this method for acetylcholine, but he too was unsuccessful. Neuberg then asked the author what pH* he had used, and was told that neutral pH had been used. Neuberg then said, "The trouble with you young people is that you do not know the literature." He quoted a paper that had appeared in the early 1890s(!) in Hoppe-Seyler's *Zeitschrift für Physiologische Chemie*: The determination of esters with the use of hydroxylamine requires an alkaline pH of 9 to 10. (Acetylcholine is an ester, whereas acetylphosphate is an acid anhydride, so the latter would react at neutral pH.)

The next morning the author informed Dr. Shlomo Hestrin of this discussion. He had been working on this problem at that time in the author's laboratory; he later became the chairman of biochemistry at the Hebrew University in Jerusalem. After two days the method worked perfectly. In the following years the method was widely used, in other laboratories as well as in the author's, both for the enzymes forming and for those splitting acetylcholine, until the method was later replaced by still more precise methods. Many important results were obtained. Hestrin, for instance, observed that the reaction of acetate with hydroxylamine is greatly accelerated by the enzyme acetylcholinesterase, the enzyme that splits acetylcholine. On the basis of several observations he suggested that the first step in the catalytic action of acetylcholinesterase may be an acetylation of the enzyme—that is, an acetyl enzyme may be formed. This suggestion was fully borne out during the following decade by a variety of experiments in several laboratories. This observation of Hestrin's initiated another important development. A large group of compounds called *organophosphates*, to which belong the famous nerve gases and many widely used insecticides, form a phosphorylated enzyme instead of an acetylated enzyme. The acetylated enzyme reacts rapidly, in less than a thousandth of a second, with water to form acetate and restored enzyme. In contrast, the phosphorylated enzyme does not react with water and becomes inactive. Although many other enzymes are also inactivated, the author attributed the fatal action specifically to the inhibition of acetylcholinesterase in view of its vital function in nerve activity. He thus postulated a "specific biochemical lesion" as the basis of the fatal action of organophosphates, whereas the overwhelming majority of investigators in the field attributed that action to a "general toxic effect." A compound may be formed that reactivates the phosphorylated acetylcholinesterase by the removal of the phosphoryl group attached. The restoration of the enzyme activity would prevent the fatal action. Such a compound may act as an antidote in organophosphate poisoning. Hestrin's experiments suggested that hydroxylamine attacks the carbon of the carbonyl (CO) in the acetyl group and thereby accelerates the reaction.

---

* pH is a symbol indicating the concentration of hydrogen ions (i.e., the acidity or alkalinity of the solution used).

Thus the author assumed that hydroxylamine may attack the phosphorus atom of the phosphoryl group and remove it from the enzyme. He suggested to I. B. Wilson that Wilson investigate whether hydroxylamine might restore the enzyme activity of phosphorylated enzyme. It did. It was the first time that an enzyme "irreversibly" inhibited by organophosphates was reactivated. However, the concentrations required were too high and the reaction time too long to use hydroxylamine as an antidote; the work showed only the possibility of developing an antidote on this basis. It was the beginning of the efforts to develop a fast-acting compound in low concentration. Within a few years an efficient antidote was developed. It is today widely used all over the world against organophosphate poisoning. It has saved many lives.

The many pertinent results of the work that was initiated by the information Neuberg recalled from a publication over a half-century old vividly illustrate the value of Neuberg's extraordinary knowledge of biochemical literature and his eagerness to help younger investigators in their problems. This story is also a timely reminder of the value of studying the literature carefully. We live in an era in which science students tend to ignore literature more than a few years old, and even senior investigators feel that publications more than 20 years old have little value.

Jewish Attitude

Neuberg's deep devotion and genuine love of Germany did not present a conflict for him in stressing his Jewish identity. Although completely unorthodox and liberal, he was proud of his Jewish heritage, of its tradition of respect for intellectual and ethical values. He knew the Bible well and was familiar with Jewish history. He was a member of an association founded in 1893 called *Central Verein deutscher Staatsbürger jüdischen Glaubens* ("German citizens of Jewish faith"). One of the chief aims of this association was an active fight against anti-Semitism, widespread among the population, particularly in the government. Emphasizing their cultural attachment and their genuine devotion to the fatherland, the members of the *Central Verein* stressed the obligation of the Jews to participate actively in the political life of the nation and to fight for complete de facto equality of Jews as citizens, wherever it was threatened. Such a struggle required special efforts in areas where there was still blatant discrimination against qualified Jews, such as in the civil service. At that period, at the turn of the century, there were still several such areas. At the universities there was considerable discrimination in the humanities and in law, relatively little in science and medicine. The *Central Verein* wanted to keep the Jewish faith from being used as a pretext for considering Jews different from other Germans, whatever their religious faith might be. Baptism as a precondition for equal rights was unacceptable.

Neuberg, like a great number of other German Jews, fully shared this attitude and these principles. The status of the Jews, particularly in Prussia, had vastly improved in the latter part of the nineteenth century. Many Jews played important roles in a variety of fields and were highly respected. But the remaining anti-Semitic forces had to be actively fought until they completely disappeared. Neuberg actively participated in this fight whenever he had an opportunity. He especially disapproved of baptism for purely opportunistic reasons, such as for promoting one's career. This was for him incompatible with the dignity of man. Neuberg was full of scorn and contempt for Jews who tried to hide their origin. He frequently expressed his feelings quite vigorously. He took the obligation of fighting anti-Semitism very seriously.

An interesting episode reflecting his attitude occurred in the late 1920s. The Hungarian minister of education was in Berlin and visited Neuberg. Under the fascist regime in Hungary at that period Jews suffered great hardships and many left or were forced to leave, among them outstanding physicists and mathematicians, as mentioned before. The minister asked Neuberg to accept two very gifted Hungarian biochemists in his institute for training. Neuberg had recently accepted two Jewish biochemists who had been forced to leave. "Your Excellency, I am sorry, but my Hungarian quota has just been filled," was Neuberg's reply. The minister understood the meaning. This is one of several episodes known to the author, which shows Neuberg's forceful and courageous attitude in his fight against anti-Semitism and anti-Semites. Chaim Weizmann was a great admirer of Neuberg and, especially after World War I, turned to him frequently for advice on his plans to build up chemistry and biochemistry in Israel. Two of Neuberg's pupils went to Israel and achieved prominence. The late Ernst Simon was with Weizmann in the Daniel Sieff Institute in Rehovot from the beginning; Jesaiah Leibowitz joined the Hebrew University in Jerusalem.

When Hitler came to power Neuberg considered the event to be an episode of short duration, as did many Jews and non-Jews. He believed that the violent anti-Semitic movement would soon be overcome, as many had been before. The liberal forces would regain power. The author met Neuberg in mid-January 1933 after Meyerhof's lecture on philosophy. Neuberg told him that it was high time for him to be promoted to *Privatdozent*. He would take the action in his own hands. He knew the author's reputation and he would gain strong support in the faculty. An appointment was arranged for January 31 for discussing details. The author was at that time actively preparing his emigration, since he considered the situation to be hopeless. When Hitler became chancellor on January 30, the author called Neuberg to inquire whether he should come in view of the new situation. The answer was short: "I will see you tomorrow morning." The next morning Neuberg forcefully expressed his views: Jews should not be intimidated; they should fight. The regime could not last long. He predicted that the violence and anti-Semitic propaganda would die down within a couple of months. He had lived through several such dangerous-looking periods. The author was asked to have his *Habilitationsschrift* ready by May. Neuberg was not the only one who held this view. In fact, the author was surprised about the indifference among many Jewish and liberal non-Jewish colleagues, how few took the threat seriously or recognized the catastrophic situation created.

In 1934 Neuberg was forced to resign as director of his institute. He soon had to give up the editorship of the famous *Biochemische Zeitschrift*, which he had founded in 1906 and which had greatly stimulated the growth of biochemistry. He had been its editor for three decades. He left Germany and came to Israel. For reasons explained before he left Israel and arrived in the United States in 1940. But he remained attached to Israel and followed with great interest and pride the magnificent development of the Weizmann Institute.

F. GUSTAV EMBDEN (1874–1933)

Gustav Embden was one of the brilliant and outstanding leaders in biochemistry during the period described in this essay. His name is inseparably associated with the elucidation of the metabolic pathway of glycolysis, the scheme generally referred to as the Embden–Meyerhof pathway. He made a great number of pioneering contributions to this problem, which was one of the central and most passionately pursued fields of that era. Under the influence of his teacher Franz Hofmeister and his friend Albrecht Bethe his interest leaned strongly toward the physiological aspects and significance of his investigations.

He was a superb, dedicated, and inspiring teacher. His lectures were of a remarkable lucidity. His warm personality and his magnanimity attracted many investigators in Germany as well as from abroad and many leaders of biochemistry of the following generation were trained in his laboratory. It became, under his dynamic leadership, one of the great centers of biochemistry in Germany. In spite of his untimely and sudden death at the age of 58, in the midst of a most creative period, his impact on the development of biochemistry in this century was profound and has been generally recognized.

Family Background and Career

Embden was born in Hamburg as the son of a prominent Jewish lawyer. His mother was not Jewish. His brother was a famous neurologist. He married a student of his, Johanna Fellner. It was a most happy marriage. Since his wife was not Jewish, his children were in Nazi terminology one-quarter Jewish. His son was drafted into the army and killed in World War II.

Embden studied medicine and finished his studies in 1899. After a few years, part of them spent in Zurich, he joined Franz Hofmeister at the University of Strasbourg in 1903. This move had an important influence on his scientific thinking and directed his main interest toward physiological chemistry. In 1904 he accepted an offer from Carl von Noorden to become a chemical assistant and to direct his laboratory in the City Hospital in Frankfort on the Main. Under his leadership the laboratory developed into a very active center. After the foundation of the University of Frankfort, in 1914, he became *Ordinarius* of physiological chemistry, a position he kept until his death in 1933. There he met again his good friend Albrecht Bethe, the father of the physicist Hans Bethe, who became *Ordinarius* of physiology. Embden maintained his close contacts with the university clinics, which were gradually extended and were considered to be among the most modern institutes in Germany.

Scientific Contributions

Embden was at first mainly interested in chemical problems. One of his earliest contributions was the isolation of cystine, an amino acid named by Berzelius, from split products of proteins. This finding was independently and at about the same time achieved by K. T. Moerner. It was known that cystine contained sulfur and could be readily reduced to cysteine.* Embden elucidated its exact structure. Both amino acids play an important role in protein chemistry.

However, under the influence of the powerful and dynamic personality of Franz Hofmeister and his friend Bethe, his interests turned toward physiological chemistry and particularly toward the processes of intermediary metabolism. He carried out these investigations on perfused isolated organs, particularly on liver, and he improved the method used in many ways. These investigations later became the basis of his work on carbohydrate metabolism.

Among the pertinent contributions of this early period are the results he obtained on perfused livers of dogs after removal of the pancreas, which, as was shown in 1889 by Josef von Mering (1849–1908) and Oscar Minkowski (1858–1931), produces diabetes. Embden discovered the formation of acetone,

---

* Cysteine is an amino acid containing a sulfhydryl group (CH_2-SH); on reduction a disulfide group ($-S-S-$) is formed.

$$CH_3 - \underset{\underset{O}{\parallel}}{C} - CH_3$$

in the liver of those diabetic dogs. Acetone may accumulate under certain conditions in the blood by decarboxylation of acetoacetic acid,

$$CH_3 - C \overset{O}{\diagup} CH_2 - COOH$$

and a few other compounds. Such accumulations are usually, although not quite correctly, called ketone bodies. The mechanism of their formation was explained many decades later, after a vast amount of information had accumulated about intermediary metabolism. Embden found that fatty acids with even numbers of carbon atoms (C_4, C_6, C_8, etc.) did form acetone, whereas those with odd numbers (C_3, C_5, C_7, etc.) did not. Odd-numbered fatty acids, on the other hand, gave rise to the formation of glucose. The reverse reaction, the formation of fatty acids from glucose, remained obscure and a challenge to biochemists for many decades. Another important observation of that period was made by Embden and Oppenheimer, in 1912. They found that pyruvic acid in perfused liver may act as precursor to acetoacetate; they suggested that glucose, via lactic and pyruvic acid, may lead to the formation of acetoacetic acid.

Subsequently, Embden became increasingly interested in intermediary carbohydrate metabolism. He found in perfused liver the formation of lactic acid from glycogen or sugar. His observations led him to the assumption of the cyclic character of carbohydrate metabolism. As was pointed out before, Meyerhof's success, in 1925–1926, in extracting the glycolytic enzymes from muscle and obtaining them in solution was a dramatic turning point that decidedly helped the exploration of the glycolytic pathway. It is interesting in this context that Embden in 1912 prepared pressed muscle juice with the hope of finding some clues to the intermediary carbohydrate metabolism in solution. Both men were inspired by the revolutionary effect that Buchner's success in extracting the enzyme(s) of yeast cells had on the exploration of alcoholic fermentation. The enzymatic theory had been enthusiastically endorsed by Embden's teacher Hofmeister. Embden hoped to find in the solution prepared from muscle enzyme processes of glycolysis equivalent to those observed in alcohol fermentation in Buchner's extracts.

He and his associate Fritz Laquer found a rapid lactic acid formation, but there was no equivalent decrease of glycogen or sugar as observed in the experiments with perfused liver. The two investigators concluded that there must be another still unknown precursor, which was referred to as "lactacidogen." In addition to the lactic acid formed there was an approximately equimolecular amount of phosphoric acid. These observations suggested that this precursor might be hexose diphosphate, the ester discovered by Harden and Young. Indeed, on addition of hexose diphosphate both lactic acid and phosphoric acid were increased, thus supporting the assumption proposed. But there was still a long way to go before glycolysis could be understood. These experiments were carried out in the years before the outbreak of World War I and were consequently soon disrupted for several years.

After the war Embden took up his experiments on lactacidogen under various experimental conditions. On the basis of several observations he began to realize that there might be some connection between the formation of phosphoric acid and muscular activity. This physiological function increasingly became one of his major areas of interest. It may be recalled that it was at about the same period (i.e., in 1918) that Meyerhof initiated his investigations on the chemical reactions taking place during muscular contraction with the specific aim of providing an explanation of the heat changes measured by A. V. Hill. There was thus a certain amount of similarity in the problems studied, which led to some overlapping in some areas, although the approaches and some basic concepts were quite different. Occasionally there were some contradictory results. Frequently they were based on differences in the experimental conditions. More frequently, there were differences of interpretations, which is quite natural considering the ignorance prevailing in this field in the early 1920s. Many of these differences became resolved as a result of the rapid increase of information that accumulated in the following two decades, an era during which muscular contraction became one of the central problems of biology and biochemistry. Some more detailed comments on these differences will be made later.

Embden's experiments led him to the conclusion that muscular contraction was accompanied by a reversible formation of phosphoric acid. He considered lactacidogen to be the source of this reversible formation of phosphoric acid. It was only after the discovery of creatine phosphate, in 1926, that this problem was taken up in Meyerhof's laboratory. It was demonstrated by the author, in 1927, that the hydrolysis of creatine phosphate paralleled muscular contraction and was the source of increased formation of phosphoric acid. A certain fraction of this compound was very rapidly resynthesized immediately following relaxation (Nachmansohn, 1928/29 I). Embden, however, did not give up his belief that a phosphorylated intermediate of carbohydrate metabolism yielded, in the course of the same reaction, lactic and phosphoric acid, both contributing to what he assumed to be an acidification of the muscle. One of the main factors of Embden's conviction was the observation, reported by several investigators, that phosphoric acid in urine excretion greatly increases following muscular activity. As was learned later, this increased excretion was derived from the phosphate released by the creatine phosphate hydrolysis, probably because phosphate diffuses from the muscle into the blood faster than creatine phosphate. Moreover, at that time Embden still assumed an acidification of the muscle. But it must be realized that little was then known about the physicochemical constants and properties of phosphate esters and the degree of acidification of phosphoric acids. Only later was it found that these carbohydrate esters, under the conditions prevailing in the muscle, are stronger acids than the phosphates resulting from hydrolysis. When a carbohydrate ester is split, alkalinization results. Under physiological conditions (i.e., close to neutrality), creatine phosphate does not act as a buffer, but when it is hydrolyzed it becomes a potent buffer; since more creatine phosphate may be split than lactic acid, depending on the conditions of the experiment, an alkalization results, not an acidification.

The lactacidogen theory was finally abandoned in 1931, when Carl and Gerty Cori confirmed that the increase of inorganic phosphate was, under the conditions of Embden's experiments, mostly derived from ATP, the newly discovered compound of Lohmann, and very little, if any, from hexose diphosphate.

On the basis of his investigations on lactacidogen, Embden developed the concept that the ability of muscle to build up phosphorylated compounds depends on the vital conditions of the muscle. According to his concept the processes that determine all basic vital functions are the changing conditions of colloidal proteins; they are energy-bound interactions of the exo- and endo-thermic chemical processes. They permit the simultaneous existence of assimilation/dissimilation of the living cell. The energy of contraction is primarily provided by colloidal processes, which are then reversed by chemical reactions, in particular by lactic acid; these chemical reactions permit the restoration of the original colloidal state. This general concept, however, did not interfere with Embden's main interest, the exploration of the chemical reactions; they were the focus of his experimental work.

In the course of his studies of phosphorus-containing compounds, Embden discovered the presence in muscle of adenylic acid (i.e., adenosine-monophosphate). Two years later Embden and Gerhard Schmidt reported that the adenylic acid isolated from rabbit muscle differed from that of yeast. Schmidt then discovered an enzyme that converts adenylic acid into inosinic acid by the deamination of adenine, thereby releasing ammonium. Embden found that ammonium is formed during contraction, but is rapidly synthesized again to adenine. He therefore attributed to the ammonium a role in the process of muscular contraction. Although the ammonium formation was confirmed by Parnas and by the author, working at that time in Meyerhof's laboratory, no rapid disappearance was found. Shortly afterward, in 1929, Lohmann discovered ATP and it was soon established that adenylic acid occurs in muscle in the form of adenosine triphosphate.

The discovery of creatine phosphate and of ATP drastically changed, within a few years, all the concepts of the chemical processes providing the energy for muscular contraction and clarified the situation, as previously described. Many ideas and assumptions had to be modified or discarded. This is not surprising. In the early 1920s the chemical reactions associated with muscular contraction were still virgin country. Many important factors were unknown in this extremely complex process. Many ideas proposed were necessarily more or less speculative in their nature.

However, Embden in the last year of his life, together with his associates H. J. Deuticke, H. Jost, G. Kraft, and E. Lehnartz, proposed a scheme of glycolysis that must be considered as the crowning of his lifework; it reveals the strength of his intuition and his imaginative genius. Although three-carbon compounds (triose phosphates) had been repeatedly proposed in alcoholic fermentation, it was Embden and his associates who proposed, in 1932–1933, a complete scheme of the glycolytic pathway. In short, their scheme suggested that fructose diphosphate, the Harden–Young ester, is split in glyceraldehyde-3-phosphate and dihydroxyacetone phosphate.* These products were proposed to yield 3-phosphoglycerol and 3-phosphoglyceric acid; it was suggested that the latter was converted into pyruvic acid and phosphate. In a second reaction pyruvic acid was reduced to lactic acid, and phosphoglycerol was oxidized to glyceraldehyde-3-phosphate. This scheme provided a tentative pathway for the conversion of hexose into lactic acid. (For details see Deuticke, 1933; Lehnartz, 1933.)

Embden unfortunately did not see the triumph of his scheme, since he died shortly afterward. It remained to Meyerhof and his associates to verify experimentally in the next five years the correctness of the proposed scheme, step by step, and to isolate and purify the enzymes cata-

---

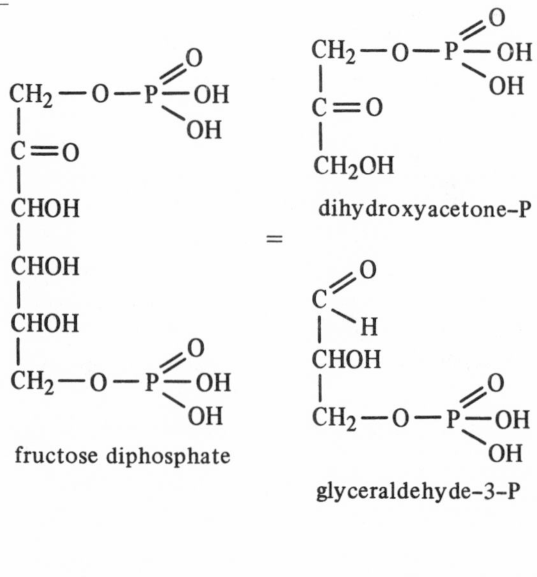

*

$$CH_2-O-P\overset{O}{\underset{OH}{\lessgtr}}OH$$
$$|$$
$$C=O$$
$$|$$
$$CHOH$$
$$|$$
$$CHOH$$
$$|$$
$$CHOH$$
$$|$$
$$CH_2-O-P\overset{O}{\underset{OH}{\lessgtr}}OH$$

fructose diphosphate

=

$$CH_2-O-P\overset{O}{\underset{OH}{\lessgtr}}OH$$
$$|$$
$$C=O$$
$$|$$
$$CH_2OH$$

dihydroxyacetone–P

$$C\overset{O}{\underset{H}{\lessgtr}}$$
$$|$$
$$CHOH$$
$$|$$
$$CH_2-O-P\overset{O}{\underset{OH}{\lessgtr}}OH$$

glyceraldehyde–3–P

lyzing the single reactions. The scheme required only a few minor corrections. Some important contributions were made by other investigators, as has been mentioned before, and by the end of the 1930s glycolysis had been almost completely elucidated. Biochemists interested in the exciting history of this episode are referred to the competent description by Florkin (1975) in his book, *A History of Biochemistry*. But the short outline given may suffice to show the justification of referring to the scheme as the Embden–Meyerhof pathway of glycolysis. This contribution is a lasting tribute to Embden's brilliance.

The Relationship between Embden and Meyerhof

Frequently one encounters views about the antagonism between Embden and Meyerhof that do not quite correspond to reality. It seems appropriate to analyze this relationship briefly in a broader perspective. From the end of World War I to the death of Embden—a period of 15 years—both were deeply and passionately engaged in research on the chemical reactions that take place during muscular contraction, in particular the pathways of glycolysis. It was almost impossible for their work not to overlap, and for them not to report minor differences in results, particularly in view of the complexity of biological materials. For instance, in Meyerhof's laboratory increased lactic acid formation was found to parallel the tension developed (i.e., contraction), but in Embden's laboratory the increased lactic acid formation did not precisely coincide with contraction but outlasted it. Meyerhof interpreted Embden's data as the result of overstimulation; Embden considered the data as an indication that lactic acid was not directly involved in the elementary process, which had never been claimed, as Meyerhof and Lohmann had stressed. These and a few other experimental differences led to some controversies, occasionally producing irritation. Since controversies are disliked in Anglo-Saxon countries, these differences were blown up out of proportion and interpreted as a strong antagonism between these two men.

One difference of opinion concerned the question of whether lactic acid was close to the elementary contractile process, as Meyerhof preferred to believe, or whether phosphate was, as suggested by Embden. Since at that time the elementary process was completely unknown, both views were purely hypothetical assumptions, and the difference of opinion had little significance. The first step toward finding a possible molecular mechanism was the brilliant discovery of Engelhardt and Ljubimova in 1939 on the interaction between ATP and myosin. The molecular mechanism was not to be elucidated until the last two decades, long after both men had passed away. The advances started with the observation of Hugh Huxley and his associates: using electron microscopy they obtained the striking result, that during contraction myosin and actin filaments slide past each other. Thus the purely speculative and hypothetical assumptions of the earlier period were not of great scientific importance. They were never taken too seriously, and Meyerhof's and Lohmann's statement in 1926 clearly indicated their own reservations. Later events resulted in a completely different picture that nobody could have foreseen.

When phosphocreatine was discovered, in 1926, Meyerhof still considered lactic acid as the main source of energy, but left open the relationship with the elementary process. Only after Lundsgaard's observations of a direct relationship between phosphocreatine breakdown, tension, and heat produced in a muscle that was poisoned by iodoacetate, in which no lactic acid was found, did Meyerhof and Lundsgaard begin to consider that phosphocreatine might be directly involved in the contractile process. But both had considerable doubts and soon abandoned this assumption. Not until 1934 did Meyerhof and Lohmann come out strongly in favor of ATP being the substance directly responsible for the contractile process. These developments have been mentioned to show how the views on the primary reaction fluctuated for a long time. There is no question that minor differences of experimental observations were greatly exaggerated, probably more so by outsiders than by the investigators concerned. Both men had great respect for each other. This is borne out by the moving remarks made by Meyerhof about Embden's great achievements and the expression of his admiration on the occasion of the celebration of his sixty-fifth birthday, when the *Festschrift* edited by the author was presented to him.

However, although these small differences should be seen in proper perspective and have completely lost their interest in view of the vast amount of information that has accumulated since that time, there were indeed fundamental differences between Meyerhof's and Embden's general approaches and basic concepts. The author visited Embden in Frankfort on the Main in his laboratory in 1930. He was received by Embden with great cordiality and they had an intensive discussion about various aspects of muscular contraction. The discussion went on for four hours. At the end of the discussion the large table in Embden's office was virtually covered with books and journals that Embden had picked up from the bookshelves to bring across his various viewpoints. The author was very familiar with Meyerhof's concepts, having worked in his laboratory for many years, during which time he and Meyerhof had had innumerable discussions about many aspects of their work. There is no substitute for such personal discussions. Many ideas and convictions are not printed in full detail and cannot readily be expressed in print in view of their subtle nature, but in a free personal discussion the thinking of a person frequently becomes quite evident.

The discussion with Embden was for the author a great and fascinating experience. He greatly admired Embden's competence and superb knowledge of chemistry, the lucidity with which he expressed his .views, his genuine enthusiasm for the problem, his complete frankness in speaking to a young colleague (25 years younger), treating him as an equal and listening carefully to the questions raised. This attitude was all the more remarkable since in Germany, at that time, there was usually a wide gap between an *Ordinarius* and a young postdoctoral fellow. The atmosphere prevailing in Dahlem and in the Haber colloquia was a great exception.

Thanks to this discussion the author realized how relatively irrelevant the details were as far as the chemical data were concerned. The real gap between Embden and Meyerhof was caused by a difference in fundamental concepts. For Embden the elementary mechanism of contraction was a "colloidal-chemical" process. The essential function of the chemical reactions was to restore the energy spent in the contractile process. He repeatedly expressed his skepticism about paying too much attention to the heat exchanges measured. Many details may pass unnoticed, since rapid small changes in opposite directions (positive and negative) may cancel each other. The elucidation of the sequence of the metabolic pathways during muscular contraction was for him the central problem, which required solid chemical knowledge and insight combined with the imaginative mind of the investigator.

Meyerhof, in contrast, did not speculate much about the nature of the elementary process. The heat changes were for him what the lighthouse was for the captain of a ship in a stormy sea, as he expressed it in his Nobel Prize lecture. He was fully aware that corrections were to be expected with the continuous improvements of techniques and experimental conditions, and this has been borne out by the history of the results of measurements of heat changes. But nevertheless, the heat changes were an important indicator for the biochemist and he could not ignore them. He had to provide an explanation of heat changes by the analysis of chemical reactions. These explanations were obviously always subject to change with increase of knowledge. The second equally important difference was Meyerhof's emphasis on the energy released by each single reaction involved, which would provide essential clues to the sequence of energy transformations. His epoch-making lecture to the Philosophical Society in Kiel in 1913 has been discussed, as well as his revolutionary discovery of two types of phosphate derivatives, one type releasing high energy, the other releasing low energy, established during 1926 and 1933.

It must be stressed in this context that Meyerhof was 10 years younger than Embden and that it was just during that period, at the turn of this century and in its first decade, that thermodynamics began its rapid development especially under the leadership of Walther Nernst and Fritz Haber. Its general implications and its significance for many scientific fields became increasingly apparent. In 1912 Meyerhof had spent a year with Georg Bredig, a friend of Haber, studying physicochemistry. Bioenergetics was an essential and crucial factor in his thinking. His collaboration with Otto Warburg on oxidation (oxygen being the ultimate source of energy in all human cellular functions) not only attracted him to cellular biology and chemistry, but increasingly directed his interest toward bioenergetics in general, of which he was the founder and the uncontested pioneer.

It may be mentioned that among the two schools there was always a great mutual respect. The author personally had many pleasant experiences with Embden's collaborators. Friendly feelings prevailed, based on mutual scientific interest, undisturbed by differences of experimental observations or theoretical views.

G. GENERAL GROWTH OF GERMAN BIOCHEMISTRY

In the preceding sections the elucidation of the metabolic pathways of fermentation and glycolysis was described. The great number of enzymes involved in these two processes were isolated and purified, and quite a few were crystallized. Much information accumulated about enzymes, coenzymes, and the mechanism of oxidation and dehydrogenation. Since three of the most brilliant leaders (Neuberg, Warburg, and Meyerhof) worked in the Kaiser Wilhelm Institutes, Dahlem became one of the most outstanding centers in this special field, especially in the 1920s. In order to get a balanced view one must realize that the progress achieved in Dahlem represents only a small section in the gigantic field of biochemistry, even today still rapidly and continually expanding. It is true that the three leaders in Dahlem and their schools initiated one of the most dynamic trends by their notions and concepts, by providing biochemists with a great number of advanced methods and invaluable tools. They thus had an enormous influence on the progress made in many other areas of biochemistry. Even during the period described in Chapter IV several other outstanding scientists—Embden, Wieland, and Windaus for example—worked on these problems in Germany. Research also went on in other European countries and in the United States, in the laboratories of Dorothy Needham (1896–   ) in Cambridge, Jacob K. Parnas (1873–1949) in Lvov (at that time Poland; now part of the U.S.S.R.), Hans von Euler (1873–1964) in Sweden, and Carl F. and Gerti Cori in the United States, to name but a few. Several of them and others have been mentioned before. Many other areas of biochemistry also advanced very quickly. The achievements of Wieland in steroid chemistry have been mentioned. Steroids are, however, a very large group of compounds that have a great diversity of biological activities. Their study therefore attracted a vast number of investigators. As mentioned before, they are related to the bile salts and to many hormones, in particular to sex hormones; some of them are crucial in the female sex cycle. Some steroids play a role in the metabolism of carbohydrates, fats, and proteins. One of the most important steroids is cholesterol, which is abundant in the brain; it is also present in cell membranes, where it apparently plays an important but not yet explored role. Although cholesterol was discovered in the eighteenth century, its structure was elucidated only in the 1930s by several investigators, among them Wieland and Adolf Windaus. The report of J. D. Bernal, in 1932, of the X-ray diffraction pattern of ergosterol (a closely related compound) permitted researchers to establish the structure of cholesterol with still greater precision. The brilliant achievement of the synthesis of cholesterol by F. Lynen and K. Bloch has been mentioned before.

Adolf Windaus (1876–1959) was one of the great pioneers in the field of steroids. Born in Berlin, he first studied medicine, but under the influtnce of Emil Fischer he soon turned to chemistry. The topic of his *Habilitationsschrift* in Freiburg was cholesterol. He became *Ordinarius* in Göttingen, in 1915, and kept his position until his retirement. Windaus succeeded in elucidating the structure of several vitamins belonging to the steroids, especially of the B and D series; he also synthesized quite a few of them. He became famous for the preparation of vitamin D_2, the vitamin that prevents rickets. He obtained the active form by ultraviolet radiation of ergosterol. The achievement virtually wiped out rickets in Europe. Among his many other contributions may be mentioned the elucidation of the structure of thiamine (vitamin B_1), simultaneously with Robert R. Williams in the United States. Karl Lohmann, in Meyerhof's laboratory, found thiamine pyrophosphate to be a coenzyme. As was shown by Ochoa, in collaboration with R. A. Peters (1889–   ) it is required for the oxidation of pyruvic acid. Windaus also made important studies on steroid hormones. For his many brilliant achievements in the field of steroids, he received the Nobel Prize in 1928. He was deeply shocked and depressed by the destruction of the University of Göttingen under the Nazi regime as well as by their persecution of Jews in general.

Another great center was created by Franz Knoop (1875–1946) in Tübingen. He and his school, and especially Carl Martius, made many important contributions to intermediary metabolism. Obviously it is impossible even to mention the many accomplishments of biochemistry of that era in Germany as well as in Europe, the United States and elsewhere. But it is important to realize that the knowledge of the structure of an ever increasing number of cell constituents and the great advances in physicochemistry in these decades both were essential for the rapid growth of biochemistry.

H. REFLECTIONS ON THE ROOTS AND FRUITS OF THE
COLLABORATION BETWEEN GERMAN
AND JEWISH SCIENTISTS

The question arises as to the factors that favored the extraordinary success and the great achievements that resulted from the close collaboration and association between German and Jewish scientists during the era and in the fields described, or in other fields, such as medicine; a history of the latter field is being prepared under the sponsorship of the Leo Baeck Institute. An in-depth analysis would require the collaboration of historians, sociologists, psychologists, and scientists in different fields. In the opinion of the author, a simple answer does not exist, and differences of opinion are to be expected. But at least some of the factors that were important may be pointed out.

Science, because it is international in character, is particularly favorable for collaboration independent of nationality, race, or creed, much more so than the humanities, law and other fields. Atoms, molecules, cellular functions, astronomy, and mechanics do not differ anywhere on our planet, nor are they dependent on an era or tradition. The question raised may therefore be limited to the special case of why just in Germany in the half century before Hitler collaboration between German and Jewish scientists led to a momentous success hardly matched previously anywhere else.

Some of the external factors have been described and may be only briefly recalled. After the foundation of the *Kaiserreich* in 1870 the German leadership realized that military strength alone does not represent real power. Industry, agriculture, and natural resources form an integral part of the real power of a nation. Since French and English industry were at that time, in the middle of the nineteenth century, ahead of that of Germany, one potentially great source of power was the development of a strong German industry that far surpassed that of France and England. Science and science-based technology offered an almost unlimited source for this aim. Into the development of science, the German government poured funds of an order of magnitude previously unknown. New universities, institutes, and attractive positions mushroomed. The success of this effort has been described. Manpower, or rather brainpower, was required on an unprecedented scale and was recruited wherever gifted and creative people could be found. The extraordinary conditions in the Kaiser Wilhelm Institutes have been described. The number of Jews in Germany had grown considerably, and quite a large fraction belonged to the upper middle class; some of the richest families in the country were Jewish. The cultural and intellectual atmosphere in many of these families stimulated the children to choose creative careers, in the arts and sciences. In France and England the number of Jews was extremely small compared to that in Germany, and the possibilities for scientific careers were very limited in view of the few positions available. The tremendous empires of France and Britain attracted talented Jews to build up large fortunes in many ways, whereas the few German colonies were poor and had little to offer. As Frédéric Joliot-Curie once told the author, the empires of France and England hindered the development of science in these two countries. Most industrialists and bankers had little interest in and understanding of the value of science and technology and were unwilling to support them financially; nor did their governments see any compelling reason for making special efforts for science. There were many other ways of accumulating wealth and power. Exploitation, at times extremely ruthless, of large, rich colonies such as India and China in the British empire offered many opportunities for enterprising and talented people to become very rich, sometimes in an easy way.

Anti-Semitism in the Wilhelmian era blocked Jews, with few exceptions, from government and the military establishment. Even in the universities, in which positions were civil service jobs, anti-Semitic feelings were strong in some fields in the humanities and particularly in law. But in medicine and especially in science, a liberal spirit prevailed. Moreover, the anti-Semitism of Kaiser Wilhelm II was not as uncompromising and fanatic as that of the Nazis. It did not include such fields as science. He had seen how much Emil Rathenau had contributed to German wealth, power, and prestige by the A.E.G. (Allgemeine Elektrizitäts-Gesellschaft). He had seen how outstanding Jewish scientists and industrialists had contributed to the amazing growth of the German chemical and pharmaceutical industry. It was not difficult for Walther Nernst, Emil Fischer, Max Planck, and many other liberal leaders in science to convince the kaiser of the importance of having the most brilliant and creative scientific minds working for the glory of Germany, even if they were Jews. In the Weimar Republic the conditions for Jews further greatly improved in most fields.

In addition to these external factors, there were others that favored the close collaboration and many genuine friendships between German and Jewish scientists. A person's abilities and creative forces are formed not only by genetic endowment but also by education, experiences, and the impact of the environment. Many if not most of the Jewish scientists in Germany had the same kind of education as their German colleagues; many of them had attended humanistic gymnasiums. They were deeply influenced by the same literature, by Goethe, Schiller, and Hölderlin; they loved Bach, Mozart, and Beethoven and went to the same concerts, operas, and theaters. They spent their vacations on the German lakes and seas, in the forests and mountains. A large fraction of the Jewish scientists came from more or less completely assimilated homes. Most of them had little or no connection with the Jewish community and knew little about

Jewish rites and traditions. They loved Germany, to which they gave their main loyalty. They were Jews because their ancestors were Jews, in their view a simple historical fact with little meaning. There were some differences in emphasis, but with little consequence. Some of them had experienced the manifestation of anti-Semitic feelings on several occasions, but the majority considered anti-Semitism as a remainder of the Middle Ages that had lost much of its momentum since the era of Enlightenment and was bound to disappear sooner or later with the progress of civilization. Moreover, most scientists, German and Jewish, were little interested in politics. All their passion and love and all their time were devoted to science, art, and nature. Among German scientists of that era, at the turn of the century, were many extremely liberal elements. Scientists such as Max Planck, Emil Fisher, Walther Nernst, Max von Laue, Heinrich Wieland, and David Hilbert had not only no prejudice against Jews, but some of them, such as Wieland and Hilbert, left the church. The reason was frequently not that they were not religious, but that the church was not sufficiently liberal and free of prejudice. The combination of these many factors favored close ties and created strong bonds between the scientists of the two ethnic groups. Continuous exchange of views and mutual stimulation were a natural result. All these elements do not explain the huge number of real scientific giants, the explosive rise of German science during the relatively short period of about half a century. This is, however, not a unique phenomenon. History offers several examples, where certain fields flourished to an amazing degree and produced many geniuses in certain places in a relatively short time, such as in Athens during the Periclean age or in Florence during the Renaissance in the fourteenth and fifteenth centuries.

The most intriguing and interesting problem is the question of whether the two ethnic groups had specific gifts that were complementary, and whether they stimulated each other in some specific way. In art, differences of national character are manifest, although the origin does not affect in the slightest their international effects. In some fields specific national traits are easy to identify. The typical Russian or German or French character of Tolstoi and Dostoevski, of Goethe and Hölderlin, of Balzac and Maupassant is not just an expression of an era, determined by prevailing conditions; it is an artistic manifestation of specific national characteristics. The fundamental themes may be similar—love and hatred, human conflicts and relationships, friendships, lust for power or wealth, fight for freedom and against tyranny, and so on. In fact, they have not changed fundamentally, except in their form, from the time of the ancient Greek poets. Therefore we still read Homer and Sophocles today with interest and enjoyment. There are some fields of art in which the manifestations of national character are more subtle and more difficult to recognize, but they are there.

In science, however, the role of ethnic or national differences is much more open to question. The problems are identical, at least in the basic fields. Does there, nevertheless, exist a difference of approach, of analysis, based on genetic endowment, heritage, tradition of thinking? This is a question to which an answer appears extremely difficult. Willstätter stated that good scientific work should end with a question mark and not with an exclamation mark. The question raised here is much more complex in its nature. In this respect the discussion will close with a question mark. But whatever the basis of the great achievements of German-Jewish scientific collaboration has been, both Germans and Jews have every reason to look back with great pride on this glorious chapter of their history.

V

Worldwide Effects on Biochemistry Due to Nazi Persecution

The influx of many brilliant refugees from Nazi Germany led to a great resurgence in atomic physics in the United States, England, and France, as described in Chapter II. However only about 10% of the 1200 or so scientists who left Germany were physicists. A large variety of other fields also benefited from the immigration of so many qualified scientists. The growth of biochemistry in particular was greatly stimulated. As a biochemist who left Germany immediately after the Nazis took over the government, the author has witnessed these effects most closely, especially since many of the scientists were colleagues and friends; thus he is more familiar with the details here than in other branches of science. In the first half of this century the most intellectually exciting area in science was atomic physics, but from the 1940s on it has been biochemistry. The spectacular rise of biochemistry and some of its special branches widely referred to as molecular biology has changed our knowledge in an absolutely fascinating way and has dramatically changed our insight into the understanding of the living organism. Biochemistry has penetrated almost all branches of the biological sciences: anatomy, physiology, pharmacology, immunochemistry, and others. As every layman knows, these advances have revolutionized medicine and agriculture and we are in the midst of continuous new and striking progress. The insights provided by molecular genetics and evolution of life have also begun to influence our philosophical outlook, as mentioned before.

The author had therefore originally planned to devote a relatively extended chapter to this subject. Two factors motivated him to abandon his plan. As stressed before, this book is not intended to compete with the many excellent and detailed treatises available in the fields mentioned.

It tries to present some selected highlights as illustrations of the particular aspect expressed by the title, reflections on an era and on topics in which the author was personally and passionately involved. However, the size of the work has already far exceeded the original intention. Second, Professor G. Semenza of the *Eidgenössische Technische Hochschule* in Zurich has fortunately initiated a large project that will cover much of the history and the advances of biochemistry in this generation and will of necessity reveal the effects of the scientific migration. Many outstanding leaders, among them Sir Hans Krebs, Severo Ochoa, Feodor Lynen, Eugene Kennedy, Albert Lehninger, Britton Chance, and Melvin Calvin, will give their life stories in this book. Therefore, the effects of Nazi persecution on the development of biochemistry after 1933 will be only very briefly summarized and the profiles of only a few of the many brilliant leaders who had received their decisive scientific formation in Germany will be presented, just as examples.

It may be useful to exemplify the rapidity of the advances made by describing at least briefly a few milestones in a field in order to convey to the reader a meaningful picture of the revolution which has taken place since the turn of the century. The history of protein chemistry offers a suitable subject. Emil Fischer, one of the most eminent chemists of this century, received the Nobel Prize in 1902 for his great achievements in the chemistry of carbohydrates and purines. Only in 1899 did he become interested in the structure of proteins and this work was just in the beginning stages. Proteins play a crucial role in virtually all biological processes. Enzymes, discussed in Chapter IV, are proteins. At present more than 1000 enzymes have been isolated and purified. But the catalytic function is only one of a huge number of others. Proteins form the contractile substances of muscle (myosin, actin, tropomyosin, etc.); they transport oxygen (hemoglobin in erythrocytes, myoglobin in muscle); they are essential for the actions of genes; they are the target of most biologically active substances; antibodies are proteins; they form the structural element of many tissues (e.g., collagen in connective tissue); some hormones are proteins. These examples may suffice to indicate their many roles.

The structural units of all proteins are amino acids. When Fischer started his studies, 12 amino acids were known to be cleavage products of proteins. He introduced new methods for the separation of amino acids from protein hydrolysates and identified three more amino acids. But neither these nor later methods were fully adequate. A really satisfactory method of analysis and separation was made available only after World War II by the introduction of chromatography. Today the number of amino acids from which proteins are formed is known to be 20. Fischer also tried to synthesize amino acid chains to form proteins. Amino acids are linked by peptide bonds,

$$-\overset{\displaystyle \underset{|}{\text{H}}}{\underset{\displaystyle \underset{\displaystyle \text{O}}{\|}}{\text{C}}}-\text{N}-$$

as discovered simultaneously by Hofmeister and Fischer, and form poly-peptides. Fischer succeeded in synthesizing polypeptide chains formed by 18 amino acids; their molecular weight would amount to about 1500. At the time some scientists estimated the molecular weight of proteins to be about 12,000. Fischer was skeptical as to the solidity of these estimates. As late as 1916 he considered the molecular weight of proteins to be not higher than 5000. Fischer's work on proteins in the last two decades of his life was instrumental for the subsequent developments in the field. He was fully aware of the paramount importance of proteins for cellular functions, but he also realized the difficulties involved in analyzing them and explaining their properties and functions. In his Faraday lecture at the Royal Society in London, in 1907, he emphasized that no matter how important the knowledge of their chemical structure might be, this infor-mation alone would be inadequate; only a close collaboration between chemistry and biology could provide a better understanding of their prop-erties and function in biological processes, as actually envisaged by Liebig, Dumas, and other leaders in the nineteenth century (described by Florkin, 1977). Today his knowledge of proteins may appear primitive, but that applies to almost all spheres of scientific activities in view of the rapid progress made in this century.

Soon it became evident that Fischer's estimates of the molecular weights were much too low. The first step was made by Sören P. L. Sörensen (1868–1939), an outstanding physicochemist in the Carlsberg Laboratories in Copenhagen, which was destined to become a great center under sev-eral outstanding leaders, among them Kaj Linderström-Lang (1896–1959). Applying physicochemical methods, Sörensen obtained results that sug-gested a molecular weight for crystalline egg albumin of 34,000 and for serum albumin of 45,000. Still higher values were found for other pro-teins by Gilbert S. Adair (1896–   ) in England and Edwin Cohn (1892–1953) and his associates at Harvard, among them John T. Edsall, who be-came one of the most eminent leaders in protein chemistry. A decisive turning point in our knowledge of molecular weights was reached when Theodor Svedberg (1884–1971) introduced the ultracentrifuge in Sweden in 1925. By using high speed ultracentrifugation with 50,000 rotations per minute in combination with most ingenious optical methods, he succeeded in determining within a decade the molecular weights of a large number of proteins; the values ranged from about 18,000 to 6,700,000 (for hemo-cyanin, an oxygen carrier found in snails; later measurements found the molecular weight to be 2,800,000). The instrument has been greatly improved, especially the optical methods; the rotation speed may reach 75,000 per minute. Thus Svedberg's results proved beyond any doubt that proteins are very large macromolecules. This has been confirmed by additional physical methods, many of them used in polymer chemistry in general.

Another aspect of protein molecules was revealed by Niels J. Bjerrum (1879–1958). He demonstrated that they are not uncharged neutral molecules as was assumed, but *Zwitterions*, or dipolar ions (of the structure $+NH_3CH_2COO-$), having positive and negative charges. Their charge depends on the pH; at a certain pH, called the *isoelectric point*, the number of positive and negative charges is equal and therefore the net charge is zero. Bjerrum's work on the electrolyte nature of proteins was greatly extended by Edwin Cohn and his associates at Harvard. A great amount of data on the acid–base behavior of proteins was accumulated. It was firmly established that proteins are macromolecular polyelectrolytes. This property was used by Arne W. K. Tiselius (1902–1971) for the development of a moving-boundary electrophoresis, an instrument formed by a U-shaped tube. Proteins are placed at the bottom, at a definite pH, and move in an electric field. Since the charges of the proteins are different, depending on the pH, this method permits an elegant separation of the proteins. One of the striking results of this method was the observation that when it was applied to the separation of proteins, the globulin fraction contained several components; some of them, the so-called γ-globulins, are antibodies. In the last two decades many new and simple techniques have been developed for the separation of proteins on the basis of such characteristics as size, solubility, charge, and binding affinity. Among these techniques may be mentioned chromatography, the use of Sephadex columns and ion-exchange resins, and polyacrylamide gel electrophoresis. They have become tools that form part of the daily routine in every laboratory working with proteins. For the separation and purification of enzyme proteins a variety of procedures and a combination of them were used for several decades. The most elegant, efficient, and now generally applied procedure is affinity chromatography: A compound that specifically binds the enzyme is attached to a Sepharose column. When the protein mixture passes through such a column, the enzyme is retained, whereas the other proteins are removed. Subsequently the enzyme is removed with salt or other solutions and may be obtained in pure form and with a high yield. The method was introduced in 1968 by P. Cuatrecasas, M. Wilcheck, and C. B. Anfinsen (Cuatrecasas et al., 1968).

Another landmark in protein chemistry was the determination of the amino acid sequence in insulin by Frederick Sanger (1918–   ). Insulin is a peptide hormone of a molecular weight of about 6000, formed by two polypeptide chains. The amino acid sequence is well defined. The determination of amino acid sequences has been greatly facilitated by ingenious methods developed by William H. Stein (1911–   ) and Stanford Moore (1913–   ). Today the amino acid sequences of more than 100 proteins have been determined. Each protein has a unique and precisely defined amino acid sequence that is genetically determined. The knowledge of these sequences in proteins has opened the way for the understanding of many essential aspects of their properties and has formed the basis for much of the important progress achieved in the last decades.

In the 1950s many successful efforts were made by a great number of outstanding biochemists to obtain information about the amino acid groups at the active site of the enzyme proteins—that is, at the site involved in the catalytic process—and to deduce from the observations the exact mechanism of enzymatic catalysis. Extremely valuable information was obtained about many aspects of the reaction mechanisms, although some of the ideas had to be corrected and modified when in the following decade the three-dimensional structures of proteins and enzymes were elucidated by means of X-ray crystallography. Many new notions emerged from these efforts to analyze the mechanism of enzyme catalysis. One example may be mentioned. Emil Fischer had proposed the "key-lock" concept for the interaction of enzyme and substrate: Assuming a rigid structure for the enzyme, he suggested that the enzyme specificity was due to the fitting of the substrate to the configuration of the enzyme as a key fits into a lock, absorbing the substrate to a specialized catalytic group. By the new developments it became apparent that this view did not provide satisfactory explanations in many cases. Daniel Koshland (1920–   ) and his associates introduced the notion of an "induced fit." On the basis of a large amount of brilliant experimental work, they proposed that the substrate induces conformational changes of the protein on the binding to the enzyme, thereby promoting the catalytic reaction. This concept was borne out in many ways.

A fascinating step was the concept of the regular molecular form (configuration) of the polypeptide chains elaborated by Linus Pauling (1901–   ) and Robert B. Corey (1897–1971) and their associates in the early 1950s after studies stretching over a period of more than a decade. They recorded first the X-ray diffraction pattern of crystals of amino acids and simple di- and tripeptides. From their observations they deduced the precise three-dimensional structure of each molecule and, in particular, of the peptide bond. It turned out that the carbon-to-nitrogen bond of the peptide linkage has a double bond character and cannot rotate freely. The four atoms of the peptide bond and an additional carbon lie in a single plane as a result of resonance stabilization. Thus the backbone of a polypeptide chain may be pictured as consisting of a series of rigid planes. For example, in a polypeptide chain formed by 100 amino acids, there are about 300 single bonds in the backbone (not counting side chains). However, only 200 single bonds of the chain have complete freedom of rotation. Using models to study the ways in which a peptide chain could fold in view of the constraint due to the peptide bonds, they concluded that the simplest arrangement is a helical structure. The most common form is the α-helix (where α stands for the right-hand direction in which the chain is coiled). There are about 3.6 amino acids per turn. In such a coil a single turn would extend about 5.4 angstroms (1 angstrom $= 10^{-8}$ cm) along the axis. This helical arrangement of the polypeptide chains is favored because it permits the formation of hydrogen bonds between carbonyl and amido groups,

$$\geq C = O \cdots HN \leq$$

between the successive turns of the helix. Each peptide bond of the chain takes part in the hydrogen bonding, thereby promoting a maximal bond strength and contributing to stability. Other conformations exist, but it is impossible to discuss all the details; they can be found in every biochemistry textbook. There are also nonhelical parts in polypeptide chains, but the helical form is essential for their functions.

The ingenious concept of the conformation of the polypeptide chains was soon confirmed in a variety of ways. The most conclusive evidence was derived from X-ray diffraction analysis. The diffraction effects of X-rays were studied by von Laue in 1912. The wavelength of X-rays is shorter than the distance between the atoms in a crystal. Shortly after von Laue's discoveries, William H. Bragg (1862–1942) and his son William L. Bragg (1890–1971) decided to use X-rays for determining crystal structure. In 1913 they succeeded in describing the structures of simple inorganic crystals, such as NaCl. In the 1920s they started the analysis of organic compounds, and in 1929 Kathleen Lonsdale (1903–1971) succeeded in determining the structure of hexamethylenebenzene. In the 1930s John D. Bernal (1901–1971) and a number of brilliant associates, among them Dorothy Crowfoot Hodgkin (1910–   ), started to apply X-ray analysis to larger molecules. It was not until the early 1960s that clear pictures of the three-dimensional structure of larger protein molecules were obtained: of myoglobin by John E. Kendrew (1917–   ) and of hemoglobin by Max Perutz (1914–   ). Myoglobin is formed by a single polypeptide chain; it has a molecular weight of about 17,000. Hemoglobin is formed by four polypeptide chains—two pairs of identical chains (two α and two β chains). Its molecular weight is about 67,000. In the following decade the three-dimensional structures of a large number of proteins and enzymes were established, and the number grows steadily.

Knowledge of three-dimensional structures has opened a new chapter in protein chemistry. It became possible to study the catalytic mechanism of enzymes on a completely new basis. Through the use of a variety of additional methods—optical methods such as optical rotatory dispersion and circular dichroism, and a series of chemical and kinetic determinations—our understanding of the catalytic mechanisms has dramatically advanced. A decisive role in this progress was played by Eigen's fast-reaction measurements. Many of the existing concepts, such as the helical structure of polypeptide chains, or the conformational changes during activity, have been conclusively confirmed.

One more of the many advances of protein chemistry may be briefly mentioned. Most enzymes are formed by subunits, and the number of subunits may vary considerably. They may also have different functions: Some carry the active site, whereas others have a regulatory function. For the sites that are not the active (catalytic) ones, Jacques Monod (1910–1976) proposed the term *allosteric* sites. The way different subunits cooperate has been investigated in many laboratories and models of cooperation have been proposed, particularly by Monod, Jean-Pierre Changeux (1936–   ), and J. Wyman Jr., in Paris and by Koshland and his associates in Berkeley. The complexity of proteins with several subunits was made strikingly apparent by intensive studies of J.-P. Changeux, J. C. Gerhart, and H. K. Schachmann on the enzyme aspartate transcarbamylase, and of E. R. Stadtman on the enzyme glutamine synthetase. The latter enzyme is formed by 12 subunits each having a molecular weight of about 50,000.

During the 80 years since Emil Fischer started his experiments on proteins the progress made has been of a magnitude surprising even for the experts in the field. Several large volumes of treatises on proteins exist today, and there are equally large ones on enzymes, which are also proteins. Just a few aspects of this progress have been touched upon here. But proteins, including enzymes, form only one section of biochemistry. Equally spectacular progress was seen in many other areas, such as intermediary metabolism; the structure, properties, and functions of carbohydrates; lipids; hormones; vitamins; nucleic acids and genetics; photosynthesis; immunochemistry; and biomembranes. In addition, there was a remarkably strong cross-fertilization between these fields. It is difficult even for a biochemist and impossible for a layman to appreciate fully the profound changes in our knowledge and the insight gained in the understanding of the living organism. It required a huge amount of ingenuity and brainpower to obtain these stupendous achievements. Moreover, biochemistry was decisively helped, particularly after World War II, by the knowledge of electronic structure of atoms and molecules and the highly refined instruments and methods developed, as mentioned before. The few pages devoted to some of the aspects of the progress in protein chemistry are intended only to help the reader get at least a rough idea of what happened in the short space of less than a century in one field of biological science.

However, this information may facilitate the understanding of the situation faced by the refugee biochemists—Jews, half- or quarter-Jews, non-Jews with Jewish wives or non-Jews who refused to live under the Nazi regime. Their field was in a phase of rapid and profound transformation with many outstanding leaders and active centers existing all over the world. It shows the international spirit of the scientific community that many active centers were eager to accept the refugee colleagues. In fact, there was a widespread realization of the great opportunities that the fast-advancing field of biochemistry offered and that it might gain by accepting highly qualified colleagues. It was not easy in view of the widespread economic depression of the 1930s, but great efforts were made to overcome the financial difficulties and to permit the scientists to continue their research. The many contributions to biochemistry by the refugees justified the expectations. A full description would require a volume as large as this book. For the two reasons mentioned in the beginning of this chapter, the author decided to give just a few examples in a very abbreviated form and to describe the contributions of only four biochemists, randomly selected.

A. HANS A. KREBS (1900–   )

Hans A. Krebs was one of Otto Warburg's most brilliant pupils. He worked in Warburg's laboratory from 1926 to 1930, and Warburg influenced his scientific thinking more than any other of his teachers. Krebs' fame in the history of biochemistry is based on his establishment of the cyclic nature of intermediary metabolic reactions. This notion was an important landmark in biochemical thought; it provided a decisive clue to the understanding of metabolic pathways and in the following decades dominated the studies of intermediary metabolism. It is one of the notions that decisively stimulated both the experimental work and our concepts of cellular reaction sequences.

The first cycle discovered by Krebs in 1932 was the urea cycle (Krebs and Henseleit, 1932). Urea (NH_2—CO—NH_2) is the end product of many nitrogen-containing compounds of the body. Most of it is formed in the liver, although small amounts are formed in other organs. Albrecht Kossel (1853–1927) and Henry D. Dakin (1882–1952) had found, in 1904, that the enzyme arginase forms urea from arginine. Krebs investigated the problem and obtained quantitative data and information about other factors influencing the rate of formation, using Warburg's tissue slice technique. When liver slices were incubated with ammonium salt, urea was formed (consistent with the equation $2NH_3 + CO_2 = NH_2$—CO—NH_2). Several compounds tested did not affect the reaction of arginase. Unexpectedly ornithine was found to promote urea synthesis in a catalytic way; that is, small amounts of ornithine added had a very strong effect on the rate and the amounts of urea formed. On the basis of the chemical structures it seemed reasonable to assume that citrulline was an intermediary compound between ornithine and arginine; this was borne out by experiment. Thus in the presence of arginase the reaction is not a one-way synthetic system, but a cyclic process in which ornithine acts as the catalyst: ornithine → citrulline → arginine (releasing urea by the action of arginase) → ornithine. This sequence is usually written in the form of a cycle. It was the first cyclic reaction unequivocally demonstrated.

Only a few years later, in 1937, Krebs established a second cycle: the citric acid cycle, also referred to as the tricarboxylic acid cycle or Krebs cycle. The citric acid cycle is the final common pathway for the oxidation of the end products of carbohydrate, protein, and lipid metabolism. It is the main source of energy in higher organisms, as well as aerobic microorganisms and plants. Its elucidation was therefore of primary importance for our knowledge of how the intermediary metabolism uses the fuel introduced into the organism to provide energy for biosynthetic processes. Oxidation has been intensively studied since the beginning of the century. An important step was the findings of Albert Szent-Györgyi (1893–   ) in 1934 and 1935, that four-carbon dicarboxylic acids (succinic, fumaric, malic, and oxalacetic acids) have a catalytic effect on oxidation. In 1937 Krebs found that citric acid (a six-carbon tricarboxylic acid) acts as a catalyst of oxidation in the same way as succinic acid. In the same year Carl Martius (1906–   ) and Franz Knoop (1875–1946), studying the fate of citric acid in oxidation in liver, found that α-ketoglutarate was a product of citrate oxidation; they suggested that *cis*-aconitate (by dehydration) is formed, followed by the formation of isocitrate (by hydration) as intermediary steps. In the same year Krebs and Johnson (1937), in studies aimed at linking carbohydrate metabolism with oxidation, demonstrated a decisive step, namely the formation of citrate from oxalacetate and pyruvate. On the basis of this observation they proposed the original citric acid cycle: citric acid → *cis*-aconitate → isocitric acid → α-ketoglutaric acid → succinate → fumaric acid → malic acid → oxalacetic acid → citric acid. (This sequence is generally presented in the form of a cycle.)

Like all advances in science, no matter how fundamental and important, the continuous progress continuously adds new information. Both the urea and the citric acid cycle have been subsequently extended. The additional steps in the urea cycle, for instance, are the transfer of a carbamoyl group to ornithine to form citrulline. An enzyme (arginosuccinate synthetase) then catalyzes the condensation of citrulline and aspartate to arginosuccinate, which is then split into arginine and fumarate. The nitrogen added in the conversion of citrulline into arginine does not come from ammonia, but from aspartic acid, as was demonstrated in 1954 by Sarah Ratner (1903–   ). Through fumarate and aspartate the urea cycle is linked to the citric acid cycle. Similarly the citric acid cycle has been extended and many details elucidated by several investigators, particularly by Severo Ochoa (1905–   ) and Feodor Lynen. However, the fundamental steps of the two cycles still stand. Moreover, the introduction of the notion of the cyclic character of cell reactions is now solidly established by experiment. Meyerhof had envisaged, as mentioned before, the possibility of the cyclic nature of carbohydrate metabolism, but the experiments were performed under inadequate conditions and later substituted by the Cori cycle.

The notion had a catalytic effect on the advances in intermediary metabolism. In the following decades a vast number of cycles in intermediary metabolism were established. Thus the introduction of the concept of the cyclic nature of cellular reactions stands as a milestone in the history of biochemistry.

In the more than 50 years of his intensive research activities Krebs has enriched biochemistry in many ways. For reasons mentioned at the beginning of this chapter a detailed description is impossible; the work of Krebs and his personality will be described in a book by Professor Frederic L. Holmes, which is in preparation (personal communication).

Krebs was an excellent teacher. His lucid and penetrating mind attracted many gifted students and he trained a whole generation in biochemistry; his pupils came from all over the world. One of his pupils and friends, Hans L. Kornberg, a brilliant investigator who also came originally from Germany, has vividly described Krebs' qualifications as a teacher (Kornberg, 1968). Krebs is an able and thoughtful organizer and administrator. This became particularly apparent when under his chairmanship a new institute of biochemistry was built in Oxford. Although retired, he is still fully active and productive and spends much time in his laboratory.

Krebs was born in Hildesheim, the son of Georg Krebs, an ear, nose, and throat surgeon in that city. His mother was Alma, *née* Davidson. His father came from Gleiwitz in Silesia. He was the only child. His ancestors, as far as one can trace them, lived in Silesia since about 1760. His mother belonged to the Jewish families who had lived there since the early eighteenth century and enjoyed relative freedom and many privileges. The families of Otto Meyerhof and Carl Neuberg belonged to this small group and there were many intermarriages, as mentioned before. After having attended the gymnasium in Hildesheim, Krebs studied medicine at the universities of Göttingen, Freiberg, and Berlin. He received his M.D. degree in 1925. He spent about a year in Peter Rona's laboratory, which was the chemistry department of the Institute of Pathology at the University of Berlin. It was the only department at the university where the new trends in biochemistry were taught and where many well-known biochemists were trained (among them Rudolf Schoenheimer, Fritz Lipmann, Karl Meyer, Ernst B. Chain, and Hans H. Weber). Krebs joined Otto Warburg in 1926 and stayed with him until 1930, when he returned to clinical work. The investigations on the urea cycle were carried out in the medical clinic of S. J. Thannhauser at the University of Freiberg.

In 1933 he left Germany and joined Sir Frederick Gowland Hopkins at the Institute of Biochemistry in Cambridge. In 1935 he was appointed lecturer in pharmacology at the University of Sheffield and later became professor there. In 1954 he was appointed Whitley Professor of Biochemistry at the University of Oxford. He received the Nobel Prize for his work on the citric acid cycle in 1953. In addition to numerous honors, he was knighted in 1958. After his retirement he became the head of the Metabolic Research Laboratory at the Nuffield Department of Clinical Medicine in Oxford. In 1938 he married Margaret Fieldhouse of Wickersley, Yorkshire. They have two sons, Paul and John, and a daughter Helen.

Coming from a family that had lived for two centuries in Germany, it is not surprising that Krebs had no ties to his Jewish heritage. However, in discussions with the author over many years—they worked at the same places for a period of about six years, first in Goldscheider's clinic, then in Rona's laboratory, and finally in the Kaiser Wilhelm Institute of Biology—Krebs showed more real interest in the Zionist movement, its aims and roots than most other Jewish colleagues. The rise of the Nazis to power led to a strong reaction. He identified himself completely with the Jews. After his move to England he seriously began to consider joining the Hebrew University in Jerusalem. He had several talks with Chaim Weizmann, who had at that time great plans to build a great research center in Rehovot, where he hoped to attract at least some of the illustrious scientists who were forced to leave Germany. The opening ceremony of the first institute of chemistry in 1934, presided over by Willstätter, was mentioned before. Weizmann was most anxious to have a second building for biochemistry, since these two fields have many common interests and seemed to be an appropriate nucleus for the further development of the center. Weizmann invited Krebs and the author to visit him in Rehovot in the spring of 1936 to discuss the building of a biochemistry institute in which a group of biochemists was interested. Krebs was deeply impressed by the enthusiastic spirit prevailing in Palestine, the remarkable atmosphere in the *kibbutzim*, the great achievements of the Jewish community in a short period of time under the most trying circumstances. Shortly after the visit serious riots started with the Arabs. The plans were first postponed and finally abandoned in view of the increasing threats of war in Europe and the strong pro-Arab policy of the British government.

After the war Krebs became very active and helpful in building up the Hebrew University in Jerusalem. He visited Israel many times, frequently with Margaret. Through his great scientific competence and his broad experience in organizing research and administration of university affairs, he made many invaluable contributions to the development of the university. He received the *Doctor honoris causa* of the university in 1960. He also became an honorary fellow of the Weizmann Institute in 1972, considered to be the highest scientific honor in Israel.

Among his many personal qualities may be mentioned his absolute reliability, his sound and fair evaluation of people, his spirit of cooperation, his still youthful devotion and enthusiasm for science. He is a most faithful friend. He is genuinely modest and dislikes pomposity. He is thoughtful and rarely acts impulsively or without serious reflection. But he has the courage to present his views, *suaviter in modo, fortiter in re.*

B. SEVERO OCHOA (1905–   )

Severo Ochoa was considered by Meyerhof to be his most brilliant pupil, as Meyerhof told the author on several occasions. He was deeply impressed by Ochoa's lucid and penetrating mind, by his rapid grasp of the essentials of a problem, by his personality, originality, and intuition. He predicted that Ochoa would become one of the truly great leaders in biochemistry of this generation because of his enthusiasm, his charisma, and his many leadership qualities. It was for Meyerhof a source of deep satisfaction and great pride to follow Ochoa's brilliant achievements in the 1940s and the rapid rise of the recognition, admiration, and affection of his colleagues. Meyerhof's great and genuine affection for Ochoa and his wife Carmen, and their friendship, has been described before. Ochoa always recognized the deep impact that his beloved and admired teacher Meyerhof had on his thinking. Francisco Grande and Carlos Asensio (1976) in their article, "Severo Ochoa and the Development of Biochemistry," quote Ochoa's statement: "Meyerhof was the teacher who most contributed to my formation and the most influential in directing my life work."

Meyerhof unfortunately did not live long enough to see Ochoa's great triumphs and brilliant achievements in molecular biology. Ochoa received, in 1959, the Nobel Prize for the discovery of an enzyme that makes RNA (ribonucleic acid). He shared the prize with his friend and one-time pupil Arthur Kornberg (1918–   ), who shortly afterward and independently discovered the enzyme that makes DNA (deoxyribonucleic acid). The two nucleic acids, DNA and RNA—symbols with which every high-school student is familiar—are the essential cell components for the expression and transmission of genetic information.

The establishment of the tridimensional structure of DNA by F. Crick (1916–   ), J. Watson (1928–   ), and M. Wilkins (1916–   ), for which they received the Nobel Prize in 1962, was the crowning of the investigations of many illustrious scientists who for a century had worked on the exploration of the structure of this giant molecule. However, two types of macromolecules—proteins as well as nucleic acids—are equally essential for life. Without the former virtually no reaction takes place in living cells. Proteins take part in building the nucleic acids and in carrying out the instructions encoded in DNA by a complex molecular mechanism; nucleic acids control the synthesis of the proteins. Ochoa's discovery of the enzyme polynucleotide phosphorylase, an enzyme that makes RNA compounds, and Kornberg's discovery of the enzyme that synthesizes DNA present a major breakthrough in the history of genetics. Their ingenious achievements opened the way to the analysis of the dynamic features and of the molecular events in the genetic mechanisms.

The biological function of the two nucleic acids is different. DNA is the main component of the chromosomes in the cell nucleus and plays an essential role in cell growth and in the transmission of hereditary characteristics. RNA is mostly located in the ribosomes and in the cytoplasmic fluid and is essential for protein synthesis (ribosomes are submicroscopic organelles that are either free or attached to the membranes of the endoplasmic reticulum). Polynucleotide phosphorylase catalyzes the synthesis of high-molecular-weight polyribonucleotides from nucleotide diphosphates with release of orthophosphate. The reaction requires magnesium ions and is reversible. It can be formulated by the equation

$$n\text{X} - \text{R} - \text{P} - \text{P} \overset{\text{Mg}^{2+}}{\rightleftharpoons} (\text{X} - \text{R} - \text{P})_n + n\text{P}$$

where R stands for ribose, P—P for pyrophosphate, P for orthophosphate, and X for one or more bases, including among others adenine, hypoxanthine, guanine, uracil, or cytosine. In the reverse direction the enzyme cleaves polyribonucleotides to yield ribonucleoside diphosphates.

The discovery of polynucleotide phosphorylase started intense activity on a large number of problems in the laboratory of Ochoa, who had attracted a group of highly qualified collaborators from all over the world. One of the most fundamental questions involved how the four-letter language of DNA and RNA is translated into the 20-letter language of proteins; that is, how can the four bases of polynucleotides direct the synthesis of proteins? Most proteins consist of a large number of 20 different amino acids in varying sequences; the amino acid sequence of each protein is unique for that particular protein. The set of instructions contained in DNA, that expressed through the manufacture of a given protein, is known as the genetic code. Cells contain three kinds of RNA: ribosomal (rRNA), transfer (tRNA), and messenger RNA (mRNA). mRNA is the template for protein synthesis; it itself is synthesized in the nucleus on DNA as a template. tRNA carries amino acids in an activated form to the ribosomes for assembly of the polypeptide chain with an amino acid sequence that is determined by the sequence of nucleotide bases in mRNA. The ribosomes, made up of various kinds of rRNA and a great number of protein subunits, function together with mRNA as the protein assembly line. Without going into details, it may be mentioned that a group of three adjacent bases (called a codon) specifies an amino acid.

Ochoa's group started their work on the genetic code in 1960. Before anything was published Marshall Nirenberg (1927–   ) announced at the International Congress of Biochemistry in Moscow, in 1961, that polyuridylic acid functioned as mRNA in a cell-free protein-synthesizing system and led to the synthesis of polyphenylalanine. It was the first public announcement of a breakthrough in the understanding of the genetic code and it produced great excitement. A month later Ochoa and Nirenberg reported on their results at a meeting of the New York Academy of Medicine, at which Ochoa presented the codons for 11 amino acids. In the next few years the studies of the two groups resulted in the elucidation of the main elements of the genetic code. This pioneer work was further strengthened by Gobind Khorana (1922–   ), Charles Yanofsky (1925–   ), and many others. Under the inspiring leadership of Ochoa, his group greatly extended the knowledge of the mechanisms controlling the expression of the genetic message, including the mechanism of replication of some RNA viruses, and various aspects of protein synthesis. It is impossible in this context to mention all the brilliant collaborators who helped Ochoa in his work, or the many other brilliant scientists all over the world who contributed to the field of molecular genetics.

Although Ochoa's fame is closely associated with his contributions to molecular biology, his achievements in the field of intermediary metabolism are of equal brilliance. In the 1940s he significantly extended the knowledge of the Krebs citric acid cycle by the elucidation of several essential steps. He isolated and purified several enzymes of the cycle and thereby succeeded in clarifying the mechanism. The oxidation of pyruvic acid, which plays a central role in intermediary metabolism, requires thiamine pyrophosphate, as was shown by Ochoa in collaboration with R. A. Peters (1869–   ). In the reaction pyruvic acid loses CO_2 to form acetyl CoA (CoA stands for "coenzyme for acetylation"). Lynen had just isolated acetyl CoA and Ochoa demonstrated that it reacts directly with oxalacetic acid to form citric acid, clarifying one of the most obscure steps of the cycle. Ochoa, in collaboration with his associate Joseph Stern, isolated the enzyme citrate synthetase, which catalyzes the formation of acetyl CoA and oxalacetate. The enzyme was crystallized by Ochoa's group; it was the first enzyme of the citric acid cycle to be crystallized. Another step of the citric acid cycle was clarified when Ochoa studied the enzyme isocitric dehydrogenase; it catalyzes the oxidation of isocitric acid, which is then decarboxylated and forms α-ketoglutaric acid. Another important enzyme was discovered by Ochoa, Alan Mehler, and Arthur Kornberg in 1948 and was referred to as "malic enzyme." Harland Wood and Chester Werkman had made the important discovery that some microorganisms utilize carbon dioxide for the synthesis of compounds related to those in the Krebs cycle. The mechanism of this "Wood–Werkman reaction" remained unexplained. The malic enzyme fixes carbon dioxide to pyruvic acid, producing malic acid. The reaction proved to be reversible, producing pyruvic acid from malic acid. The enzyme, like other enzymes discovered later, serves to replenish intermediates of the citric acid cycle; it is of importance in the intermediary metabolism of animal and bacterial cells. Ochoa was also the first to estimate the amount of ATP formed by the citric acid cycle from glucose. He found that 36 molecules of ATP are formed from 1 molecule of glucose oxidized. This strikingly demonstrated the efficiency of the main function of the citric acid cycle, the supply of energy to the cell, since the energy stored in ATP is the main source of energy for a vast number of cellular reactions. Ochoa made many other important contributions to fatty acid metabolism, photosynthesis, and other processes, but the few examples mentioned may suffice to illustrate his important role in the development of biochemistry.

In addition to the Nobel Prize, Ochoa has received many distinctions. He is a member of many leading academies and has a great number of honorary doctoral degrees. He was president of the International Union of Biochemistry. He is still actively engaged in research and takes part in many international meetings and symposia all over the world; he gives lectures and plays an active and important role. Carmen is a faithful companion on all his trips and few couples enjoy such a high degree of popularity and affection as Severo and Carmen do.

The admiration and affection of Ochoa's many friends, pupils, and colleagues from all over the world found a forceful expression on the occasion of his seventieth birthday, celebrated by symposia of an extremely high standard and by magnificient festivities, in Barcelona and Madrid, in September 1975. Many of the most celebrated biochemists of this generation, among them 10 biochemists who are Nobel laureates, took part in this stirring event, to pay tribute to his inspiring leadership, his dynamic personality, and his great contributions. A book entitled *Reflections on Biochemistry* (Kornberg et al., 1976) was published in his honor and all the publications of his laboratory were reprinted and presented to him in three volumes, each about 1000 pages thick. An Institute of Molecular Biology carrying his name was inaugurated on this occasion in the presence of the then prince and princess of Spain, the present king and queen.

Severo and Carmen Ochoa are passionate lovers of music and art. They have been regular subscribers to the opera; they love concerts. Art and music, and visits to museums, cathedrals, and historic monuments are for Severo not just a pleasant relaxation—they are deep emotional experiences and form an integral part of his life. The author has made many extended trips with the couple, to Greece, Israel, Sicily, Italy, and Spain; a short summary is published in the article, "Highlights of a Friendship," in *Reflections on Biochemistry*. Both Severo and Carmen have always been strong and devoted friends and supporters of Palestine and, after the establishment of the state, of Israel. They have repeatedly visited Israel and have many good friends in the Weizmann Institute. They regularly attend the annual Weizmann Dinner in New York.

Ochoa was born in Luarca, Spain. He is the son of Severo Ochoa, a lawyer and businessman, and Carmen de Albornoz. Educated at Malaga College, he received his B.A. degree in 1921. His interest in biology was greatly stimulated by the famous neurobiologist Santiago Ramón y Cajal. He went to the medical school at the University of Madrid and obtained his M.D. in 1929. He immediately joined Otto Meyerhof at the Kaiser Wilhelm Institute of Biology in Berlin–Dahlem. He stayed with Meyerhof until 1931, when he returned to Madrid as lecturer in physiology and biochemistry. In 1931 he married Carmen Garcia Cobian. The Spanish Civil War ended all possibilities of scientific work and Severo and Carmen decided to leave Spain. They went in 1936 to Meyerhof in Heidelberg and moved to England in 1937. There Ochoa worked with Sir Rudolph Peters in Oxford. But again, the outbreak of the war made scientific work almost impossible. Ochoa went in 1941 to the United States. There he joined Carl and Gerty Cori at Washington University in St. Louis. In 1942 he moved to New York and joined New York University. He became professor of biochemistry and chairman of the department in 1954 and stayed there until his retirement. He then joined the Roche Institute of Molecular Biology in Nutley, New Jersey, where he is still fully active and highly productive.

C. RUDOLF SCHOENHEIMER (1898–1941)

In spite of the tragically short period of his scientific career, Rudolf Schoenheimer's pioneer work and ideas left an impact on biochemistry, biology, and medicine so profound and an imprint so far-reaching and permanent that he must be considered as one of the foremost leaders in the history of biochemistry. His genius was early recognized and admired by many colleagues during his lifetime, and the subsequent developments continued to demonstrate even more strikingly the real revolution he had initiated.

Of his many achievements, two stand out as real milestones with which his name will be forever associated in history. First is the brilliantly conceived and ingeniously executed use of isotopes as tools for the analysis of many important, previously unapproachable problems of intermediary metabolism. The use of isotopes has become indispensable and almost routine in a wide variety of fields of biology and biochemistry. Second is the unequivocal evidence that all cell constituents, small or large, even proteins and other macromolecules, are in a dynamic state; that is, they are not in a static condition, but are continuously broken down and rebuilt. This notion had, and still has, a profound effect on many problems and on biological thinking in general. There were some biologists and biochemists who had assumed this idea on the basis of theoretical considerations, but it was the classical contribution of Schoenheimer, in collaboration with his associates David Rittenberg and Sarah Ratner, to bring the clear and unequivocal evidence for this fundamental notion. A superb presentation of this concept and the basis on which it was built may be found in Schoenheimer's Dunham lectures (1942), which were given shortly after his death. He had prepared the draft, but after his death Clarke, Rittenberg, and Ratner revised it and the lectures were presented in this form.

Schoenheimer was born in Berlin. He attended the humanistic gymnasium and studied medicine at the University of Berlin, where he received his M.D. degree in 1922. He spent a year at the Moabit Hospital in Berlin. There he became interested in biochemistry in general and particularly in cholesterol in connection with the problem of atherosclerosis. In order to obtain a solid training in organic chemistry he joined the Department of Chemistry of Karl Thomas in Leipzig and spend three years there. He then spent half a year in Rona's laboratory to fill the gaps in his knowledge of physicochemistry and biochemistry. In 1926 he joined the brilliant pathologist Ludwig Aschoff, who had a marked influence on his scientific development. He eventually became the head of the division of physiological chemistry in Aschoff's institute. He again took up his studies on cholesterol. He belongs to the group of great pioneers who contributed to the elucidation of the structure and function of this compound. In 1933 he was forced to leave Germany and went to the United States where he spent a year at the University of Chicago in 1930–1931. On the invitation of Hans T. Clarke he joined the biochemistry department of Columbia University Medical School.

At first he worked on fatty acids, a field of intensive investigation for decades by Hofmeister, Embden, Knoop, Dakin, and many others. Knoop had used marked substances, but they had different physical and chemical properties than the unmarked ones and were inadequate for answering the questions raised.

In 1932 Harold Urey, at Columbia's chemistry department, had discovered deuterium, the isotope of hydrogen. Schoenheimer had encountered the idea of using isotopes in Freiburg through Georg de Hevesy (1885–1966). Hevesy had applied radioactive isotopes as tracers in botanical studies in 1923 to observe the distribution of lead in bean plants. But at that time few elements were available that could be used as isotopes in biological research. Urey's deuterium provided an ideal opportunity to label organic compounds without changing their chemical properties. Clarke and Urey arranged for David Rittenberg (1906–1970), a physicochemist from Urey's laboratory, competent in the determination of stable isotopes, to join Schoenheimer in Clarke's laboratory. From their association developed the whole field in which stable isotopes were used as labels for the investigations of many important reactions in intermediary metabolism, which were inaccessible with ordinary compounds. Very soon pertinent new and exciting information resulted from this work in experiments on fatty acids. The data indicated that a rapid interchange took place between components of the diet and those of the tissues. The reactions involved not only the replacement of chemically identical fatty acids but rapid transformations, saturation and desaturation of the fatty acids, their degradation and elongation, and their reduction to alcohols.

Soon afterward, in 1935, a stable isotope of nitrogen, ^{15}N, became available. Schoenheimer and his colleagues applied this isotope to studies of amino acid and protein metabolism. Amino acids were synthesized from isotopic ammonia and added to the diet of adult rats—that is, with tissues in nitrogen equilibrium. Many exciting results were obtained in a relatively short period. Other observations, such as the Krebs urea cycle or the transamination catalyzed by transaminase, as shown by A. E. Braunstein and his associates, were confirmed by this new means. The pathways of the formation of creatine and hippuric acid were elucidated. Most important, the amino acids were found to be rapidly and extensively incorporated in tissue proteins. These investigations, performed with Rittenberg and Ratner, as mentioned before, were first presented by Schoenheimer in his Harvey lecture (Schoenheimer, 1937). They led to the evolution of the notion of the dynamic state of body constituents in the form presented in the Dunham lectures.

At the time of his death Schoenheimer was surrounded by a large group of highly qualified associates who were able to continue and to extend his magnificent pioneer work. David Rittenberg and Sarah Ratner have been mentioned; among many others were G. L. Foster, Konrad Bloch, Hans de Witt Stetten, A. S. Keston, Karl Bernhard, Heinrich Waelsch, and Samuel Graff. In the 1940s they trained a whole new generation in the use of isotopes. This was one of the important features by which the laboratory became one of the leading centers in biochemistry. An isotopic radioactive carbon (^{14}C) with a very long lifetime was prepared in 1940 by Samuel Ruben (1913–1943) and Martin D. Kamen (1913–   ), but it became available only after World War II. Its ready availability and ease of determination (first with the Geiger counter and later by scintillating spectrometers), made its use widespread and opened completely new approaches to a vast number of various fields and problems previously considered to be beyond experimental approach.

Schoenheimer had an extraordinary, imaginative, and original mind. He had a deep and passionate devotion and love for science. These qualities, combined with a rare enthusiasm, with a personal warmth and understanding for his collaborators, attracted a great number of highly talented and gifted investigators. He had an unusual ability to inspire and lead a large group. He was a forceful and dynamic personality with a wide range of interests. Whenever an idea attracted his attention he became completely absorbed by it and his mind worked with a rare and amazing intensity to realize its implications or to find a solution. Unfortunately he suffered from recurrent deep depressions. But otherwise he was a cheerful, gay, and fascinating fellow and at parties he usually became the center of attention. His suicide, at the age of 43, at the pinnacle of his creative activities came as a terrible shock not only to all his collaborators, to Hans Clarke personally, to Columbia University, but to the whole scientific community, which realized that it had lost one of its truly great and promising leaders.

Schoenheimer's family had lived for generations in Germany and was deeply attached to its culture. It had lost its interest in the Jewish community and Jewish heritage. His father, Dr. Hugo Schoenheimer, was born in Leipzig in 1867; he was a physician in Berlin and married Gertrud, *née* Edel, the daughter of a physician. The father was passionately interested in German literature. Rudolf married Salome Glücksohn, an embryologist, in Freiburg in 1932.

In spite of his family background, Rudolf and his older brother Fritz, who died in 1976 at the age of almost 80 in New York, vigorously reacted to the growing anti-Semitism in Germany by a proud identification with their Jewish heritage, as did quite a few people of the younger generation. In their school years they joined the Jewish youth organization, *Ivriah*. After World War I, in which Rudolf participated actively as a soldier, he reacted very sharply to the explosive and violent anti-Semitic movement in Germany. He became firmly convinced of the need to find a solution by establishing a Jewish homeland where Jews could develop their creative abilities in dignity and freedom, and he joined the Zionist Student Organization. He visited Palestine in 1926. Although at that time there was no possibility of developing his scientific plans there, he remained a faithful and devoted supporter of the movement for a Jewish homeland. A few weeks before his death he spent a week in Woods Hole, where he was invited to give a lecture. In a long discussion about the future of Palestine after the war he told the author that he had decided to visit Palestine after the war as soon as it was feasible, to see whether and how he could help build up science in that country.

D. ERNST B. CHAIN (1906–   )

In the whole history of medicine few achievements can match that of Ernst B. Chain. His success in preparing the first efficient antibiotic, penicillin, produced a real revolution in medicine. The use of penicillin, some of its chemical modifications, and the many other types of antibiotics developed as a result of Chain's success has virtually wiped out the most dreaded bacterial epidemics, such as plague, cholera, typhoid, bacterial pneumonia, and many frequently fatal streptococcal infections, such as scarlet fever. History is full of reports of how whole countries, or even parts of a continent, were devastated by epidemics. Chain's contribution initiated a new era in medicine. The number of people whose lives were saved by antibiotics in the past few decades is probably several times as large as that of the people killed during World War II (55 million). The genius of Paul Ehrlich, the father of chemotherapy, searching for compounds capable of destroying pathogenic bacteria without hurting the host organism, could not have foreseen the almost incredible efficiency and the wide scope of antibiotics in the fight against bacterial infections. In Ehrlich's time biochemistry was, of course, in its infancy. Chain is not only an excellent and competent biochemist; in addition he has a rare combination of abilities; not only does he explore the chemical properties of biologically active and important compounds, but he makes every effort to use his discoveries for the benefit of medicine. Thus the success with penicillin was not incidental. It was the outcome of his fundamental approach to his research. The proceedings of a symposium organized by the Royal Society in honor of Chain's birthday have been published in a book edited by D. A. Hems (Hems, 1977). It could not have a more appropriate title: *Biologically Active Substances—Exploration and Exploitation.* Chain has made many important contributions to the exploration of biochemical problems, but it has always been his aim, supported by his enthusiasm and dynamism, his vision and ingenuity, to exploit the new knowledge for practical applications.

Chain was born in Berlin. His father, Dr. Michael Chain, was a chemist and industrialist. He had immigrated to Germany from Russia. Ernst attended the *Luisengymnasium.* He was interested in chemistry in his school years—an interest that was probably stimulated by visits to his father's laboratory and factory. He also became an accomplished piano player. When he finished the gymnasium he was not sure whether to study music or chemistry, but finally decided for the latter. He studied chemistry at the University of Berlin and received his Ph.D. in 1930. Biological problems were always his primary interest and he joined Rona's laboratory, where he spent almost three years and worked on enzymes, particularly on esterases. When Hitler became chancellor Chain immediately left Germany and went to England. After a short stay at the University College Medical School in London, where he found the research facilities inadequate for his research plans, he went to Cambridge and joined Sir Frederick Gowland Hopkins, for whose personality and scientific stature he had a great admiration. There he worked on the inhibition of alcoholic fermentation and glycolysis by snake venoms in cell-free extracts and found that the inhibition was caused by the destruction of a coenzyme by hydrolysis. In 1935 he was invited by Sir Howard Florey to join the Sir William Dunn Institute of Pathology in Oxford and to build up a biochemical laboratory. Florey had no biochemical training, but he fully realized the great importance of biochemistry for experimental pathology.

In Oxford Chain explored the action of lysozyme, an enzyme with bacteriolytic action occurring in tears, nasal secretions, and egg white. In 1938 Chain and Florey became interested in investigating antibacterial substances produced by microorganisms. This problem was of special interest to Chain, since at that time a war with Nazi Germany became a serious threat and Chain hoped to prepare something useful for the treatment of infections, the cause of many deaths in wars. Antagonisms between bacteria—that is, the production of substances by one kind of bacteria toxic for others—have long been known. Louis Pasteur had observed in 1877 that anthrax bacilli were destroyed by other bacteria. Other antagonisms were observed and some attempts were made to isolate these substances, but without success. In 1929 Sir Alexander Fleming (1881–1955) had observed the bactericidal action by a substance produced by a mold, *Penicillium notatum*; he called it *penicillin*. He hoped it could be isolated and would be useful, but he could not work on this problem since he was not a chemist. Attempts at isolation by chemists were unsuccessful.

For several reasons Chain selected penicillin as a promising source for isolating and purifying a substance with the desired bactericidal activity. By dialysis experiments he had shown that penicillin was not a protein, but a substance of low molecular weight. A great variety of other properties had to be established before it was possible to proceed with purification, but here his competence and experience with modern methods of biochemistry proved to be decisive. The work required, moreover, the help of a group of chemists, pathologists, and microbiologists. By his enthusiasm and his leadership qualities he attracted enough qualified people for the project, with the strong support and encouragement of Florey. The potency of the penicillin preparation increased steadily with the use of a variety of steps. Experiments on mice proved convincingly the high degree of efficiency that the preparation had reached. In 1941 penicillin was first applied to humans. But unfortunately not enough material was available, since the Medical Research Council turned down the request for the culture media required for growing the molds. Florey flew to the United States, and thanks to the strong and generous support of A. N. Richards he received large amounts of culture media. It became possible to prepare the necessary amounts of pure and crystalline penicillin. The results were fabulous. Previously hopeless cases were cured in no time. The triumph of medicine in the fight against bacterial infections was so obvious and striking that recognition came quickly: in 1945 Chain, Fleming, and Florey received the Nobel Prize.

The dramatic story of penicillin did not develop smoothly. Many difficulties had to be overcome. The laboratory suffered from lack of funds. One day in 1937 the funds were so drastically cut that it was impossible to buy a glass rod. An ambitious project such as finding substances for fighting infections required a minimum of equipment and materials. Here Chain's extraordinary energy, enthusiasm, willpower, and his determination to fight any obstacle became apparent. From the Rockefeller Foundation he received a grant of $5000 for a five-year period. Today this sum may seem ridiculously small, but at that time it was a large amount and crucial for success. Obviously, the funds were just a prerequisite. It was his vision and intuition, his grasp of complex biochemical problems, his indefatigable energy, and his devotion to his work that brought about the preparation of the first antibiotic with curative properties.

Chain recognized very early the importance of elucidating the chemical structure of penicillin. He started to work on this problem in collaboration with several outstanding chemists. Similar studies were later taken up independently in the United States with great intensity, with the help of many leading chemists, institutions, and industries. Knowledge of the exact structure and its properties did facilitate, as anticipated, a slight modification of the preparation, which increased the range of its applicability and presented its destruction by penicillinase, an enzyme present in bacteria. The efforts were successful and many improvements were obtained over the years.

After the war Chain had bold ideas about exploiting the success achieved with penicillin and initiating large-scale experiments, which required a pilot plant. He was frustrated by the fight with the bureaucracy and the inertia of the establishment, which was opposed to any unconventional projects and ideas. He therefore accepted an offer from Professor Domenico Marota, director of the *Istituto Superiore di Sanità* in Rome. Marotta wanted to build up a great scientific center of international reputation. He succeeded in getting the necessary funds, enabling Chain to build up a magnificent biochemistry department with modern equipment and a pilot plant for fermentations. He soon attracted a group of gifted and devoted Italian scientists and collaborators and many outstanding colleagues from all over the world spent some time in the laboratory, which offered many unique facilities. In 1948, just before leaving Oxford, Chain married Anne, *née* Beloff. She is a competent biochemist, trained by Sir Rudolph Peters and Baird Hastings, and became invaluable in superbly organizing, coordinating, and supervising the work of the large group. Much productive work in many fields of biochemistry and microbiology was accomplished during the period the Chains spent in Rome; the superb facilities combined with the large amount of brainpower brought about an extraordinary breeding ground for creative work. Chain's home became a real intellectual center in which one met interesting people from all over the world. A special additional attraction were the piano performances of Ernst, which were of professional quality. Ernst and Anne have an extremely happy family life. They have two sons, Benjamin and Daniel, and a daughter Judith. Anne is not only an extremely efficient scientist, but also a gracious hostess. Music plays a great role in their home; all three children are gifted musicians.

In the late 1950s some English scientists realized the loss their community had suffered by Chain's move to Rome. In 1958 Professor P. M. S. Blackett discussed with Chain the possibility of building at the Imperial College in London a biochemistry department with facilities comparable to those in Rome. After strong efforts the funds required, very large by English standards, were finally obtained, mainly from the Wolfsohn Foundation and the Science Research Council. The Chains moved to London in 1964. There again their scientific efforts yielded a rich harvest in many fields of biochemistry. Although now retired, Ernst continues his scientific activities with Anne, a collaboration affectionately referred to by many as the "Ernst and Anne bimolecular Chain reaction," as was recalled by Baird Hastings at Ernst's seventieth birthday party.

Because he has had the courage to voice his views—often sharply critical—frankly and forcefully, he has antagonized some people, especially bureaucrats and administration officials, but even some colleagues. Nevertheless, few will dispute his inestimable service to humanity. Few scientists have received such a large number of distinctions from all over the world—medals, honorary doctoral degrees, memberships in countless academies, and so on. He was knighted in 1969. An impressive and dignified tribute was paid him by the scientific community on the occasion of his seventieth birthday. A symposium was arranged under the auspices of the Royal Society in London. It was superbly organized by Anne with great thoughtfulness and taste. Many leaders of biochemistry attended the meeting. His friends and colleagues were the chairmen, whereas the lectures were given by his pupils. The lectures gave a vivid picture of the wide range of Ernst's interest, and the lecturers expressed in moving words their affection and their gratitude for Ernst's inspiring leadership. A special flavor was an hour of music before a festive dinner, in which Ernst played the piano, alone and with colleagues, and then his oldest son, Benjamin, also played.

In view of his Russian-Jewish ancestry it is not surprising that Chain has always been conscious and proud of his Jewish heritage, and a strong supporter of the Zionist movement. After the establishment of the State of Israel and the foundation of the Weizmann Institute in Rehovot he became deeply involved in its development. He is an honorary fellow of the institute, and a member of the board of governors, its scientific and academic advisory council, and the executive council. He has frequently played a key role in many decisions. Because of his persuasive, ebullient, and dynamic personality, his enthusiasm, and his scientific eminence, many members of the board—primarily businessmen, industrialists, and bankers

—pay great attention to his judgment and to his views. The appreciation of his many outstanding services to the institute and of his deep devotion found an eloquent expression on the occasion of the meeting in honor of his seventieth birthday: The president of the State of Israel, Professor Ephraim Katzir, and the president of the Weizmann Institute, Professor Michael Sela, attended the meeting.

The four biochemists described in Chapter IV trained many investigators, and quite a few of them became leaders in the field throughout the world. The four biochemists described in this chapter were randomly selected for reasons explained in the beginning of this chapter. However, the number of biochemists who left Germany because of the Nazi regime and who have greatly enriched the field is very large indeed. Just a few additional names of international fame will be mentioned here.

Among those who left Meyerhof's laboratory in 1933 was Hermann Blaschko. His parents were good friends of Max and Hedwig Born (see II, D). He went to England and found much recognition for his pioneering work on catecholamines. He can trace his Jewish ancestry over a period of 500 years. Gerhard Schmidt, a pupil of Embden's, left Germany in 1933 and settled in the United States. His many outstanding contributions, especially in the field of nucleic acids, have found wide recognition. Max Bergmann, a pupil of Emil Fischer and director of a Kaiser Wilhelm Institute in Dresden, left Germany in 1934 and joined the Rockefeller Institute in New York. His chief interest was protein chemistry and his achievements and those of his associates are recognized as pioneer work in this field. Konrad Bloch has already been mentioned; he shared the Nobel Prize with Feodor Lynen for the synthesis of cholesterol. Fritz Lipmann is internationally known for his many important contributions to enzyme chemistry, protein synthesis, and genetics. Edgar Lederer, who had worked with Richard Kuhn in the Kaiser Wilhelm Institute in Heidelberg, went to Paris in 1933. He became the director of a large institute in Paris and his many brilliant achievements have won international acclaim. Hans Gaffron, an eminent authority on photosynthesis and an associate of Otto Warburg, was already mentioned as one of the non-Jewish scientists who left Germany because he was unable to live under the Nazi regime. Max Delbrück, a direct descendant of Justus von Liebig, left Germany for the same reason. He received his Ph.D. with Max Born in Göttingen in theoretical physics, spent some time with Niels Bohr and Wolfgang Pauli and then joined Otto Hahn and Lise Meitner in Dahlem. He left Germany in 1937 and joined the California Institute of Technology in Pasadena. He turned to biological problems and shared the Nobel Prize in 1969 with S. Luria for their work on reproduction and genetic aspects of viruses.

But in addition one should recall that after the *Anschluss* of Austria in

1938 many Jewish scientists left that country; many of them were biochemists. Ephraim Racker is one of the leaders in oxidative phosphorylation and the biochemistry of mitochondrial membranes. Max Perutz went to Cambridge and succeeded in elucidating the tridimensional structure of hemoglobin by X-ray crystallography, a milestone in protein chemistry. He shared the Nobel Prize in 1962 with John Kendrew. The Department of Biochemistry of Columbia University Medical School has two outstanding biochemists who came from Austria: Zaccharias Dische, who has made many outstanding contributions, particularly in the field of carbohydrates, and Erwin Chargaff, whose contributions to the structure of nucleic acids were important for further progress in the field, in addition to his many other achievements. Felix Haurowitz, who had worked at the German University in Prague, settled in the United States and is an authority in protein chemistry. He had spent some time with R. Willstätter, F. Hofmeister, and A. Kossel. George Hess, a biochemist and physicochemist now at Cornell University, is well known for his work in enzyme and protein chemistry. Hans Neurath, a brilliant biochemist and physicochemist, is one of the most outstanding pioneers in exploring the catalytic mechanism of enzyme activity. Henry G. Mautner has made many important contributions to the molecular biology of excitable membranes. Several outstanding collaborators of Parnas went to different countries.

These are again just a few randomly selected examples. The list could be greatly enlarged but may suffice to emphasize how world biochemistry benefited as a result of the Nazi persecution.

E. COLLABORATION BETWEEN GERMAN
AND ISRAELI SCIENTISTS

One of the most remarkable developments of the postwar period has been the establishment of collaboration between German and Israeli scientists. It is a story that offers many fascinating political, psychological, and ethical aspects; it illustrates the power of science, its ethical and spiritual force, to initiate creative and constructive collaboration between two nations that had been separated by a barrier that seemed impossible to cross: the terrible crimes committed against the Jews by Nazi Germany, the Holocaust in which 6 million Jews perished, almost half of the world's Jewry.

Konrad Adenauer, the first chancellor of the Federal German Republic, a great statesman and a forceful personality with great vision, succeeded in rebuilding in the postwar period a devastated and demoralized country, making Germany once again a great power in Western Europe. Even before the Nazi period he had been a strong supporter of a Jewish homeland in Palestine. In the 1920s he was a member of the Pro-Palestine-Committee, a group of influential Jewish and non-Jewish personalities that included political figures belonging to various parties, from left to right. Adenauer was a fierce and vigorous opponent of Hitler. In 1933 he was immediately dismissed by the Nazis as mayor of Cologne, and he and his family found themselves in a difficult and dangerous situation. A great industrialist, Dannie Heinemann, born in North Carolina of German-Jewish descent, a friend of Adenauer, advised him to hide in a monastery, since his life was in danger; he would take care of the family. When Adenauer became chancellor, he stressed, repeatedly and forcefully, that one of his chief aims would be to try to build up relations between Germany and Israel, and world Jewry in general, for moral as well as political reasons. He realized that the monstrous crimes committed in the name of Germany could not be forgotten and that it would take a long time, requiring patience and great efforts, to restore relations leading to eventual normalization. Under the forceful leadership of Adenauer and with the support of several courageous German leaders who also were appalled by the Nazi crimes, a reparations agreement was signed in 1952 between Germany and Israel. The State of Israel received $800 million, and in addition countless individual restitutions were paid for losses suffered because of Nazi actions. These amounts went very far to strengthen the young state of Israel, which was economically, militarily, and politically in a very difficult position. Adenauer was well aware that money, no matter how much, could not atone for the Holocaust, but he hoped that this action would be recognized by Israel and the whole world as a gesture of good will, a testimony to the spirit of the new Germany, its efforts to build up a progressive society and to overcome the terrible stress on the consciousness of many Germans, especially of those who had been violently opposed to the Nazi regime and deeply upset by the savage crimes.

Chaim Weizmann, a statesman–scientist, had the vision of building a great scientific center in Israel (then Palestine) long before World War I. He was extremely familiar with the history of Germany's rise to power through science and technology. He realized very early that the greatest hope for building a strong Jewish homeland was its brainpower. As mentioned before, when many leading Jewish scientists were forced to leave Germany in 1933, he hoped to attract some of them to Israel. A small chemistry department was opened in Rehovot in 1934. Unfortunately, the deteriorating political situation in the 1930s under the British Mandate, and the lack of financial support by rich Jews, even during the Hitler period, prevented the realization of his plans. Even the small institute passed through difficult times. But after the war, deeply shaken by the Holocaust, world Jewry and many rich Jews recognized Weizmann's wisdom and statesmanship, and decided to build a huge scientific center in Weizmann's name, which would serve as a living memorial for future generations of his great services to Israel. Weizmann, who died in 1952, saw the beginning, the opening of the first new building. Within a decade a huge and magnificent scientific campus was built, with a staff of brilliant, vigorous, and deeply devoted scientists, many of them first trained at the Hebrew University in Jerusalem and in the postwar period in the great institutes in the United States, which had become the world's leading scientific center. Today the Weizman Institute is internationally recognized as one of the world's great scientific centers. Fifty to one hundred scientists from leading institutions of the United States, Great Britain, France, and other countries visit there for shorter or longer periods for lectures or research; many international symposia take place there; many scientists are invited to participate in meetings and to give lectures or to do research. The institute is a shining monument to Weizman's vision as to the Jewish renaissance, of the spiritual and moral values of Israel.

In 1956 a delegation of German scientists visited the institute. They were led by Otto Hahn, then president of the Max Planck Society; Wolfgang Gentner, professor of the Max Planck Institute of Physics in Heidelberg; and Feodor Lynen, professor at the Max Planck Institute of Biochemistry in Munich. They were deeply impressed by the high quality of the research going on there, the vigor and the enthusiasm of the people, and the superb facilities. Independent of political and moral considerations, they realized the great benefits that might be derived from a collaboration between the scientists of the Weizmann Institute and German scientists. For German science, which had severely suffered under the Nazi regime and the ravages of the war, such a collaboration would provide a great and welcome stimulus. Professor Gentner worked out a memorandum, which was signed by Otto Hahn and submitted to Adenauer.

A major role in the further developments was played by Dr. Joseph Cohn. Born in Berlin, he received his Ph.D. in Heidelberg with Alfred Weber. Weizmann was, in the 1920s, a frequent visitor to Berlin, and he had met Joseph Cohn there several times. When Cohn left Germany in 1933, he joined Weizmann and thus became closely associated with his scientific plans and efforts. He has ever since devoted all his energy to help the Weizmann Institute and has decisively contributed to its growth and particularly to the efforts to establish a scientific collaboration between the Max Planck Society, the German government, and the Weizmann Institute. Through a letter fom Heinemann he was introduced to Adenauer. Adenauer immediately recognized the political and moral implications in addition to the scientific value and promised Cohn his full support. In his historic meeting with Ben-Gurion at the Waldorf-Astoria in New York, Adenauer committed himself that the German government would make a gift of 3 million marks to the institute as a gesture of friendship, a considerable sum at that time. This was the beginning of an ever-increasing support of the Weizmann Institute by the German government, industry, and foundations, among which the Volkswagen Stiftung played a major role. A committee was established under the chairmanship of Gentner (the Gentner Committee); it meets twice a year, once in Germany and once in the Weizmann Institute, to review projects and to make recommendations. Today a substantial fraction of the budget of the institute comes from Germany.

In the Weizmann Institute there was, not surprisingly, great reluctance to collaborate with German scientists or even accept money from Germany. Too many members had lost their whole families in the Holocaust. For some every German was a Nazi. Joseph Cohn was, however, strongly supported by Amos de Shalit, Gerhard Schmidt (half German, half Jew), Aharon Katchalsky, and Michael Sela, all leading members of the institute. Scientific collaboration started slowly but grew rapidly in scope and intensity. Many symposia were arranged, and many personal friendships developed. Today collaboration between German scientists and members of the institute forms an integral part of the activities.

In the mid-1960s, when diplomatic relationships were established between Germany and Israel, the first German ambassador, Rolf Pauls, was received with extremely hostile demonstrations. He took them calmly. When the author visited him in Tel Aviv, shortly after he had become ambassador, he told him how much importance he attached to the scientific collaboration between the Max Planck Society and the Weizmann Institute. He was well informed about all details. He was firmly convinced that in the long run these activities would improve the whole atmosphere and contribute to the improvement of the political relationship. After the 1967 six-day war there was a marked shift in the attitude of many Israelis toward Germany due to the strong moral and material support by the German government and much of the population. Since then the Hebrew University has also started relationships with German universities. A particularly close relationship exists between the University of Göttingen and the Hebrew University, initiated by a visit of Professor Natan Goldblum with the strong support of Manfred Eigen. Regular symposia take place in which the faculties of both science and the humanities participate. The meetings take place alternatively in Germany and Israel. When Pauls left Israel to become ambassador in Washington he had countless good friends in Israel.

In the first decade since the start of the collaboration between the Max Planck and Weizman Institutes, Wolfgang Gentner was the strongest driving force in these efforts and, as chairman of the Gentner Committee, played an instrumental role in efficiently managing the organization and distribution of the funds available, and in selecting the most promising projects. By his genuine devotion, his excellent judgment, and by his personal qualities he gained the full confidence of the leading scientists of the Weizmann Institute. In view of his invaluable contributions to the growth of the Institute he became Honorary Fellow there and was elected a few years ago to the Board of Governors of the Weizmann Institute.

Although several other leading German scientists have been extremely active in supporting the collaboration between German and Israeli scientists, space prevents a detailed description. But at least a few remarks should be made about Manfred Eigen, who played a decisive role especially during the last decade. He is the head of the Max Planck Institute of Biophysical Chemistry in Göttingen. His teacher was Eucken, a pupil of Nernst. Eigen has revolutionized physics, chemistry and biological sciences by the development of methods which permit precise recordings of very fast reactions, previously considered to be "immeasurably" fast. These methods permit one to directly measure reactions which take place in a billionth of a second or less. They are powerful tools that made possible the analysis, and thereby the understanding, of many reactions in a vast variety of fields including the analysis of many mechanisms until then unaccessible to experimental analysis. Eigen received the Nobel Prize in 1967. Few scientists of the present generation have been as greatly honored in so many ways as Eigen. By his brilliant mind and his dynamic and charismatic personality he has decisively contributed to the furthering of the close scientific and personal relations between Israeli and German scientists. His international fame and the high esteem and status he enjoys in Germany were a great help in these efforts. He is an Honorary Fellow of the Weizmann Institute and was the first German scientist to receive an honorary doctoral degree from the Hebrew University.

To illustrate the spirit and new relationship prevailing today between German and Jewish scientists, the recent Aharon Katzir-Katchalsky Memorial meeting should be mentioned. It was arranged by the Max Planck and Weizmann Institutes and held in September 1978 partly in Göttingen, partly in Braunlage, a nearby resort in the Harz mountains. It was attended not only by a great number of German and Israeli scientists, but by many illustrious scientists from other countries. It was opened by a Mozart Piano Concerto with Manfred Eigen as soloist and the Hannover Orchestra. The scientific standard of the presentations was very high. At the farewell dinner Ephraim Katzir gave a most moving speech stressing the strength and warmth of the new bonds. At the end Eigen read some parts of Heinrich Heine's *Harzreise*.

When Adenauer visited Israel a year before his death he was greatly impressed by the great achievements of the young state, by the spirit of genuine devotion of the young people, the creative spirit visible everywhere. He ignored the hostile demonstrations of a small minority. As a statesman he was neither surprised nor hurt. He received great ovations in the Weizmann Institute and became an Honorary Fellow there. He was deeply moved and expressed his great satisfaction in the spirit uniting nations and the cooperation in science and research in the service of human progress in a peaceful world that manifests itself so magnificently in the Weizmann Institute.

Since this book is devoted to the longtime President of the Leo Baeck Institute in Jerusalem, it may be appropriate to close these few remarks about the present cordial relationships between German and Israeli scientists by quoting a story which reveals in an impressive and stirring way Leo Baeck's extraordinary personality. Lamm (1964) describes the first encounter after the war between Leo Baeck and Theodor Heuss, the President of the Federal Republic of Germany, as follows: When Leo Baeck was received by Heuss, the two men shook hands and held them tightly and silently for some time. Before Baeck was standing the first citizen of the new German democracy, with head bowed, almost pale, unable to say a word. Baeck broke the silence. "If I had to decide, Professor Heuss, I would build a gigantic memorial, a monument of gratitude to the workers of Berlin and to the furniture movers." Heuss looked up. He found only these words to say: "Why? But. . . ." Whereupon Baeck answered: "I would build this memorial because when I traveled to my office on a streetcar with the Jewish star on my breast, the Berliner workers, almost daily, put their breakfast secretly in the pockets of my coat. And to the movers, because without asking anything for it, they offered to build into the emigration vans a place for valuables, in such a way that no customs officer could find them. And you asked, Professor Heuss, for a "but." Let us not speak today about this "but," because it touches the past. Let us talk today about the present and the future and about that which moves us both."

Those who still recall the period in the early decades of this century, in which the close collaboration between German and Jewish scientists led to one of the most glorious chapters in the history of science, can only fervently hope that the new era of collaboration between German and Israeli scientists will again be extremely fruitful and lead to great achievements, for the benefit of the scientists, of the two countries, and of humanity as a whole. The creative forces of the human mind, science and art, may be the most powerful factor in the struggle to overcome a sad past.

---

* I am grateful to Professor Herken for having brought this story to my attention.

Concluding Remarks: Science and Society

The triumphs of science in the twentieth century are unmatched by those of any period in history. The insights achieved into the structure of the universe, into matter, and into many mysteries of the living organism resulted in a revolution of technology that changed the life of humanity everywhere. The recognition of Francis Bacon that knowledge is power has never before been demonstrated in such a spectacular way. Nobody could have possibly foreseen at the beginning of this century the extent of the changes that would be made in such a short period of time. Jets cross the oceans in hours; at the beginning of the century it took weeks with the fastest steamboats. Via satellites we can see on television in our living rooms the reception of Nixon in Peking or that of Sadat in the Knesset in Jerusalem; we can hear and see opera performances at La Scala or at the Royal Opera House in London. Men walk on the moon, and instruments show us on television the surface of Mars at a distance of hundreds of millions of miles. Much hard and dangerous labor has been replaced by machines. The progress continues at a rapid rate in many directions. Most people are aware of the many changes made in medicine. Widespread epidemics such as plague, cholera, typhoid, and yellow fever have been virtually eliminated. Surgery and the treatment of many internal diseases have undergone profound changes. Agriculture has increased its crops many hundredfold by the use of fertilizers, pesticides, machines, and genetic engineering.

Early in this century many scientists lived in an ivory tower. Science was the great hope of mankind. The progress of science-based technology would continuously improve living conditions and eventually eliminate poverty and misery. Even the most abstract thinkers in science were sure that their work would sooner or later lead to some useful discoveries. The vision of Archimedes, Δός μοι ποῦ στῶ καὶ κόσμον κινήσω ("Give me a place to stand and I will move the universe"), was considered to mean that science would one day be able to provide men and women with unimaginable power that could be used to improve their lives. In fact, science has to some extent fulfilled its promises. Millions of people live a longer, healthier, and happier life than ever before, although the fraction is still much too small and a much larger part of mankind should take part in these benefits.

The turning point came with the destruction of Hiroshima and Nagasaki. It was a frightful shock for an uncountable number of scientists, perhaps more than for laymen unable to realize the implications. Some scientists were already upset by the idea that it was highly advanced technology that made possible the horrors of destruction of Rotterdam and Warsaw, of Coventry and Dresden, where hundreds of thousands of men, women, and children perished in the bombings. But the atom bombs used in Japan, followed soon by the development of the hydrogen bomb, have raised the specter of the destruction of all life on this planet. As is well known, today the nuclear arsenal of the two superpowers, the United States and the USSR, is large enough to destroy the planet hundreds of times over.

Several other frightening side effects resulting from the progress of technology became apparent as threats to society. The rapid growth of industry has led to an intolerable pollution of air, of lakes and rivers, even of oceans. One of the most dangerous aspects is the population explosion, due partly to the great advances made in medicine and agriculture. A vast number of books on the impact of science on society have appeared in the last decades; many meetings and symposia have taken place in which experts have discussed the features of the various problems. Many solutions have been proposed, although the problems are so extraordinarily complex that at present nobody can provide easy answers. Many problems will require a generation (or generations) until they are solved in a satisfactory way.

The topic is obviously quite removed from the chief aim of this book and will therefore not be discussed at length. However, since it is widely recognized, and has been stressed by the author, that science in Germany (and of course in other countries) in the first few decades of this century represented one of the most glorious chapters in the history of mankind, a few brief comments on its value and in its defense seem appropriate.

Science is one of the most forceful, magnificent, and valuable expressions of the creative human mind. There are other great values that have greatly enriched human life and culture: the works of Mozart and Beethoven, Goethe and Shakespeare, Michelangelo and Leonardo da Vinci, or the philosophy of Plato and Kant. The Bible is a cornerstone of Western civilization. Science has not only increased our understanding of the universe; its philosophers have greatly contributed to our moral, ethical, and spiritual values.

In earlier times there have been movements antagonistic to science for a variety of reasons. But in the last two decades wide segments of the population, even many scientists, have become not only antagonistic but extremely hostile to science for a variety of reasons: the danger of total destruction of the planet by atom bombs, for which scientists are solely responsible; damage to the environment; and the threat of widespread famine caused by the population explosion. After World War II science in the United States expanded at an amazing rate with the financial aid of the government and the enthusiastic support of Congress and the population, all of which were deeply impressed by the progress achieved during the war. The distributing agencies, the National Institutes of Health and the National Science Foundation, established high standards, although mistakes were made, as is unavoidable in any large institution growing at a rapid rate. But in the last 15 years drastic cuts in funding were made. Basic research was particularly hard hit; many invaluable investigations were severely hampered or even completely stopped. One important factor—among others—was the hostility of the population, which in a democratic society cannot be ignored. The emotional anti-science (and anti-intellectual) forces were strong and inaccessible to reason.

For a more balanced view and a better understanding, a few problems may be mentioned. The possibility that nuclear war may destroy life on the planet cannot be denied. Since 1979 is the widely celebrated centennial of Einstein's birthday, it seems appropriate to remind the reader that Einstein was a vigorous leader in the fight against the use of nuclear weapons. The March issue of the *Bulletin of Atomic Scientists* of 1979, dedicated to Einstein, is entitled *Einstein and Peace*. It recalls that, in 1946, Einstein accepted the chairmanship of a "Committee of Atomic Scientists," a group deeply concerned with the threat to mankind posed by atomic weapons. Einstein forcefully advocated a supranational "World Government" as the only effective solution to the problem. The issue also includes an impressive article by Einstein on the moral responsibility of the scientist, written in 1952. There he stresses the great intellectual value of science as well as the extraordinary benefits drived from it. He is distressed by the creation of the threat to mankind's survival, but calls on scientists to maintain their inner freedom and independence and to fight against the misuse of power. This attitude is in line with the concern which he showed throughout his whole life in a variety of political problems and public affairs, as described in a great number of biographies. His reactions exhibit a remarkable political instinct. For instance, in contrast to many colleagues, he recognized as early as 1920 the danger of the irrational anti-Semitic forces in Germany and helped Weizmann in his efforts to create a National Jewish home in Palestine. Internationally known as an ardent pacifist, he surprised many pacifists when, after the Nazis seized power in 1933, he strongly opposed the British disarmament and urged a strong rearmament, in order to save Europe from the threat of Nazi domination.

In the more than 30 years since the end of World War II many distinguished physicists have devoted much time and effort to fighting the nuclear threat: Niels Bohr, James Franck, Max Born, Victor F. Weisskopf, Hans Bethe, Werner Heisenberg, to mention just a few of them. They have been supported by philosophers, among them Karl Jaspers and Bertrand Russell, and by many scientists and intellectual leaders in other fields. Many national and international movements have been organized and vigorously continue the fight. The argument that wars have taken place since time immemorial and that a nuclear war is therefore unavoidable is not convincing. A war with a hope of victory is basically different from a war leading to certain suicide.

On the other hand, nuclear energy is the only long-range hope for the survival of mankind. Leading physicists and biologists in the United States forcefully expressed this view in a statement published in newspapers about three years ago and signed by some 80 of the most outstanding leaders of the country. Nuclear energy drived by fission has, as mentioned before, many drawbacks. Fusion, however, is the greatest hope. The resources are unlimited; the oceans may supply energy for the entire world for millions of years. Rousseau's idyllic dreams of a primitive society are too unrealistic to require comments. A modern society without the great benefits of technology and industry is unthinkable; large amounts of power are required, more than are available at acceptable prices. But the gigantic efforts and striking advances made with lasers, magnetic fields, electron beams, and other new tools justify the hope, in the view of the experts, that it will become available. It may take 50 years or more, however, before large scale production of energy by fusion will become feasible. Until then society will have to struggle with what is available. The ingenuity of the human mind is great and may bring surprises and shorten this period, since now strong financial support for this research is forthcoming.

The population explosion, made possible by the miraculous achievements of medicine and agriculture, is a serious threat. The population in India, for example, has increased from 350 million in 1945 to 600 million. Are only the scientists to be blamed? Or perhaps also the statesmen, the leaders of nations? Is the answer to abolish medicine and permit plague, cholera, and malaria to reduce the size of the population? Or should we reduce crops by omitting pesticides and fight the explosion by famine?

The side effects of the growth of industry, which has polluted air and water and severely damaged the environment, were in some cases unpredictable, in others the result of bad management, carelessness, or greed.

where necessary safeguards were sacrificed in the name of greater profit. There are many serious and difficult problems involved. Their solution requires time and hard work. But who is in a position to find answers, if not the scientists, engineers, and experts? Certainly not the business manager, who is unfamiliar with the complexity of the problems involved. There are other factors, psychological, emotional, and social. The air pollution in the United States amounts to about 150 million tons per year; 60% is due to automobiles. How many owners of cars would be willing or able (due to special conditions) to give up their comfort in order to contribute to the fight against pollution?

The three examples mentioned may help the reader understand some of the dilemmas and problems facing scientists and their relationship to society. Scientists have provided an almost unbelievable source of power. But they cannot decide how the power is used and in many cases other people may be more qualified to make the right decision. Planck and Einstein, Rutherford and Bohr, the whole galaxy of geniuses whose work paved the way for the use of nuclear energy, and thus the production of the atom bomb, should not be blamed for its misuse, "the monstrous perversion of science" as Churchill called it. Francis Bacon, who stressed the power of knowledge, also saw its dangers. Scientists cannot replace statesmen and the other political and intellectual leaders of society. They can only advise them, provide them with the necessary information for their decision. The final decision depends on the wisdom of society and its leaders.

In our technological society science has become an increasingly important and necessary factor. We need more, not less, science to overcome the vast problems facing society, not only of the technological kind, but also the social and economic problems created by the rapid progress of technology. Mankind is in the midst of revolutionary changes of unprecedented dimensions. Science has an extremely important role to play, but it alone cannot solve the problems. However, drastic changes have taken place in the scientific community. Today it knows that the time of the ivory tower has passed, that it must share the responsibility for the results of its research. This awareness, which few scientists had before, was greatly promoted by the atom bomb, as expressed by the Franck Report. Today physicists and other scientists fight with all their power to prevent a nuclear war. The Pugwash Conferences and the *Bulletin of the Atomic Scientists* are signs of this. The efforts must be multiplied by the strong support of many more scientists and by a vast number of enlightened leaders. Many scientists work hard to explore all ways and means to fight population explosion and the problems are discussed at various international meetings. Every reader of newspapers knows of the strong fight for the protection of the environment. But the transition of scientists from people in an ivory tower to active and leading members of society, who share the responsibility of protecting society against unwanted side effects, is a slow process. The adjustment takes time and not every scientist is qualified. This latter factor is not important, since today there are hundreds of thousands of scientists. Planned collaborative efforts are required for attacking large-scale efforts, especially when the causes are well known. Chosen leaders of the scientific community in the various academies must and do act as advisers to governments. This requires great efforts and it will take much work and time to educate populations, to carry out carefully worked out programs. Planning should not be applied to basic research, without which no real progress is possible. There planning is senseless. The theory of relativity was formulated by an unknown clerk in the patent office in Bern. He would never have received a cent from any study section reviewing applications for funds. Lavoisier's explanation of oxidation, replacing the myth of phlogiston, would have shared the same fate, since all leading scientists of that era—Priestley, Cavendish, and Scheele, for example—died believing in phlogiston. A dramatic example of the failure of planned research, when applied to basic problems, is the project to develop a cure for cancer. Five billion dollars were spent in a few years, and finally one of the strongest supporters of the effort admitted failure. A cure for cancer requires basic knowledge of the genetic factors underlying the disease. This knowledge is not available. It can only be acquired by completely free and unhampered research. Complete freedom is the prerequisite of pure science, which depends on the ingenuity and imagination of scientists.

At one of the annual dinners in memory of Weizmann in New York, the key address was written by Pierre Mendès-France. (Since he had become prime minister of France the day before, the address was read in his name.) He praised the wisdom and great vision of Weizmann to have been one of the first great statesmen to recognize the importance of science for the building of a viable state is Israel, a small and poor country without natural resources. He added that in his view the greatness of future presidents will be judged according to the strength of their support for science. Weizmann was well aware, and expressed it forcefully and repeatedly, that there are great human and spiritual values other than science. It was his hope that if a great center of science in Israel became a reality, it would contribute to building bridges between science and other human values. He hoped that Jews with their inherited respect for ethical values would perhaps be particularly qualified for such a function. The collaboration between the scientists of the Max Planck Society and the Weizmann Institute mentioned before appears as a symbol of the moral and ethical power of science to establish the first bridge between nations in an extraordinarily difficult situation.

Scientists can be extremely proud of their great achievements. However, gigantic tasks are still before them on all fronts: to explore the unknown, to cope with the many unsolved problems. Their efforts may help in solving many of the ills from which humanity is suffering. Science with all its accomplishments is still at the beginning of the road, as so forcefully expressed by Einstein (p. 103). Scientists must be in the forefront of the struggle; they can never afford to relax in their efforts to explore the unknown. They should be guided by the words of Goethe:

> *Zu neuen Ufern lockt ein neuer Tag*
> ("To new shores calls a new day").

References

Anrep, B. von (1879): Über die physiologische Wirkung des Cocain. Arch. f.d. ges. Physiol. *21*, 38.

Baeyer, A. von (1915): Special volume of Die Naturwissenschaften, dedicated to Adolf von Baeyer on the occasion of his 80th birthday, Naturwissenschaften *1915*, 599, section 2. Springer, Berlin.

Beyerchen, A. D. (1977): Scientists under Hitler. Yale University Press, New Haven, Conn.

Blumberg, S., and Owens, G. (1976): Energy and Conflict. The Life and Times of Edward Teller. G. P. Putnam's Sons, New York.

Bohr, N. (1934): Atomic Theory and the Description of Nature. The University Press, Cambridge, Mass.

Bohr, N. (1949): In: Albert Einstein: Philosopher–Scientist (P. A. Schilpp, ed.), pp. 199–241. Library of Living Philosophers, Evanston, Illinois.

Bohr, N. (1958): Atomic Physics and Human Knowledge. 101 pp. Wiley, New York.

Bonhoeffer, K. F. (1934): Obituary notice. Chem. Z. *58:118*, 205–206.

Born, M. (1949): Natural Philosophy of Cause and Chance. Clarendon Press, Oxford.

Born, M. (1965): Von der Verantwortung des Naturwissenschaftlers. Nymphenburger Verlagsbuchhandlung, Munich.

Born, M. (1968): My Life and My Views. Charles Scribner's Sons, New York.

Born, M. (1969): Albert Einstein, Hedwig und Max Born. Briefwechsel 1916–1955. Nymphenburger Verlagsanstalt, Munich.

Born, M. (1975): Mein Leben. Nymphenburger Verlagsanstalt, Munich.

Broglie, L. de (1953): The Revolution in Physics. Noonday Press, New York.

Chain, E. (1946): The Chemical Structure of the Penicillins. Nobel Lecture. Nobel Lectures Physiology and Medicine 1942–1962, pp. 110–143. Elsevier, Amsterdam, London and New York.

Chamberlain, H. S. (1899): Die Grundlagen des neunzehnten Jahrhunderts. Verlagsanstalt F. Bruckmann, Munich.

Clarke, H. T. (1941): Science *94*, 553.

Clarke, H. T. (1945): Science *102*, 168.

Coates, J. E. (1939): The Haber Memorial Lecture. J. Chem. Soc., p. 1642.

Cohen, G. D. (1975): Year Book XX of the Leo Baeck Institute, pp. ix–xxxi. Secker & Warburg, London and New York.

Crick, F. (1966): Of Molecules and Men. University of Washington Press, Seattle.

Cuatrecasas, P., Wilcheck, M., and Anfinsen, C. B. (1968): Proc. Nat. Acad. Sci. U.S. *61*, 636.

Dakin, H. D., and Dudley, H. W. (1913): J. Biol. Chem. *14*, 155.

Davis, N. P. (1968): Lawrence and Oppenheimer. Simon and Schuster, New York.

Deuticke, H. J. (1933): Ergeb. Physiol. Biolog. Chem. & Exp. Pharm. (L. Asher and K. Spiro) *35*, 32.

Du Bois-Reymond, E. (1881): Über die Grenzen des Naturerkennens. Veit & Co., Leipzig.

Dubos, R. J. (1950): Louis Pasteur: Free Lance of Science. Little, Brown, Boston.

Duclaux, E. (1896): Pasteur: Histoire d'un Esprit. Sceaux, Charaire.

Ehrlich, P. (1885): Das Sauerstoff-Bedürfnis des Organismus. Eine farben-analytische Studie. August Hirschwald, Berlin.

Eigen, M. (1971): Self-organization of matter and the evolution of biological macromolecules. Naturwissenschaften *58*, 465–523.

Eigen, M. (1977): Gesetz und Zufall-Grenzen des Machbaren. In: Schicksal? Grenzen der Machbarkeit. Symposium of the Carl Friedrich von Siemens Stiftung. Deutscher Taschenbuch Verlag, Munich.

Eigen, M., and Winkler, R. (1975): Das Spiel. Naturgesetze steuern den Zufall. R. Piper & Co., Munich and Zurich.

Fermi, L. (1954): Atoms in the Family. My Life with Enrico Fermi. University of Chicago Press, Chicago.

Florkin, M. (1975): A History of Biochemistry. In: Comprehensive Biochemistry, Vol. 31 (M. Florkin and E. H. Stotz, eds.). Elsevier, Amsterdam, Oxford, and New York.

Florkin, M. (1977): A History of Biochemistry. In: Comprehensive Biochemistry, Section VI, Vols. 30–34 (F. Florkin and E. H. Stotz, eds.). Elsevier, Amsterdam, Oxford, and New York.

Fruton, J. S. (1972): Molecules and Life. Wiley–Interscience, New York.

Goran, M. (1967): The Story of Fritz Haber. University of Oklahoma Press, Norman.

Goudsmit, S. (1947): Alsos. Henry Schumann, New York.

Grande, F., and C. Asensio (1976): Biological introduction: Severo Ochoa and the development of biochemistry. In: Reflections on Biochemistry (A. Kornberg, B. L. Horecker, L. Cornudella, and J. Oro, eds.). Pergamon Press, Oxford, New York, and Toronto.

Greiling, W. (1954): Im Banne der Medizin. Paul Ehrlich. Leben und Werk. Econ, Düsseldorf.

Haber, F. (1896): Experimental Studies on the Decomposition and Combustion of Hydrocarbons. R. Oldenbourg, Munich.

Haber, F. (1898): Outline of Technical Electrochemistry on a Theoretical Basis. R. Oldenbourg, Munich.

Haber, F. (1905): Thermodynamics of Technical Gas Reactions. R. Oldenbourg, Munich.

Haber, F. (1927): Aus Leben und Beruf: Aufsätze, Reden, Vorträge. Julius Springer, Berlin.

Haber, F. (1928): Towards an appreciation of Justus von Liebig. Z. Angew. Chem. *41*, 891–897.

Haber, F. (1963): Letters to Chaim Weizmann. In: Year Book VIII of the Leo Baeck Institute, p. 103. East and West Library, London.

Hahn, O. (1966): A Scientific Autobiography. Charles Scribner's Sons, New York. (Original edition in 1962 in German entitled: Vom Radiothor zur Uranspaltung. Friedr. Vieweg & Sohn, Brunswick.)

Halban, H. von, Joliot, F., and Kowarski, L. (1939): Liberation of neutrons in the nuclear explosion of uranium. Nature (London) *143*, 470–472.

Hamburger, E. (1968): Juden im öffentlichen Leben Deutschlands. J. B. C. Mohr (Paul Siebeck), Tübingen.

Hamburger, E. (1975): Hugo Preuss: Scholar and Statesman. In: Year Book XX of the Leo Baeck Institute, pp. 179–206. Secker & Warburg, London and New York.

Hartmann, G. R. (1976): Die aktivierte Essigsäure und ihre Folgen. Walter de Gruyter, Berlin and New York.

Hartmann, H. (1953): Max Planck als Mensch und Denker. Ullstein, Berlin and Frankfort on the Main.

Hecht, S. (1947): Explaining the Atom. Viking, New York.

Heisenberg, W. (1926): Z. Phys. *39*, 499.

Heisenberg, W. (1958): Physics and philosophy. The revolution in modern science. In: World Perspectives (R. N. Anshen, ed.), Vol. 19. Harper & Brothers, New York.

Heisenberg, W. (1971): Physics and beyond. In: World Perspectives (R. N. Anshen, ed.) 247 pp. Harper & Row, New York.

Heisenberg, W. (1974): Across the frontiers. In World Perspectives (R. N. Anshen, ed.). Harper & Row, New York. Extended new German edition: Schritte ueber die Grenzen. Piper & Co., Munich.

Hems, D. A. (1977): Biologically Active Substances—Exploration and Exploitation. Wiley, Chichester and New York.

Herken, H. (1976): Paul Ehrlich, Pioneer of Chemotherapy. Address at the Dedication of the Paul Ehrlich Wing of the Institute of Biological Sciences, Weizmann Institute of Science, Rehovot, Israel.

Hermann, A. (1968): Albert Einstein/Arnold Sommerfeld Briefwechsel. Schwabe & Co., Basel and Stuttgart.

Hermann, A. (1976): Werner Heisenberg. Rowohlt, Hamburg.

Hermann, A. (1977): Die Jahrhundertwissenschaft: Werner Heisenberg. Deutsche Verlagsanstalt, Stuttgart.

Hieronimus, E. (1964): Theodor Lessing. Otto Meyerhof. Leonard Nelson. Dietrichsche Universitäts-Buchdruckerei W. Fr. Kästner, Göttingen.

Hill, A. V. (1950): A challenge to biochemists. In: Metabolism and Function (D. Nachmansohn, ed.). Biochim. Biophys. Acta *4*, 4–11.

Hill, A. V. (1965): Trails and Trials in Physiology. Edward Arnold, London.

Hoffmann, B., with the collaboration of Helen Dukas (1972): Albert Einstein, Creator and Rebel. A Plume Book. New American Library, New York and London.

Hoffman, P. (1977): The History of the German Resistance (1933–1945). MIT Press, Cambridge, Mass.

Holmes, F. L. (1974): Claude Bernard and Animal Chemistry. Harvard University Press, Cambridge, Mass.

Irving, D. (1967): The Virus House: Germany's Atomic Research and Allied Countermeasures. W. Krimber, London.

Jaenicke, J. (1935): Haber's research on the gold content of sea water. Naturwissenschaften *23*, 57.

Jordan, P. (1936): Anschauliche Quantentheorie. Julius Springer, Berlin.

Jung, C. G., and Pauli, W. (1952): In: Naturerklärung und Psyche. Studien aus dem C. G. Jung-Institut, Vol. IV, 109. Rascher, Zurich.

Kleinzeller, A., ed. (1965): Manometrische Methoden. Gustav Fischer, Jena.

Kornberg, A., B. L. Horecker, L. Cornudella, and J. Oro, eds. (1976): Reflections on Biochemistry. Pergamon Press, Oxford, New York, and Toronto.

Kornberg, H. L. (1968): Hans A. Krebs: A pathway in metabolism. Biochem. Soc. Symposia *27*, 39.

Krebs, H. A. (1970): The history of the tricarboxylic acid cycle. Perspect. Biol. Med. *14*, 154–170.

Krebs, H. A. (1972): Otto Heinrich Warburg. Biographical Memoirs of Fellows of the Royal Society, Vol. 18, 629–699.

Krebs, H. A., and Henseleit, K. (1932): Z. Physiol. Chem. *210*, 33–36.

Krebs, H. A., and Johnson, W. A. (1937): The role of citric acid in intermediary metabolism in animal tissues. Enzymologia *4*, 148–156.

Krebs, H. A., and Shelley, J. H., eds. (1975): The Creative Process in Science and Medicine. American Elsevier, New York.

Kuhn, H. G. (1965): James Franck. Biographical Memoirs of Fellows of the Royal Society, Vol. 11, 53–74.

Kupferberg, H. (1972): Three Generations of Genius. Charles Scribner's Sons, New York.

Lamm, H. (1964): Theodor Heuss—An und über Juden. Econ, Düsseldorf and Vienna.

Langmuir, I. (1919): J. Am. Chem. Soc. *41*, 868, 1543.

Laue, M. von (1934): Obituary notice. Naturwissenchaften *22*, 87.

Laurence, W. L. (1947): Dawn over Zero: The Story of the Atom Bomb. Alfred A. Knopf, New York.

Lebedev, A. von (1912): Biochem. Z. *46*, 483.

Lehnartz, E. (1933): Arch. Physiol. 7, 475.

Leibholz-Bonhoeffer, S. (1968): Vergangen, Erlebt, Überwunden. Schicksale der Familie Bonhoeffer. Johannes Kiefel Verlag, Wuppertal Barmen.

Leicester, L. M. (1974): Development of Biochemical Concepts from Ancient to Modern Times. Harvard University Press, Cambridge, Mass.

Leschnitzer, A. (1954): Saul und David. Lambert-Schneider, Heidelberg.

Leschnitzer, A. (1956): The Magic Background of Modern Anti-Semitism, International Universities Press, New York.

Lewis, G. N. (1916): The atom and the molecule. J. Am. Chem. Soc. *38*, 762.

Lewis, G. N., and Randall, M. (1923): Thermodynamics and the Free Energy of Chemical Substances. McGraw-Hill, New York and London.

Liebeschütz, H. (1962): Treitschke and Mommsen on Jewry and Judaism. In: Year Book VII of the Leo Baeck Institute, pp. 153–182. East and West Library, London.

Lipmann, F. (1975): The roots of bioenergetics. In: Energy Transformations in Biological Systems. Ciba Foundation Symposium *31*, 3–68, Elsevier, Amsterdam, Oxford, and New York.

Lohmann, K. (1934): Biochem. Z. *271*, 264.

Mendelssohn, K. (1973): The World of Walther Nernst. The Rise and Fall of German Science 1864–1941. University of Pittsburgh Press, Pittsburgh.

Meyerhof, O. (1913): Zur Energetik der Zellvorgaenge. Vandenhoeck and Ruprecht, Göttingen.

Meyerhof, O. (1925): Chemical Dynamics of Life Phenomena. Monographs Exper. Biol. Lippincott, Philadelphia and London.

Meyerhof, O (1930): Die Chemischen Vorgänge im Muskel und ihr Zusammenhang mit Arbeitsleistung und Wärmebildung. Monographien aus dem Gesamtgebiet der Physiologie der Pflanzen und Tiere, Vol. 22. 350 pp. Julius Springer, Berlin.

Meyerhof, O. (1931): Neuere Versuche zur Energetik der Muskel Kontraktion. Naturwissenschaften *46*, 923.

Meyerhof, O. (1934): Betrachtungen über die naturphilosophischen Grundlagen der Physiologie. Naturwissenschaften *20*, 311.

Meyerhof, O. (1937): Ergeb. Physiol. Biolog. Chem. & Exp. Pharm. (L. Asher and K. Spiro), *39*, 10–75

Meyerhof, O., and K. Lohmann (1926): Biochem. A. *168*, 128.

Meyerhof, O., and K. Lohmann (1931): Naturwissenschaften *19*, 575.

Meyerhof, O., and D. Nachmansohn (1928): Neue Beobachtungen über den Umsatz des "Phosphagens" im Muskel. Naturwissenschaften *16*, 726.

Monod, J. (1970): Le Hasard et la Nécessité. Editions du Seuil, Paris.

Moore, R. (1966): Niels Bohr. His Life, His Science, and the World They Changed. Alfred A. Knopf, New York.

Morse, P. M. (1976): Edward Uhler Condon. In: Biographical Memoirs, Nat. Acad. Sci., Washington, *48*, 125.

Mosse, W. E. (1976): In: Juden im Wilhelminischen Deutschland 1890–1912 (W. E. Mosse, ed.), pp. 57–114. J. C. B. Mohr (Paul Siebeck) Tübingen.

Nachmansohn, D. (1928/1929): Über den Zerfall der Kreatinphosphorsäure im Zusammenhang mit der Tätigkeit des Muskels. Biochem. Z. I, *196*, 73 (1928); II, *208*, 237 (1929); III, *213*, 262 (1929).

Nachmansohn, D. (1929): Über den Zusammenhang des Kreatinphosphorsäurezerfalls mit Muskelchronaxie und Kontraktionsgeschwindigkeit. Medizin. Klinik *42*, 1–8.

Nachmansohn, D., ed. (1950): Metabolism and Function. Elsevier, New York, Amsterdam.

Nachmansohn, D. (1956): Carl Neuberg. Rudolf Virchow Med. Soc. *25*, 75.

Nachmansohn, D. (1969): Proteins in excitable membranes. J. Gen. Physiol. *54S*, 187.

Nachmansohn, D. (1971): Proteins in bioelectricity. In: Handbook of Sensory Physiology (W. R. Loewenstein, ed.), p. 18. Springer, New York, Heidelberg, and Berlin.

Nachmansohn, D. (1972): Prefatory Chapter. Ann. Rev. of Biochemistry, 1.

Nachmansohn, D. (1976): Highlights of a friendship. In: Reflections on Biochemistry (A. Kornberg, B. L. Horecker, L. Cornudella, and J. Oro, eds.), pp. 397–408. Pergamon Press, Oxford, New York, and Toronto.

Nachmansohn, D., Ochoa, S., and Lipmann F. A. (1960): Otto Meyerhof. In: Biographical Memoirs. Nat. Acad. Sci., Washington, *34*, 153–182.

Naturwissenschaften (1928): No. 50, Haber Festschrift. Springer, Berlin.

Needham, D. (1971): Machina Carnis; The Biochemistry of Muscular Contraction in its Historical Development. Cambridge University Press, Cambridge, Mass.

Nernst, H. W. (1893): Theoretische Chemie vom Standpunkte der Avogadroschen Regel und der Thermodynamik 11.–15. Auflage, 1926, pp. 927. Ferdinand Encke, Stuttgart.

Nernst, H. W. (1903): Theoretische Chemie. Ferdinand Encke, Stuttgart.

Neuberg, C. (1911a): The Urine and Other Excretions of Man and Animals. Springer, Berlin.

Neuberg, C. (1911b): The Carbohydrates. Monograph in Biochemical Manuals, Berlin.

Neuberg, C. (1913): Der Gärungsvorgang und der Zuckerumsatz der Zelle. Gustav Fischer Verlag, Jena.

Neuberg, C. (1913): Biochem. Z. *49*, 502.

Neuberg, C., and Kerb, J. (1913): Biochem. Z. *58*, 158.

Neuberg, C., and Kerb, J. (1914): Biochem. Z. *62*, 489.

Nord, F. F. (1958): Advances in Carbohydrate Chemistry *13*, 1, Academic Press, New York.

Northrop, J. H. (1939): Crystalline Enzymes. Columbia University Press, New York.

Ochoa, S. (1975): Collected Papers, 1928–1975, in 3 Volumes, pp. 3003 (A. Sols and C. Estevez, ed.) Publications of the Ministry of Education and Science, Spain. Ruan, S. A., Madrid.

Pasteur, L. (1878): Réponse à M. Berthelot. Comp. Rend. *83*, 10.

Pauli, W. (1961): Aufsaetze und Vortraege über Physik und Erkenntnistheorie. Friedr. Vieweg & Sohn, Brunswick.

Pauling, L. (1948): The Nature of the Chemical Bond. Cornell University Press, Ithaca, N.Y.

Peters, R. A. (1954): Otto Meyerhof. Biographical Memoirs of Fellows of the Royal Society, Vol. 9, 175–200.

Planck, M. (1922): Gesammelte Reden und Aufsätze. Hirzel, Leipzig.

Planck, M. (1949): In: Scientific Autobiographies and Other Papers, p. 52. Philosophical Library, New York.

Planck, M. (1959): The New Science. Meridian Books, Greenwich Editions.

Polanyi, M. (1967): Life transcending physics and chemistry. Chem. Eng. News *54*, 66.

References 387

Rabi, I. I. (1960): My Life and Times as a Physicist. The Claremont Colleges, Claremont, Calif.

Rabinowitch, E. (1964): Obituary of James Franck and Leo Szilard. Bulletin of the Atomic Scientists, Vol. 20 October. Chicago, Ill.

Racker, E. (1972): Bioenergetics and the problem of tumor growth. Am. Scientist 60, 56–63.

Reid, C. (1970): Hilbert. Springer, New York, Heidelberg, and Berlin.

Reid, C. (1976): Courant in Göttingen and New York. Springer, New York, Heidelberg, and Berlin.

Richarz, M. (1974): Der Eintritt der Juden in die Akademischen Berufe. J. C. B. Mohr (Paul Siebeck), Tübingen.

Sambursky, S. (1965): Das physikalische Weltbild der Antike. Artemis Verlag, Zurich and Stuttgart.

Sambursky, S. (1971): Die Willensfreiheit im Wandel des physikalischen Weltbildes. Eranos-Jahrbuch XXXVIII/1969. Rhein-Verlag, Zurich.

Sambursky, S. (1975): Der Weg der Physik. Artemis, Zurich and Munich.

Sambursky, S. (1977): Naturerkenntnis und Weltbild. Artemis, Zurich and Munich.

Schoenheimer, R. (1937): The investigations of intermediary metabolism with the aid of heavy nitrogen. In: The Harvey Lectures, 1936–1937, Series XXXII, p. 122. Williams & Wilkins, Baltimore.

Schoenheimer, R. (1942): The Dynamic State of Body Constituents. Harvard University Press, Cambridge, Mass.

Schrödinger, E. (1945): What Is Life? 91 pp. Cambridge University Press, Cambridge, Mass.

Schrödinger, E. (1951): Science and Humanism. Cambridge University Press, Cambridge, Mass.

Schulin, E. (1976): Die Rathenaus. Zwei Generationen jüdischen Anteils an der industriellen Entwicklung Deutschlands. In: Juden im Wilhelminischen Deutschland 1890–1912 (W. E. Mosse, ed.), p. 115. J. C. B. Mohr (Paul Siebeck), Tübingen.

Science and the Challenges Ahead (1947): National Science Board. National Science Foundation. U.S. Government Printing Office, Washington, D.C.

Seelig, C., ed. (1954): Ideas and Opinions by Einstein. Based on *Mein Weltbild* and other sources. Crown Publishers, New York.

Smyth, H. D. (1945): Atomic Energy for Military Purposes. The Official Report on the Development of the Atomic Bomb under the Auspices of the United States Government. Princeton University Press, Princeton, N.J.

Stern, F. (1977): Gold and Iron. Bismarck, Bleichröder, and the Building of the German Empire. Alfred A. Knopf, New York.

Stern, R. (1963): Fritz Haber—Personal recollections with a prefatory note by Fritz Stern. In: Year Book VIII of the Leo Baeck Institute, p. 70. East and West Library, London, Jersualem, and New York.

Szilard, L. (1969): Reminiscences (G. Weiss Szilard and K. R. Winsor, eds.). In: The Intellectual Migration, Europe and America, 1930–1960 (D. Fleming and B. Bailyn, eds.). Harvard University Press, Cambridge, Mass.

Traube, M. (1878): Die chemische Theorie der Fermentwirkungen und der Chemismus der Respiration. Ber. Chem. Ges. 11, 1894–1992.

Vallery-Radot, R. (1900): La Vie de Pasteur. Hachette, Paris.

Van't Hoff, J. H. (1967): Imagination in Science. Springer, Berlin, Heidelberg, and New York.

Weinberg, A. M. (1977): Is Nuclear Energy Acceptable? Bulletin of the Atomic Scientists, 33, (April), p. 54, Chicago, Ill.

Weiner, C. (1969): New site for the seminar: The refugees and American physics in the thirties. In: The Intellectual Migration, Europe and America, 1930–1960 (D. Fleming and B. Bailyn, eds). Harvard University Press, Cambridge, Mass.

Willstätter, R. (1902): Abh. Ann. d. Chemie I, II, III, *317*, 204–374; with A. Bode, IV, V, VI; with Ch. Hollander, VII; with F. Ettlinger, VIII, ibid. *326*, 1–128.

Willstätter, R. (1903): Ber. d. Deutsch. Pharm. Ges. *13*, 50.

Willstätter, R. (1927): Problems and Methods in Enzyme Research. Cornell University Press, Ithaca, N.Y.

Willstätter, R. (1928): Untersuchungen über die Enzyme. Julius Springer, Berlin.

Willstätter, R. (1949): Aus meinem Leben. 463 pp. Verlag Chemie, Weinheim.

Willstätter, R., and Stoll, A. (1913): Untersuchungen über Chlorophyll. Julius Springer, Berlin.

Willstätter, R., and Stoll, A. (1918): Untersuchungen über die Assimilation der Kohlensäure. Julius Springer, Berlin.

Witkop, B. (1977): Heinrich Wieland centennial: His lifework and his legacy today. Angew. Chem. *16:9*, 569.

Witt, H. T. (1971): Coupling of quanta, electrons, fields, ions and phosphorylation in the functional membrane of photosynthesis. Quart. Rev. Biophys. *44*, 365–477.

A RUSSIAN CHILDHOOD
SOFYA KOVALEVSKAYA
Introduced, Translated, and Edited by BEATRICE STILLMAN
with an analysis of Kovalevskaya's mathematics by P.Y. KOCHINA
of the U.S.S.R. Academy of Sciences

Towards the end of her life, which ended abruptly at the age of 41, her colleagues saw her as a bizarre but fascinating phenomenon: "a mathematical lady"—as they put it.

But Sofya Kovalevskaya was much more than that, a fact that becomes evident when you read her autobiography. *A Russian Childhood* breathes with the literary authenticity of a skilled writer and a women-of-the-world.

Kovalevskaya's childhood reminiscences are a delightful piece of cultural and social history, capturing the period of the rise of radical political groups and the emancipation of serfs of the late 1800's. Ranging from minute descriptions of earliest memories to her friendship with Dostoevsky during her teenage years, the book is a thoroughly absorbing panorama that remains as alive today as it was when she wrote it.

Although she has long been recognized throughout Europe, Kovalevskaya is still relatively unknown in America. This "mathematical lady" was in fact somewhat of a renaissance woman. At thirteen she was a mathematics prodigy; as a young woman she earned her doctorate in mathematics entirely on the strength of her original contributions; she was one of the first women to hold a university professorship (at the University of Stockholm); she was awarded the Prix Bordin (comparable to the Nobel Prize) for her work in mathematical physics; and she was the first woman to become a Corresponding Member of the arch-conservative Russian Imperial Academy of Sciences. Her literary output includes a novel, two plays, a personal reminiscence of George Eliot, a critical article on M.E. Saltykov-Shchedrin, a small body of verse, and a collection of short stories, sketches, and journalism. Always ahead of her times, she remained throughout her life very much concerned with women's rights.

1978/xiii, 250 pp./7 illus./Cloth
ISBN 0-387-90348-8